Biodiversity Offsets Between Regulation and Voluntary Commitment

Marianne Darbi

Biodiversity Offsets Between Regulation and Voluntary Commitment

A Typology of Approaches Towards Environmental Compensation and No Net Loss of Biodiversity

Springer

Marianne Darbi
Institute for Landscape Planning and Nature Conservation
Hochschule Geisenheim University
Geisenheim, Germany

Leibniz Institute of Ecological Urban and Regional Development (IOER)
Dresden, Germany

ISBN 978-3-030-25593-0 ISBN 978-3-030-25594-7 (eBook)
https://doi.org/10.1007/978-3-030-25594-7

This Springer imprint is published by the registered company Springer Nature Switzerland AG.
The registered company address is: Gewerbestrasse 11, 6330 Cham, Switzerland

"Whether you like [biodiversity] offsets or not, it looks like they are here to stay. And offsets that work could be crucial to balancing development and conservation."

(Maron M, Gordon A (2015) Could well-designed offset policies actually increase loss? Decision Point Online 91/Aug 2015)

"To use the same words is not a sufficient guarantee of understanding; . . .ultimately one must have one's experiences in common"

(Nietzsche 1955, cited in Flathman RE (1992) Willful liberalism: voluntarism and individuality in political theory and practice. Cornell University Press, Ithaca, NY)

Für Opa Manfred, Oma Lotti, und Tante Hildi

Abstract

The discussion about voluntary vs. mandatory offsets has risen to prominence in the past decade, especially with the growing interest of the private sector globally, and also in Europe with the initially intended No Net Loss initiative of the EU and the envisaged introduction of a mandatory compensation scheme at the EU level. Consequently, starting from the often observed oversimplified distinction of voluntary vs. mandatory offsets, the aim of this study was to develop a refined typology with regard to the voluntariness of biodiversity offsets.

To this end, an expert- and internet-based research approach was applied. This included establishing an online personal learning network, most notably via the Biodiversity Offsets Blog. Building on the method of empirically grounded typification after Kluge (1999) and Lazarsfeld (1937), criteria for voluntariness have been derived from theoretical concepts, e.g. altruism, voluntary environmental approaches, and corporate responsibility. Together with evidence from a worldwide desk-based screening of 90 biodiversity offsets cases, this was integrated into a typology of the following seven types of biodiversity offsets:

1. Regulatory biodiversity offsets: required by law and enforced.
2. Conditional biodiversity offsets: required by financial institutions (e.g. International Finance Corporation).
3. Enabled biodiversity offsets: fostered by governments and NGOs through pilot schemes, guidance, etc.
4. Sectoral biodiversity offsets: take part in a voluntary self-commitment of a sector (e.g. mining).
5. Corporate biodiversity offsets: resulting from a voluntary self-commitment of a corporation.
6. Local biodiversity offsets: single offsets, which are most likely developed at the local level in a consensual process.
7. Altruistic biodiversity offsets: truly voluntary offsets that are driven by the altruistic motivation to make a positive impact.

The analysis and comparison of these types has shown the variety and complexity of biodiversity offsets and that huge differences exist on a global scale. Moreover, it is concluded that biodiversity offsets are highly context dependent and no "one-size-fits-all" solution to offsets exists—neither regulatory nor voluntary. The analysis of the state of the scientific knowledge and the practical evidence explored throughout this study encourage the critical analysis and exploration of the various forms of (voluntary) biodiversity offsets, in particular with regard to the evaluation of their outcome in terms of effectiveness and efficiency. This can help to contribute to an informed debate about biodiversity offsets.

References

Kluge S (1999) Empirisch begründete Typenbildung. Zur Konstruktion von Typen und Typologien in der qualitativen Sozialforschung. Leske + Budrich Verlag, Opladen

Lazarsfeld PF (1937) Some remarks on the typological procedures in social research

Keywords

Voluntary biodiversity offsets, Impact mitigation, EU No Net Loss initiative, The business case for biodiversity offsets, Voluntary environmental approaches, Corporate environmental responsibility, Empirically grounded typification

Preface

Together with climate change, biodiversity loss is one of the biggest challenges of our times that has been baptized the Anthropocene. Accordingly, anthropogenic activities have caused dramatic changes to nature and wildlife, including habitat fragmentation, sealing, deforestation, spread of invasive species, and overexploitation of natural resources.

Whatever the rationale to reverse this trend is—i.e. preserving nature for its intrinsic value vs. for its instrumental value as the basis for human wellbeing—it cannot be done without human intervention. This means that there is an increasing need for restoration-based activities to complement conventional nature conservation activities. Building on this premise, the paradigm of "no net loss" has risen to prominence in a worldwide context and particularly with respect to EU policy.

In this scope, biodiversity offsets (i.e. compensating a biodiversity loss with a biodiversity gain) have been increasingly explored to reach the no net loss goal. While they have risen from regulatory requirements in a number of countries, their increasing popularity is mainly bound to a new trend towards the promotion of voluntary biodiversity offsets, which has started in the early 2000s.

However, this development is also highly contested. Political aspirations (especially when not well-thought-through or mature enough, yet) to foster or establish the use of biodiversity offsets are witnessing criticism of "oversimplification", "commodification of nature", and "licence to trash". Aims of the private sector to include biodiversity offsets into their business activities to counterbalance negative impacts are also often criticized for their pragmatism (argued to be at the expense of environmental quality), especially with regard to the use of market-based approaches, such as habitat banking for the implementation of biodiversity offsets.

While both proponents and opponents of biodiversity offsets bring important arguments, the discussion as such has seen a lot of polemics and oversimplification due to a lack of consideration of the complexity and context dependency of the implementation of biodiversity offsets. Against this background, with this book I attempt to help introduce objectivity into this dispute by providing differentiated viewpoints based on evidence from around the world.[1]

To this end, I have developed a typology of biodiversity offsets. Seven types (including four subtypes) have been identified. These types are described and analysed both in general and using practical examples in the form of illustrative case studies for each type.

The presentation of these case studies is structured and characterized by criteria for voluntariness that have been identified from an analysis of various theoretical concepts. On the one hand, this includes more general theories, i.e. altruism, economic theories, and willingness to pay. On the other hand, the theoretical concept of voluntariness is transferred and examined as to its application in the scope of nature conservation and care for the environment, i.e. with regard to voluntary environmental approaches and corporate (environmental) responsibility.

Furthermore, the typology is underpinned by a broad explorative, worldwide empirical base of 72 practical examples that are assigned to the seven types.

As a result, global tendencies and conclusions become visible, e.g. regarding the role of different stakeholders, notably the role of government bodies beyond the conventional understanding in terms of regulating and enforcing authorities.

In addition, differences are shown with regard to the scale of the development impacts, the sectors/types of developments addressed, and the governance of their implementation.

More generally, the typology highlights the context dependency of biodiversity offsets and helps identify and focus on the major characteristics and challenges in a given context.

Thus, the results of this study lead to a plea for a differentiated, context-sensitive consideration of the various forms of biodiversity offsets, based on practical evidence, an outcome-oriented evaluation, as well as discussion of standards and roles of different stakeholders. This means acknowledging the role of governmental authorities beyond regulation and enforcement as well as the pivotal role of certain stakeholders such as lending institutions.

Put short and simple: Biodiversity offsets, whatever they may be driven by, are not going to be the one solution to halt biodiversity loss. But they deserve consideration as one counterbalancing tool to be embedded into human development activities with impacts on nature.

Clearly, biodiversity offsets should not be seen as the prerequisite for legitimating decisions in favour of a certain infrastructure development, but rather as a means to address the consequences of biodiversity loss that has been decided and proven to be unavoidable.[2]

[1]Please note that data has been collected in the period from 2010 to 2016 and references published later could not be included (with a few exceptions).

[2]What is deemed to be an unavoidable biodiversity impact or loss would require an in-depth political and societal debate.

Consequently, biodiversity offsets should be thoroughly used, considering their specific context and limitations, but also in relation to other instruments. In doing so, they can help making "no net loss" of biodiversity a central achievement of the Anthropocene.

Geisenheim, Germany Marianne Darbi

Acknowledgements

This book would not have existed (and not been ready by now) without the help of many people—my professional and private network and "comfort zone". It is impossible to mention all of them—from the colleagues who brought me food in the last stressful weeks to the nice librarians who always cheered me up to my close friends who sent me frequent motivation messages. I owe all a wholehearted THANK YOU.

First and foremost, I want to thank my supervisor Wolfgang Wende who supported me throughout this whole journey. Support is probably a too weak word to express the difficulties but also the achievements we have managed together—with Wolfgang always being positive, encouraging, and never failing to have a backup plan. I have learned from him that excellence and pragmatism can fruitfully unite in science.

Second, I would like to express my thanks to Martine Maron. Having seen her role in the academic offsets debate and the international focus of this study, I feel honoured she agreed to be my second supervisor.

This work would not have been possible without the support of the Leibniz Institute of Ecological Urban and Regional Development where I have been working and doing research for the past 10 years. The IOER has always provided a fruitful and flexible work environment (in particular in reconciling work and family life). My individual thanks go to Gerd Lintz for being an attentive and problem-solving doctoral studies commissioner and to the director of the institute Bernhard Müller for supporting this project and pushing towards its finalization (in particular by granting a final 6 months fostering). Furthermore, I would like to thank the following colleagues: Andrea Wendebaum, Anja Heller, Evi Rockstroh, Heike Timmel, Christina Kraatz, Sabine Witschas, Sandra Lysakowska, Erhard Busch, and Jen Washco.

Most importantly, this work largely builds on my personal learning network on biodiversity offsets. Altogether this sums up to at least several hundreds of people from all around the world whom I can only thank generically here: everyone involved in the Business and Biodiversity Offsets Programme (above all Kerry ten Kate, Patrick Maguire, and Amrei von Hase from the BBOP secretariat), my LinkedIn network, the readers of and contributors to my Biodiversity Offsets Blog, and all the other experts whom I have met over the years at various occasions and who shared their knowledge and insights, sent me their papers, and answered my various questions. Nevertheless, I would like to thank a few people individually: Alan Key, Jared Hardner, and Julia Baker for their contribution to the case studies; Divya Narain for encouraging me to write about the business case for biodiversity offsets; and James Brown and Naazia Ebrahim for writing guest posts on the Biodiversity Offsets Blog that have been incorporated into this book.

For granting permissions to use figures I thank The Biodiversity Consultancy, Renato Orsato, and Kerry ten Kate.

Finally, I would like to give a big thank you and a hug to all my close friends who have gone through this with me—from encouraging me to pursue my PhD over inspirational discussions to last-minute proofreading. I am happy to have you in my life: Kathrin, Sanne, Jule, Luli, Juliane, Reimund, and Nelya (no priority in the order).

Last but not least, I would like to thank my family (continued in German).

Letztendlich verdanke ich es meiner Familie, der extremen Belastung, gleichzeitig beruflich Fuß zu fassen, diese Forschungsarbeit zu erstellen und eine Familie zu gründen, standgehalten zu haben. Ihr habt mich nicht nur zu der Person gemacht die ich bin, sondern Ihr seid auch meine Wohlfühlzone und mein Sicherheitsnetz. Ihr habt mich meinen Weg gehen lassen und mich doch stets begleitet.

Ich danke meinen Eltern, Großeltern und Tante Hildi für alles, vor allem den nervlichen Beistand und die Strapazen, die Ihr scheinbar ohne große Anstrengungen auf Euch genommen habt (wie zum Beispiel spontane mehrwöchige Besuche, um Euch um Kinder und Haushalt zu kümmern!), um mich zu entlasten und da zu sein, wenn ich es am nötigsten brauchte.

Daher widme ich dieses Buch meiner Oma Lotti und Tante Hildi, sowie dem Andenken an meinen Opa Manfred, der seit ich denken kann unerschütterlich an mich geglaubt hat, aber die Fertigstellung dieser Arbeit nicht mehr miterleben konnte. Danke.

Und schlussendlich Dank an meine Mädchen, Mona, Maya und Maria—Ihr seid meine Motivation und Inspiration, nicht aufzugeben, das Schöne in dieser Welt zu sehen und sie für Euch ein bißchen besser zu machen.

Contents

List of Figures

List of Tables

1 Biodiversity Offsets and No Net Loss: Introduction, Problem Statement, and Research Questions

Abstract

In light of ongoing global biodiversity loss, there is an increasing need for restoration-based activities to complement conventional nature conservation activities. Building on this premise, the paradigm of "no net loss" has risen to prominence in a worldwide context and particularly with respect to EU policy. In this scope, biodiversity offsets are increasingly explored and promoted to reach the no net loss goal.

Biodiversity offsets are the last step of a sequence called the "Mitigation Hierarchy" (first avoid, then minimize, and then finally restore/offset negative impacts). They are understood as "measurable conservation outcomes" that are designed to counterbalance the "significant residual adverse biodiversity impacts" from planned projects or developments, e.g., infrastructure, housing, or land-use change. Thus, offsets imply an exchange of a biodiversity loss in one location for a biodiversity gain in another location.

While biodiversity offsets have risen from regulatory requirements in a number of countries, their increasing popularity is mainly bound to a new trend toward the promotion of voluntary biodiversity offsets, which has started in the early 2000s.

Building on the growing controversy about voluntary vs. regulatory biodiversity offsets, and the complexity and context dependency of their implementation, this study attempts to develop a refined typology with regard to the voluntariness of biodiversity offsets. This shall provide the basis for an informed debate in light of a differentiated critical analysis of biodiversity offsets, the different options as to how they can be delivered in practice, and their outcomes.

Keywords

Biodiversity offsets · No net loss · Mitigation hierarchy · Voluntary biodiversity offsets · Market-based instruments · Ecological compensation · The business case for biodiversity offsets · License to trash

M. Darbi, *Biodiversity Offsets Between Regulation and Voluntary Commitment*,
https://doi.org/10.1007/978-3-030-25594-7_1

1.1 Introduction to the Context for Biodiversity Offsets: From Biodiversity Loss to No Net Loss of Biodiversity

Halting Biodiversity Loss

The natural world is being destroyed at an increasing pace. Many species and their habitats, as well as ecosystems as a whole, are threatened by human activities (European Commission 2015a, b). A broad consensus exists among scientists, politicians, businesses, and civil society that biodiversity loss is one of the biggest challenges that we are facing. This is a problem not only because of the intrinsic value of nature but also for the goods and services that it provides for human well-being, for example, through the provision of food and water, by offering natural protection from floods and storms, and by regulating the climate, i.e., because humans rely on a healthy environment as the basis for their existence (Millennium Ecosystem Assessment 2005, CBD Secretariat 2006; Darbi 2010).

Consequently, politicians and decision makers have declared that actions need to be taken to halt this global loss urgently and immediately. The problem as such is not new. But beginning with the Ecosummit in Rio de Janeiro in 1992 the world's leaders decided that it was time for a paradigm shift (CBD Secretariat n.d.). The result was the Convention on Biological Diversity (CBD), which since then has been ratified by 168 states and the European Union (United Nations 1992, CBD Secretariat n.d.). It gives a comprehensive definition of biological diversity, including the diversity of species and habitats and genetic diversity (see Fig. 1.1), and combines the preservation of biodiversity with the sustainable use of the natural resources.

Many states have elaborated national biodiversity strategies and action plans to enforce the goals of the CBD at the national or regional level (CBD Secretariat n.d.). For instance, the European Union has adopted its 2020 Biodiversity Strategy. The strategy includes a commitment to halt "the loss of biodiversity and the degradation

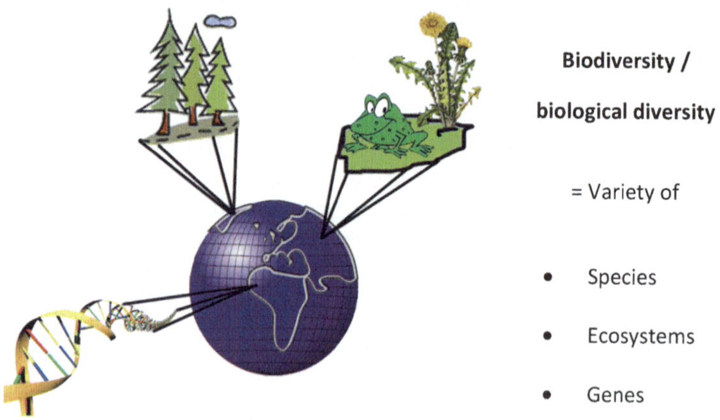

Fig. 1.1 Definition of biological diversity. Source: Author

of ecosystem services in the EU by 2020, and restoring them in so far as possible" (European Commission 2011).

The No Net Loss Commitment

As previous commitments to halt biodiversity loss have not proven effective (European Commission 2015a, b; CBD Secretariat 2006) and the loss of biodiversity is ongoing, especially through accelerated urbanization and infrastructure development (Millennium Ecosystem Assessment 2005; McKinney 2006; Quétier and Lavorel 2011), two trends are important to note with regard to policy change, in particular in the European Union (cf. European Commission 2011; Wende et al. 2018b, c):

1. A stronger focus on restoration as part of the concept of "no net loss."
2. The promotion of "new and innovative instruments and strategies" to achieve "no net loss."

Both center on the concept of "no net loss" as a key element. But what is the difference between "no loss" and "no net loss"? While "no loss" would technically mean that no loss and consequently no development shall be admissible at all (which in practice will be practicable only in rare circumstances), the term "net" in "no net loss" refers to the overall balance of loss and gain. This means that all environmental losses need to be counterbalanced by environmental gains (which may only be possible through restoration, e.g., planting trees for cutting trees) to maintain the status quo in terms of "no net loss." This principle is illustrated in Fig. 1.2 (cf. Figs. 1.3 and 1.4). It is important to note that losses and gains are measured against a baseline or counterfactual that can be assumed to be either static (a) or dynamic (b) over time.

As has been noted, the first observation is a stronger focus on restoration (or compensation) to counterbalance the loss, in addition to continued conservation efforts that alone cannot achieve the goal of halting loss (cf. European Commission 2011; Darbi 2010).

The second observation on the biodiversity policy of the European Union is the exploration and promotion of "new and innovative instruments and strategies to achieve 'No Net Loss' of biodiversity" (European Commission 2011). In light of "new" and "innovative" approaches, the actual discussion focuses on a set of keywords that can be grouped under the umbrella of "market based instruments," e.g., "Natural Capital," "Payments for Ecosystem Services," "Tradable Development Rights," "Tradable Permits," "Habitat Banking," or more general the "restoration economy" (cf. European Commission 2011, 2012a, b; Commission of the European Communities 2007; EEA 2006; Wissel and Wätzold 2010; Drechsler and Hartig 2011).

Biodiversity Offsets Between Market and Conservation

Biodiversity Offsets are often also assigned to this group of instruments because they imply a kind of "trade" between environmental losses and environmental gains as

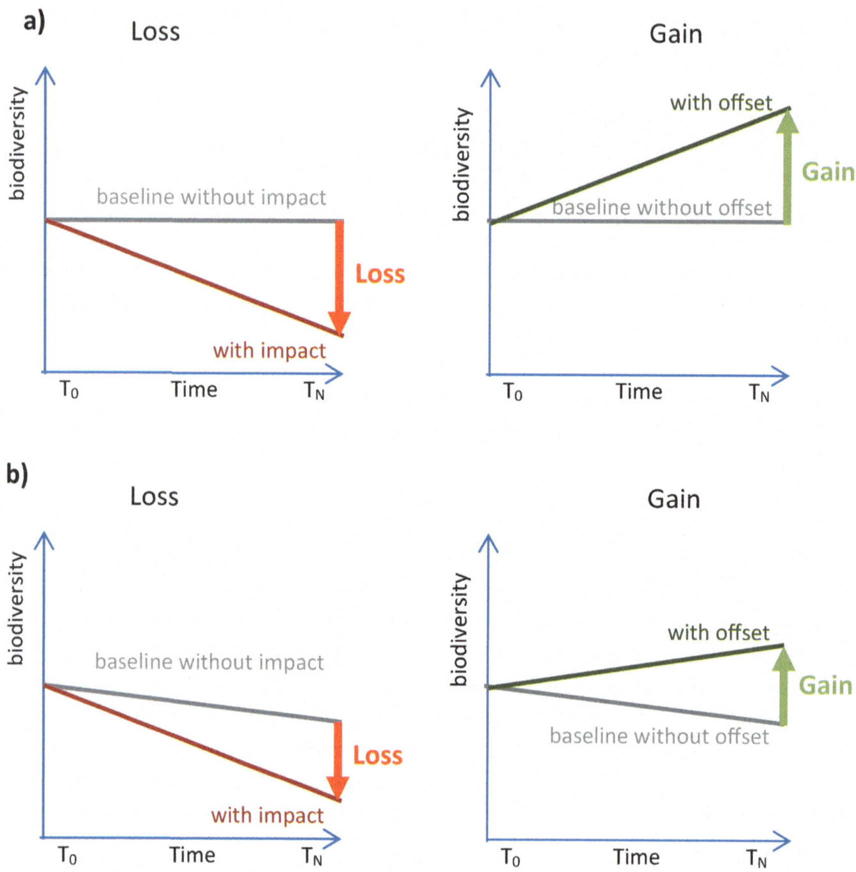

Fig. 1.2 Measuring no net loss against the static or dynamic baseline. (**a**) Measuring no net loss against a static baseline. (**b**) Measuring no net loss against a dynamic baseline. Source: Author (cf. Gibbons et al. 2015)

shown in Fig. 1.3. Thus, Trezise (2015) concludes that biodiversity offsets "are as much an economic instrument as they a biodiversity conservation tool."

Nevertheless, while the term "Biodiversity Offsets" is relatively new and mostly used with reference to market-based instruments, the underlying concept is much older (Darbi et al. 2010; Benabou 2014).

Compensation for negative environmental impacts has been institutionalized in different schemes under the environmental legislation of countries like the United States (Wetland Mitigation after Clean Water Act and Conservation Banking, cf. Bayon et al. 2012; Salzmann and Ruhl 2002; Ambrose 2000), Germany (Impact Mitigation Regulation after Federal Nature Protection Act, cf. Köppel et al. 2004; Wende et al. 2018d), Canada (Fish habitat compensation after the Fisheries Act, cf. Quigley and Harper 2006), Brazil ("forest set-aside offsets" after the Forest Code and "Project developers' offsets" after the SNUC Act, cf. De Olivieira and Filho n.d.),

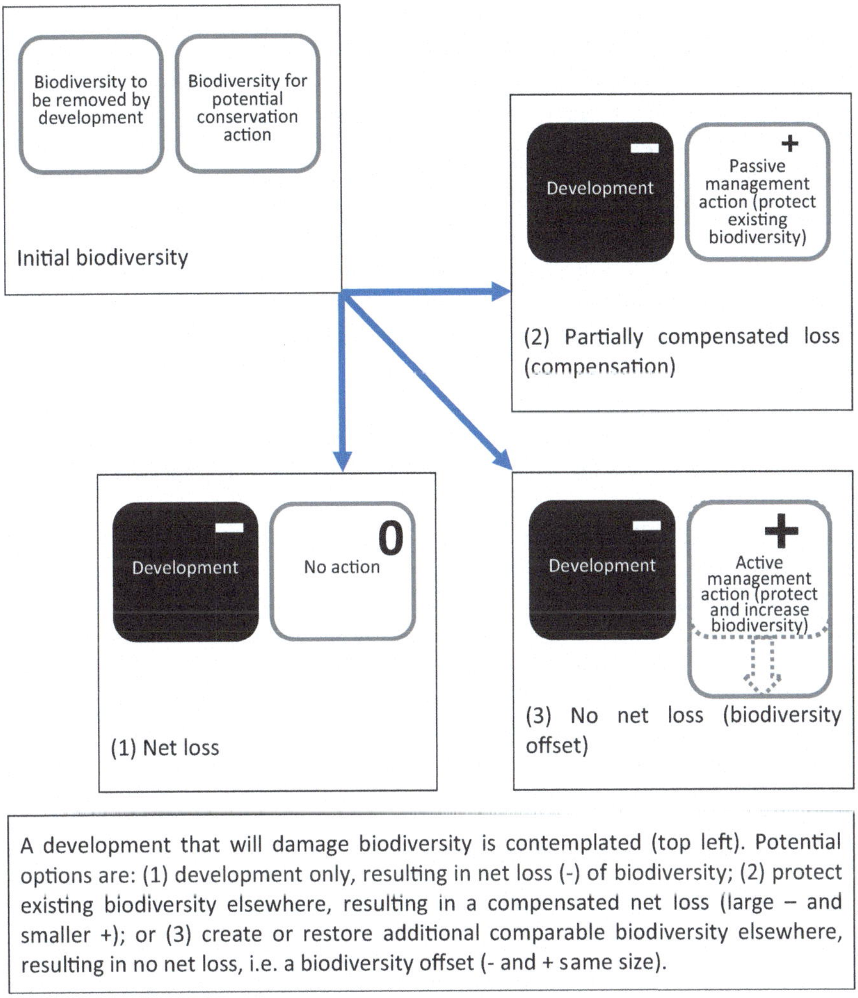

Fig. 1.3 Schematic of the offsetting principle for development impacts (It must be noted that this schematic is based on a static frame of reference or baseline at the time the impact occurs, i.e., without consideration of potential negative (biodiversity loss) or positive (biodiversity gain) development of the environmental baseline over time.). Source: Based on Bull et al. (2013)

and Australia (Biobanking in New South Wales and Bush Broker in Victoria, cf. Miller et al. 2015). Furthermore, compensation approaches in the context of infrastructure planning exist, e.g., in the Netherlands, Finland, and Sweden (cf. Rundcrantz and Skärbäck 2003; Cuperus et al. 1999, 2001; Madsen et al. 2010; Darbi et al. 2010; van Teeffelen 2018; Wende 2018).

What Are the Implications for the Present Study?
In light of ongoing biodiversity loss, there is an increasing need for restoration based activities to complement conventional nature conservation activities. Building on this premise, the paradigm of "no net loss" has risen to prominence in a worldwide context and particularly with respect to EU policy. In this scope, biodiversity offsets are increasingly explored and promoted to reach the no net loss goal.

Biodiversity offsets are a tool for compensation for environmental impacts rooted in compensation schemes under the environmental legislation of countries like the United States, Germany, Brazil, and Australia.

1.2 Introduction to the Concept of Biodiversity Offsets

Emergence of Biodiversity Offsets as a Specific Form of Environmental Compensation
Compensation approaches for impacts on biological diversity exist in numerous countries worldwide and liability for damages is stipulated under various sectoral laws, e.g., environment, mining, forests, waste, and water (Darbi et al. 2010). The "Polluter Pays Principle" is widely recognized (Madsen et al. 2010; Manuel 2013; Morandeau and Vilaysack 2012; Darbi 2010; Peters et al. 2003). According to this, generally the project proponent is liable for the damages caused by development, and has, therefore, to put in place appropriate compensation measures (cf. De Sadeleer 2015).

In this context, the project proponent (hereafter also a developer) is understood as the responsible party of a development project that upon execution and operation may (or may not) have impact on the natural environment. The term development project implies that only the impact from planned developments is meant here, and not the potential harmful outcomes of hazards and disasters (e.g., oil spill). In this sense, development projects (hereinafter also development or project) include, for example, the construction or altering of rail and road infrastructure as well as built infrastructure for housing and industry, mining or excavation of mineral resources, changes in land use like deforestation and plowing up of grassland (cf. Rhineland-Palatinate (Germany), Hufen et al. 2011).

Unsurprisingly, the situation on the ground is very heterogeneous and a variety of approaches and terms exists. Especially the term "compensation" has many different meanings and connotations in a global context as previous research (cf. Darbi et al. 2010) has shown. It may even be misleading when used to refer to financial compensation for social inequalities rather than for environmental degradation. Thus, more and more "biodiversity offsets" or "biodiversity offsetting" (used synonymously hereafter) is becoming the dominant term (see below: Promotion and popularity of biodiversity offsets as a common framework).

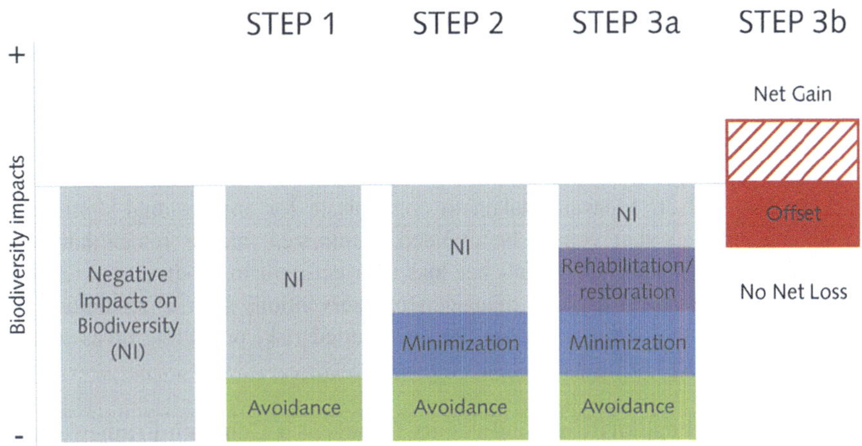

Fig. 1.4 Biodiversity Offsets as part of the Mitigation Hierarchy. Source: Darbi (2010), adapted from Rio Tinto and Western Australia EPA

Definition of Biodiversity Offsets

Biodiversity Offsets are defined as "measurable conservation outcomes" that are designed to counterbalance the "significant residual adverse biodiversity impacts" on the environment from projects or development (BBOP 2012a). Biodiversity offsets are the last step of a sequence, called the "mitigation hierarchy" (see Fig. 1.4).

According to this, developers are asked to first **avoid** and second **minimize** the **negative impacts** of their projects, then **rehabilitate/restore** as far as possible, before ultimately applying an **offset**.

Biodiversity offsets as part of the mitigation hierarchy target the goal "to achieve no net loss and preferably a net gain of biodiversity on the ground with respect to species composition, habitat structure, ecosystem function and people's use and cultural values associated with biodiversity" (BBOP 2009a). Briefly, this means that negative impacts on biological diversity need to be compensated or counterbalanced by conservation or restoration measures in order to ensure no net loss of biodiversity. Thus, offsets imply an exchange of a biodiversity loss in one location with a biodiversity gain in another location, or as James Trezise (2015) puts it in metaphoric words: biodiversity offsets "operate like a barter system. Someone will take an apple, provided that they give you an apple or plant some apple trees in return."

As has been noted, biodiversity offsets are an integral part of the mitigation hierarchy and their success in terms of no net loss relies on the adherence to the mitigation hierarchy, i.e., proper implementation of the precedent steps in the order outlined in Fig. 1.4 and as follows (BBOP 2012a):

- **Step 1—Avoidance**: Measures taken to avoid creating impacts from the outset, such as careful spatial or temporal placement of elements of infrastructure, in order to completely avoid impacts on certain components of biodiversity.

- **Step 2—Minimization**: Measures taken to reduce the duration, intensity, and/or extent of impacts (including direct, indirect, and cumulative impacts, as appropriate) that cannot be completely avoided, as far as is practically feasible.
- **Step 3a—Rehabilitation/restoration**: Measures taken to rehabilitate degraded ecosystems or restore cleared ecosystems following exposure to impacts that cannot be completely avoided and/or minimized.
- **Step 3b—Offset**: Measures taken to compensate for any residual significant adverse impacts that cannot be avoided, minimized, and/or rehabilitated or restored, in order to achieve no net loss or a net gain in biodiversity. Offsets can take the form of positive management interventions such as restoration of degraded habitat, arrested degradation, or averted risk, protecting areas where there is imminent or projected loss of biodiversity.

Promotion and Popularity of Biodiversity Offsets as a Common Framework
One reason for the popularity of biodiversity offsets is inherent to the term itself: "biodiversity" is probably one of the most prominent vogue terms of the twenty-first century (after "sustainability" at the end of the twentieth century).

Simply typing "biodiversity" into the Google search engine delivers more than forty million results, and every day several hundreds or thousands of new information sources are being produced and added (as of August 2014).

While the Google search hits for "biodiversity offset" (including both "biodiversity offsets" and "biodiversity offsetting") can in no way be compared to the ones for "biodiversity," they still deliver a remarkable number (more than 25,000)—seeing that it is a very specific and complex concept. What can be seen from Fig. 1.5 is that

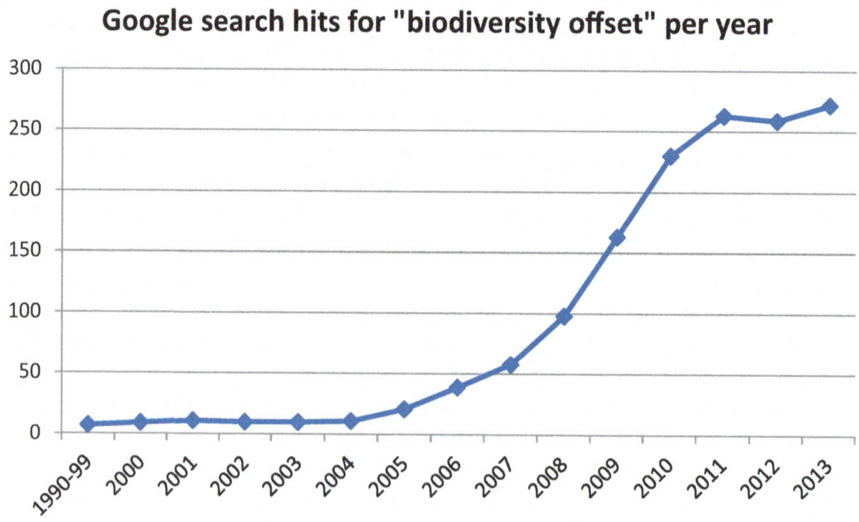

Fig. 1.5 Google search hits for "biodiversity offset" per year (websites retrieved by Google search engine). Source: Author, building on web search using the Google search engine

in the past decade the information that is made available and shared on the Internet has significantly increased.

The simple answer to the question why biodiversity offsets have become popular would be that, similar to the biodiversity "hype," the interest in biodiversity offsets is a result of growing awareness and the insight that biodiversity is being lost at an unprecedented rate (CBD Secretariat 2006; European Commission 2015a, b). The International Year of Biodiversity 2010 and the following United Nations Decade of Biodiversity (2011–2020) brought this fact to the broad public debate.

While it may be true that this did, in fact, influence the promotion of biodiversity offsets worldwide, the (abstract) awareness of ongoing biodiversity loss falls short of providing an incentive for concrete action.

Instead, the success (in terms of growing interest, not necessarily in terms of environmental quality) of biodiversity offsets is mainly a systemic one. The term "biodiversity offsets" provides a common framework (cf. Darbi 2014a). Under this term, a multitude of compensation approaches is grouped (e.g., compensation, restoration, remediation, habitat banks, impact mitigation, environmental offsets, biodiversity banking, and wetland mitigation). In the past, these were mostly isolated and an exchange was hindered through the high level of complexity and context dependence (the Brazilian Amazon cannot be compared to a wetland in the US or the African savannah).

Now Bush Broker in Australia, compensation agencies in Germany, and Wetland Mitigation Banks in the United States (to name only some of the most common) all refer to one common point of reference: the concept of biodiversity offsets.

Biodiversity offsets have been increasingly promoted and their number continues to grow (Calvet et al. 2015; Evans 2015; Maron and Gordon 2015; Miller et al. 2015; Maron et al. 2010, 2015a; Curran 2014; Evans et al. 2013; Brown et al. 2013). "Billions of dollars are spent each year on planning and implementing offsets" (Maron et al. 2015a).

They have been translated into different contexts worldwide (Darbi et al. 2010). In countries with existing offsetting schemes this has opened up a new global perspective on biodiversity offsets, e.g., in the United States and Australia (cf. Evans et al. 2013; Maron et al. 2015a–c). It has also helped to foster enforcement, e.g., in France (cf. Quétier et al. 2014; Morandeau and Vilaysack 2012). In other countries, the use of biodiversity offsets is being trialed, ranging from single pilots to whole campaigns to introduce offsets as a legally binding instrument, e.g., in the United Kingdom (see Sect. 4.1.3, cf. Defra 2011, 2013; Dyck et al. 2013; The Natural Capital Initiative 2010).

Voluntary Biodiversity Offsets/Emergence of the Business Case for Biodiversity Offsets

Biodiversity offsets "have arisen from regulatory requirements in a number of countries (e.g. USA, Brazil and in Europe)" (Doswald et al. 2012; cf. Wende et al. 2018a). These regulatory schemes, e.g., Environmental Impact Assessment, US Wetland Mitigation and other (see Sect. 4.1, Darbi et al. 2010), have served as a source of inspiration for the development of new toolkits and standards for the "the business case" for biodiversity offsets (Doswald et al. 2012), for example, by the Business and Biodiversity Offsets Program (BBOP 2010, 2012a–d, 2018), the International Council on Mining and Metals (ICMM 2005a, b) and the International Finance Corporation (IFC 2012). This resulted in the proliferation and promotion of "voluntary offsets" as opposed to the existing "regulatory biodiversity offsets" (cf. Benabou 2014; Doswald et al. 2012; Darbi 2010; Bull et al. 2013).

> A growing number of businesses, including financial institutions, are undertaking voluntary biodiversity offsets as part of their environmental or risk management strategies [...] the motivations for financial institutions to engage are clear, demonstrating leadership, gaining an ability to influence the regulatory process, reducing operational risk exposure, and taking advantage of new business development opportunities. (Doswald et al. 2012)

A reason for businesses and governments to get involved in this process was growing public pressure to manage the environmental impact of their activities. Howard (2007) stresses that environmental management has become an indispensable task—not only for "global players". At the same time biodiversity offsets provide a chance or a step toward developing a broader "biodiversity market": "the business case for biodiversity offsets" (cf. BBOP n.d., Benabou 2014, see Sect. 4.2).

> Central to the actors' concerns was what to do in countries where offsets are not legally mandated—how to build up voluntary offsets robust enough to save companies from potential adverse campaigning by local communities and national and international NGOs for poor biodiversity management (ten Kate et al. 2004). It thus took a decade or so (from the mid 1990s to the mid-2000s) for this conservation instrument to become institutionalized in the rapidly growing field of "business and biodiversity", through a triple process of (1) alignment of corporate executives, senior actors in conservation organizations, and bureaucrats, (2) articulation through multiple "dialogues" and publications, and (3) orchestration in major conservation meetings. (Benabou 2014)

This has been promoted for the first time on a global scale by the Business and Biodiversity Offsets Program of the international non-profit Forest Trends Association (Benabou 2014). The BBOP platform has been in place since 2004. It has engaged business, NGOs, administration and academia in discussing various aspects of biodiversity offsets (most prominently the mitigation hierarchy and the appropriateness of biodiversity offsets, including the "No Go" option for development), set up a community of practice and most importantly developed quality standards and an implementation toolkit for biodiversity offsets and tested these on biodiversity offset pilots with business partners around the world (e.g., Rio Tinto, Anglo American, and Newmont) (BBOP 2009a, b).

What Are the Implications for the Present Study?
Biodiversity offsets are the last step of a sequence, called the "Mitigation Hierarchy" (first avoid, then minimize, and then finally restore/offset negative impacts). They are understood as "measurable conservation outcomes" that are designed to counterbalance the "significant residual adverse biodiversity impacts" from planned projects or developments, e.g. infrastructure, housing or land use change. Thus, offsets imply an exchange of a biodiversity loss in one location with a biodiversity gain in another location.

While biodiversity offsets have risen from regulatory requirements in a number of countries, their increasing popularity is mainly bound to a new trend toward the promotion of voluntary biodiversity offsets as opposed to the existing regulatory biodiversity offsets, which has started in the early 2000s. Fostered at a global scale in particular by the Business and Biodiversity Offsets Program (BBOP), this implied the establishment of the "business case for biodiversity offsets," based on the argumentation that the negative impacts from a business' operations present a business risk and in reaction to this, biodiversity offsets can turn it into a business opportunity, e.g., in terms of license to operate and improved stakeholder relations.

1.3 Problems of Biodiversity Offsets, Research Gaps, and Problem Statement

Biodiversity Offsets on the Research Agenda
Biodiversity offsets are increasingly becoming the subject of research projects, masters courses, and dissertations. They are no longer only presented in gray literature (e.g., ten Kate et al. 2004; ten Kate 2005; Bishop et al. 2006; BBOP 2007; Treweek et al. 2009), but feature prominently in scholarly literature, including monographs (e.g., Bayon et al. 2012; Darbi et al. 2010; Levrel et al. 2015; Wende et al. 2018a), book chapters (e.g., Quétier et al. 2015b, c) and numerous articles in peer-reviewed journals (e.g., Maron et al. 2015a, b; Miller et al. 2015; Villarroya et al. 2014; Rainey et al. 2015; Quétier et al. 2014; Albrecht et al. 2014; Bull et al. 2013; Gardner et al. 2013; Pilgrim et al. 2013; Maron et al. 2012; Quétier and Lavorel 2011; Kiesecker et al. 2015; McKenney and Kiesecker 2010; Bekessy et al. 2010; Moilanen et al. 2009; Norton 2009; Burgin 2008).

From 2008 on, a growing number of research projects emerged to examine "what is out there on the ground" (e.g., Madsen et al. 2010, 2011; Darbi et al. 2010; Tucker et al. 2013; eftec, IEEP et al. 2010; PricewaterhouseCoopers LLP 2010; Morandeau and Vilaysack 2012; Escorcio Bezerra 2007; Lopez Arbeláez and Quintero Sagre 2015; ICMM IUCN 2012; ICH GHK, Bio Intelligence Service 2013; Treweek et al. 2009; Enetjärn et al. 2015; Doswald et al. 2012). This exploration phase is ongoing.

Research and practical work thus far has focused on two aspects: (1) the legitimation and setting of boundaries of appropriateness for biodiversity offsets, including how Biodiversity Offsets fit into the mitigation hierarchy (Miller et al. 2015; Evans 2015; Herbert 2015) and (2) the technical details of the implementation of biodiversity offsets (getting implementation right, methods and standards, metrics, loss–gain calculation, timing, habitat banking, etc.) (e.g. Bull 2015; ICH GHK, Bio Intelligence Service 2013; Treweek et al. 2009; IUCN 2014; Aiama et al. 2015; Evans et al. 2013; Bull et al. 2013; Gibbons et al. 2015; Bruggeman et al. 2005; Moilanen et al. 2009; Parkes et al. 2003).

Doswald et al. (2012) note that "[d]espite the experience and benefits generated, engaging in offset activities raises a number of challenges". Bull et al. (2013) list several of these theoretical challenges associated with biodiversity offsets that have been discussed in the literature as shown in Table 1.1, e.g., with regard to different methodologies (currency and equivalences) and temporal aspects (longevity and time lag).

Notably, the concept of No Net Loss remains a core challenge in terms of whether it requires restoration measures to achieve its goals and whether both restoration and protection measures are admissible. However, preference for the former or the latter depends on the specific context and conditions and the related frame of reference or counterfactual. While Bull et al. (2013, see Fig. 1.2) recommend measuring No Net Loss against a dynamic baseline (i.e., in practice often a negative counterfactual assuming ongoing future biodiversity decline) that allows for protection measures. By contrast, when measuring No Net Loss against a static baseline, only restoration measures would qualify to reach the No Net Loss goal. It is thus crucial to always be very clear against which baseline No Net Loss is measured.

Furthermore, the outcome and effectiveness of biodiversity offsets to reach No Net Loss is contested, e.g., Garrard (2015) notes that "[a]n increasing body of evidence suggests that biodiversity offset policies will struggle to achieve goals of no net loss, let alone net gain."

"How much compensation is enough?" asks Michael Curran (2014). Bull and Brownlie (2015) add that in fact "[t]he transition from No Net Loss to a Net Gain of biodiversity is far from trivial." This appraisal is backed by four arguments: "(1) the two principles represent different underlying conservation philosophies; (2) ecological uncertainties make it difficult to know where the threshold between No Net Loss and Net Gain lies; (3) different frames of reference are more or less appropriate in evaluating the ecological outcomes, depending on the principle chosen; and (4) stakeholder expectations differ considerably under the two principles" (Bull and Brownlie 2015).

In fact, several authors (Brown et al., 2013; Quigley and Harper 2006; Evans et al. 2013) have observed poor performance of offsets and a lack of transparency and evaluation (Miller et al. 2015; Edgar et al. 2005; Madsen et al. 2010; Evans et al. 2013). From this arises the need to evaluate the outcome of biodiversity offsets in their specific policy context (Evans 2015; Evans et al. 2013).

Table 1.1 A summary of the main theoretical challenges for biodiversity offsets (including design recommendations)

Problem	Description	Relevant research	Design recommendations
Currency	Choosing metrics for measuring biodiversity	Humphries et al. (1995); Salzman and Ruhl (2000); Godden and Vernon (2003); McCarthy et al. (2004); Burgin (2008); Lipton et al. (2008); BBOP (2009a); Norton (2009); Walker et al. (2009); McKenney and Kiesecker (2010); Temple et al. (2010); Treweek et al. (2010)	Use multiple or compound metrics; incorporate measure of ecological function as well as biodiversity
No net loss	Defining requirements for demonstrating no net loss of biodiversity	Gibbons and Lindenmayer (2007); BBOP (2009a); Gorrod and Keith (2009); Bekessy et al. (2010); McKenney and Kiesecker (2010); Gordon et al. (2011)	Measure no net loss against a dynamic baseline, incorporating trends; state whether no net loss is at project or landscape level; consider discounting rate (e.g., Dunford et al. 2004)
Equivalence	Demonstrating equivalence between biodiversity losses and gains	Godden and Vernon (2003); Bruggeman et al. (2005, 2009); Gibbons and Lindenmayer (2007); Lipton et al. (2008); Norton (2009); McKenney and Kiesecker (2010); Burrows et al. (2011); Quétier and Lavorel (2011)	Do not allow "out of kind" trading unless "trading up" from losses that have little or no conservation value
Longevity	Defining how long offset schemes should endure	Morris et al. (2006); Gibbons and Lindenmayer (2007); BBOP (2009a); McKenney and Kiesecker (2010); Pouzols et al. (2012)	Offsets should last at least as long as the impacts of development; offsets should be adaptively managed for change
Time lag	Deciding whether to allow a temporal gap between development and offset gains	Morris et al. (2006); Gibbons and Lindenmayer (2007); Moilanen et al. (2009); Norton (2009); Bekessy et al. (2010); McKenney and Kiesecker (2010); Drechsler and Hartig (2011); Gordon et al. (2011); Maron et al. (2012)	Require offsets to be delivered through biodiversity banking mechanisms

(continued)

Table 1.1 (continued)

Problem	Description	Relevant research	Design recommendations
Uncertainty	Managing for uncertainties throughout the offset process	Fox and Nino-Murcia (2005); Moilanen et al. (2009); Norton (2009); Treweek et al. (2010); Maron et al. (2012); Pouzols et al. (2012)	Development of a framework for uncertainty in offsets is a research requirement
Reversibility	Defining how reversible development impacts must be	Godden and Vernon (2003); BBOP (2012a)	Define reversibility; require all biodiversity losses to be reversible
Thresholds	Defining threshold biodiversity values beyond which offsets are not acceptable	Morris et al. (2006); Gibbons and Lindenmayer (2007); BBOP (2009a, 2012a); Norton (2009)	Define explicit thresholds for impacts that cannot be offset

Source: Bull et al. (2013)

Framing Biodiversity Offsets as Part of a "Neoliberal Conservation" Paradigm
As can be concluded from the above, most research centers around the question of applicability: In which contexts? Under which conditions? Which limits exist? But this does not cover the more fundamental question as to whether biodiversity offsets as such are acceptable for sustainable development.

The question of appropriateness is now increasingly being brought up in academic debate (cf. Sect. 9.1) and " [t]here have been prolific, widespread and valid criticisms of the theory, methodology and mode of implementation of offsets by both natural and social scientists in recent years" (Benabou 2014; cf. Bekessy et al. 2010; Garrard et al. 2015; Walker et al. 2009; Sullivan 2013; Robertson 2004; Moilanen et al. 2009; Maron et al. 2012; Curran 2014; Gibbons and Lindenmayer 2007; Spash 2015). Hence, it is not only an issue of technical feasibility but also of moral responsibility, of how we value and see our environment. Does the support of biodiversity offsets per se imply an anthropocentric "pro development" judgment that will inevitably contribute to biodiversity decline rather than halting the loss?

> Contemporary debates on appropriate biodiversity conservation policy have seen the emergence and consolidation of a "neoliberal conservation" paradigm (Igoe and Brockington 2007; Igoe et al. 2010; Büscher et al. 2012) based on the premise that the target of halting biodiversity loss will only be achieved by subordinating environmental conservation to capitalist logics and associated market mechanisms, for instance, through commodification and trading of ecosystem services (McAfee 2012). Biodiversity offsets, I suggest, draw much of their legitimacy from their recent alignment with this dominant discourse on sustainability. (Benabou 2014)

Construing biodiversity offsets from a market framing and economic logic has led to the assumption that offsets are perverted in that they depend on substantial (construction) development and nature lost (Benabou 2014). This means

Fig. 1.6 Putting a price on nature. Source: Second forum on the natural commons 2014

biodiversity offsets might be used as a "logic to legitimise, rather than prevent, ongoing habitat destruction [and] ... a means of commodifying habitat for exchange" (Spash 2015).

Moreover, this claimed "commodification of nature" (see Fig. 1.6) has also risen to a growing public controversy and concerns about biodiversity offsets (Monbiot 2012, 2014; Dearden 2013; Bull et al. 2013; Benabou 2014; Darbi 2014b).

The "license to trash" criticism, i.e., allowing the continuation of projects at the potential expense of biodiversity (Howarth 2013; Mathiesen 2013; Monbiot 2012) has resulted in various writings (e.g., by George Monbiot in his column in the Guardian, see Monbiot 2012, 2014), events (e.g., Forum on the Natural Commons, see Verpoest 2014) and campaigning by several nature conservation NGOs, e.g., FERN and Friends of the Earth (FERN 2014a, b; FERN and Friends of the Earth 2014), and a recent mockumentary[1] (Counter balance 2014).

On the other side, proponents for biodiversity offsets argue that "by putting a price on ecosystems, developers will be discouraged to use a site that involves high

[1]A fictional film done in a parodic documentary style.

biodiversity compensation costs. Should work go ahead nonetheless, the developers will have to finance conservation" (Benabou 2014).

In placing biodiversity offsets in this line of argumentation (i.e., valuing or commodifying nature as natural capital), the distinction between opponents and proponents in the current discussion comes down to a question of framing: There are on the one side the opponents (e.g., Spash 2015; Monbiot 2012, 2014; O'Neill 2014), who argue from an ethical and moral (or even philosophical) perspective that biodiversity offsets cannot be measured and valued in economic terms and by its very nature is irreplaceable and thus not offsettable. On the other side the proponents of biodiversity take a more pragmatic view claiming that biodiversity offsets are inevitable whenever development takes place (e.g., Scharks and Masuda 2016; Quétier et al. 2015a; Edwards 2015; The Wildlife Trusts n.d.).

However, this debate reduces biodiversity offsets to the question of monetization of biodiversity and misses a more fundamental point: the question of the admissibility of projects that generate negative environmental impacts and the responsibility to mitigate these impacts (polluter pays principle).

Research Gaps and Problem Statement of the Study

Biodiversity offsets have matured as regards their scope, definition, and growing use worldwide. The concept of biodiversity offsets has been refined particularly in terms of technical and methodological quality standards (e.g., Evans et al. 2013; Gibbons et al. 2015; Bull et al. 2013, 2014; Parkes et al. 2003; Bruggeman et al. 2005; Moilanen et al. 2009).

However, as has been noted, a number of critical points persist from which research demand derives. This refers mainly to the lack of a sound evidence base of documented offset cases from around the world and the evaluation of the outcomes of biodiversity offsets in light of the different contexts under which they arise (Maron et al. 2010; Edgar et al. 2005; Madsen et al. 2010; Miller et al. 2015; Evans 2015; Evans et al. 2013; Curran et al. 2014).

This differentiated, context-dependent evaluation (i.e., evaluation of biodiversity offsets in their specific policy and market context as well as societal anchorage, cf. Evans 2015) is of particular importance because "[t]here is no 'standard' approach to offsetting, and many different approaches have been developed in multiple jurisdictions" (Miller et al. 2015).

> Biodiversity offsetting is a simple phrase for what in reality is a very complex idea (The Wildlife Trusts n.d.).

Given not only the complexity of biodiversity offsets as a concept in light of achieving the goal of no net loss/net gain, but also the plurality of their implementation, a generalized judgment of biodiversity offsets seems inappropriate, that is, it is necessary to take into consideration the differences in offsets between the developing and developed world, pristine and transformed (natural and cultural) landscapes, varying legal and policy traditions and anchorage as well as different policy discourses. Research that takes into account this complexity and these differences is still rare. Maron et al. (2015b) note the necessity of counterfactuals for biodiversity offsets, while Pilgrim and Bennun (2014) and Maron et al. (2015c) examine the relation of offsets and protected areas. Evans (2015), Evans et al. (2013), and McKenney and Kiesecker (2010) focus on policy frameworks and policy evaluations of offsets.

The growing controversy on biodiversity offsets has sometimes led to unqualified and unjustified generalizations in that the failure or shortcomings of single offset cases have been scaled up to a failure or shortcomings of offsets in general, e.g., the suggested offsets in Smithy Wood and Alvecote Wood in the United Kingdom, where the planned destruction of forests (i.e., old-growth habitat) was considered in light of offsetting possibilities and in response have been instrumentalized to demonstrate that biodiversity offsetting "is a flawed concept" (Walters 2014) and enables what "might have been unthinkable until recently", i.e., "[i]f you trash a piece of land here you can replace its value by creating some habitat elsewhere" (Monbiot 2014).

Much of the concerns have risen from what has evolved as "the business case for biodiversity offsets" and scrutiny exists as to why a business would voluntarily implement offsets (Benabou 2014; BBOP n.d.; Howard 2007). Thus, both the outcomes and motivations for voluntary biodiversity offsets are being questioned (Benabou 2014).

Despite this scrutiny, a detailed analysis of voluntary biodiversity offsets (or "the business case") beyond "various sets of narrative strategies that serve the business and biodiversity community in justifying their actions and legitimizing their views on the biodiversity agenda" is lacking (Benabou 2014). This is problematic because the business case in itself (as opposed to mandatory/regulatory biodiversity offsets) is varied as a growing number of offset cases shows.

A result from the more implementation-oriented research on biodiversity offsets, attempts to generate a typology of biodiversity offsets focused on procedural aspects of the implementation, i.e., the distinction between "single" biodiversity offsets, "aggregated" biodiversity offsets and "banking" of offsets (Brownlie et al. 2009).

> [V]ery little has been said in the academic literature about what biodiversity offsetting could cover in such voluntary situations. In what context and on what grounds have these corporate commitments emerged? How has the concept of biodiversity offsets been configured and circulated to gain traction among corporate leaders? Why, despite these developments, does the implementation of voluntary biodiversity offsets on the ground remain limited? (Benabou 2014)

In this regard, Benabou (2014) presents the first comprehensive analysis of voluntary biodiversity offsets in terms of their development, context, involved parties, and drivers. She suggests a classification according to four different drivers:

1. Regulation
2. Finance
3. Business risk management
4. Business opportunities

While this is a valuable first attempt, it does not go far enough and provides the potential for even further differentiation. In particular, the following aspects and problems make a case for further exploration and structuring of voluntary biodiversity offsets (in terms of a typology, underpinned by practical evidence):

- The discussion about voluntary vs. mandatory offsets has risen to prominence with the consultation of the European Commission on its (initially) planned no net loss initiative (see Sect. 4.1.3) and the envisaged introduction of a mandatory scheme (i.e., through an EU directive or regulation). While the scope of the no net loss initiative included consideration of different aspects (e.g., which drivers of biodiversity loss need to be addressed primarily) and potential tools, the discussion centered very much around the question whether biodiversity offsets shall be mandated by law or carried out on a voluntary basis. Due to concerns over the appropriateness of biodiversity offsets and the fear that these might increase habitat destruction "a significant number of respondents were against the inclusion of offsetting/compensation in any future EU No Net Loss initiative and that the majority of respondents were against a binding EU legislation on offsetting/compensation" (European Commission 2015c). In contrast to these generalized fears, due to the different levels of control and the possibility of sanctions in voluntary biodiversity offsets, many experts now see a strong regulatory system (i.e., mandatory offsets) as the preferred way forward (Albrecht et al. 2014; Tucker et al. 2013; Herbert 2015; Darbi et al. 2016; BBOP 2014). There are clearly many discrepancies and open questions in the debate about mandatory vs. voluntary biodiversity offsets, which may only be solved based on a deeper understanding (in terms of definition, scope, drivers, involved parties, governance, and implementation) of each of them as well as their relation and distinctiveness toward each other.
- As has been noted, offset schemes are increasingly emerging in different contexts worldwide. This involves numerous questions, in particular, regarding the role of government: Is there a need to regulate? To provide guidance? Should government be actively involved or leave offsets to the business sector?
- In addition to the role of government, a number of key actors are involved in offsetting, e.g., a trend can be observed of nature conservation NGOs partnering with business (Benabou 2014) as well as the rising influence of financial institutes, e.g., the World Bank or the International Finance Corporation (Benabou 2014). Examining the role of these different actors can help to analyze the drivers and context conditions of biodiversity offset implementation (e.g., is there a sort of "new global players" or governance schemes replacing or complementing government schemes?).
- As has been discussed above, on the one hand, there is growing controversy about biodiversity offsets that extends to the public sphere. On the other hand, voluntary biodiversity offsets clearly target local stakeholders in terms of "license to operate" (ten Kate et al. 2004; Howard 2007; Benabou 2014). It is thus crucial to analyze (voluntary) biodiversity offsets with regard to their societal embedding

and the role of (local) stakeholders—both from the perspective of business and the local population.

- The increasingly international exchange of biodiversity offset evidence and experience clearly presents a source for learning and improvement for practitioners. However, in light of the mentioned complexity and variety of offsetting practices, a refined typology might help increase transparency, facilitate comparability, identify similar mechanisms, and typical pitfalls that could be avoided, etc.
- For academia, undoubtedly, offsets are a prospering new research field. Increasingly there is a need to find substructures for this field to further refine research on this topic and to build theory and generalizations from the growing evidence base. This can then also provide the base for practice and policy advice and recommendations in view of best practices and standards.
- Summarizing, a refined typology of biodiversity offsets helps to structure and classify the increasingly emerging evidence for a broad number of stakeholders (practice, policy, research, wider society, business, etc.).

These described problem settings illustrate that a more sophisticated description of biodiversity offsets in terms of a classification or typology can increase transparency of offset evidence worldwide and add to an informed debate that goes beyond nonconstructive oversimplification or polemics of "offsets are good" vs. "offsets are bad" (Evans 2014) and help to inform the topical discussion of regulatory vs. voluntary biodiversity offsets.

Building on the aforementioned discussion of the challenges of biodiversity offsets and associated research gaps, Table 1.2 summarizes some (not exclusive) research issues and highlights where the need for further research persists. Issues in bold will be approached by this study (not exhaustive, see Sect. 1.4). The focus lies on no. 5, Mandatory vs. Voluntary Biodiversity Offsets.

Table 1.2 Summary of identified research gaps/need for further research

No.	Research issues	Research gaps/need for further research
1.	Concept of "No Net Loss" (NNL)	• Examine the complexity of NNL: Under which premises? How does it work? What are the limits? • Analyze the controversy of NNL: What are the reasons for and against NNL? Who are the stakeholders? What are their roles? • Clarity concerning the frames of reference/ counterfactuals against which NNL is measured. • Transition from no net loss to net gain.
2.	Concept of "biodiversity offsets" (BO)	• Further develop definition and characterization of BO: What are the criteria for BO? How can the outcome of BO be measured/evaluated? • Analyze the (societal) debate on the appropriateness of the concept of BO: What are the risks and challenges? What are the opportunities? What are the alternatives?

(continued)

Table 1.2 (continued)

No.	Research issues	Research gaps/need for further research
3.	**Drivers of biodiversity offsets**	• What are the driving forces? • What are the constraints? • Analyze the role of different actors: Who are the actors? What is their role? Which motivations?
4.	**Context dependence of biodiversity offsets**	• Examine different levels of context dependence, from concept (BO schemes) to site level (every BO is unique) • Description and comparison of offset schemes with regard to the specific context: What are the natural features? What are the political, economic, and sociocultural conditions?
5.	**Mandatory vs. voluntary biodiversity offsets**	• Disambiguation: What do "voluntary" and "mandatory" mean in the context of BO? • Examine the practical feasibility of this distinction. • Comparison and critical analysis of the outcomes.
6.	Biodiversity offsets as market-based instruments	• Comparison and critical analysis of market- and nonmarket-based (natural compensation) approaches to implement BO: What are the outcomes? What are the limits? What are the advantages? What are the risks? • Examine the relation of markets and regulation: How much regulation is possible/needed? What are the limits to government control?
7.	Business case for biodiversity offsets	• Examine incentives and motivations for the business case • Evaluate the outcomes for business and nature/ biodiversity
8.	Biodiversity offsets and EIA	• At a worldwide scale (and with to some extent scattered BO schemes) Impact Assessment is often a vehicle for BO. • Examine the role of EIA for BO.

Source: Author

What Are the Implications for the Present Study?

As biodiversity offsets have matured in terms of scope, definition, and growing uptake worldwide, they have also become a prospering new research field. A number of challenges associated with them have been widely discussed in the literature, in particular regarding technical and methodological quality standards.

Nevertheless, a number of critical points persist from which further research demand derives. This refers particularly to the evaluation of the outcomes of biodiversity offsets in light of their complexity and different contexts and conditions under which they arise. Seen the (growing) plurality of their implementation, a generalized judgment of biodiversity offsets seems no longer appropriate.

(continued)

This calls for a more sophisticated description of biodiversity offsets in terms of classification or typology, with special emphasis on the growing controversy about voluntary vs. mandatory biodiversity offsets. In particular, the scrutiny on businesses implementing voluntary biodiversity offsets, the role of government in newly emerging offsets schemes, the discussion on the envisaged introduction of a mandatory offset scheme in the scope of the EU no net loss initiative, and the implications and lessons learnt for practitioners/consultancies need be examined.

1.4 Aim and Research Design

Premises and Aim of the Study

As has been said, the implementation and outcome of biodiversity offsets are highly dependent on the conditions and context. Worldwide very different approaches to compensation for impacts on biological diversity exist. These primarily differ with regard to the basic conditions or general framework for the implementation of biodiversity offsets, notably:

1. The legal framework, i.e., whether an obligation is fixed in the law that negative impairments have to be compensated for.
2. The enforcement and control of these legal provisions.
3. The kind of development and related impacts, in particular with regard to intensity and duration.
4. The affected ecosystem, i.e., its characteristic, rarity, and vulnerability.
5. The different stakeholders/developers and their motivations and corporate policies (local business, global corporations, voluntary negotiated agreements).
6. The sector or line of business of the developer (e.g., mining, settlement development, infrastructure, and transportation) and associated standards, current/best practices, etc.
7. The social context, i.e., the affected population with its requirements and needs (e.g., by becoming active in negotiation processes).

These context factors (not exhaustive) illustrate the complexity of biodiversity offsets in practice. As a result, generalizations (of what biodiversity offsets are and how they are to be applied) fall short of being applicable in practice and thus can only serve as a very restricted theoretical model for the understanding of biodiversity offsets.

Therefore, the present study aims to build a more elaborate structuring and description of the concept of biodiversity offsets (obviously this also cannot be exhaustive). This is done by suggesting a typology of biodiversity offsets with regard to the degree of voluntariness of the implementation.

Voluntariness was chosen because it is inherent to the majority of the above-named conditions (no. 1, 2, 5–7). It is therefore assumed that voluntariness is a key factor that influences the implementation and outcome of biodiversity offsets. This is one of the five premises of this study (no. 4, see Fig. 1.8):

1. Biodiversity offsets are important in practice because a growing number are being implemented.
2. Biodiversity offsets are gaining importance as a field of research.
3. Biodiversity offsets are complex and context dependent: they are shaped by many factors.
4. Voluntariness is an important factor in the implementation of biodiversity offsets.
5. So far, mandatory and voluntary biodiversity offsets have been distinguished.

Up to now, these two types of offsets have been used mostly literally (see Chap. 3, Voluntariness) and not further underpinned. However, as previous research in the scope of this work has found, the distinction of two types of biodiversity offsets (mandatory vs. voluntary) is too simplistic (cf. Benabou 2014).

If one looks at mandatory and voluntary biodiversity offsets as antipodes, a cursory examination of the field makes it obvious that this "black-or-white" distinction does not work. Biodiversity offsets are far more complex than that, as will be shown in Chaps. 5 and 6.

First, the literal meaning of mandatory vs. voluntary needs to be clarified. From a scientific point of view, this dichotomy is not precise, because neither "mandatory" nor "voluntary" biodiversity offsets have been defined properly and the terms "voluntary" and "mandatory" have many connotations. As a result, in practice it is hardly feasible to draw a strict line between mandatory and voluntary offsets. Thus, from a practical point of view, the dichotomy is not applicable, because the limits between both have become blurred: It can be observed that "voluntary" (colloquially referred to in the absence of law) biodiversity offsets are not seldom driven by some kind of obligation or incentive. And "mandatory" biodiversity offsets (while theoretically being compulsory by law) are not always enforced and properly implemented into practice.

This leads to the second observation that mandatory and voluntary offsets are not two distinct categories but represent the two ends of a continuum. All offset cases range in between.

This makes clear that instead of the distinction of two classes of biodiversity offsets (mandatory vs. voluntary) a more refined typology is needed that takes into account the complexity of biodiversity.

Therefore, the aim of this study is to develop a refined typology to structure and better understand the conditions for the implementation of biodiversity offsets (see Fig. 1.7).

From a scientific point of view the typology and systematization shall help to structure and understand the complexity of biodiversity offsets. From a practical

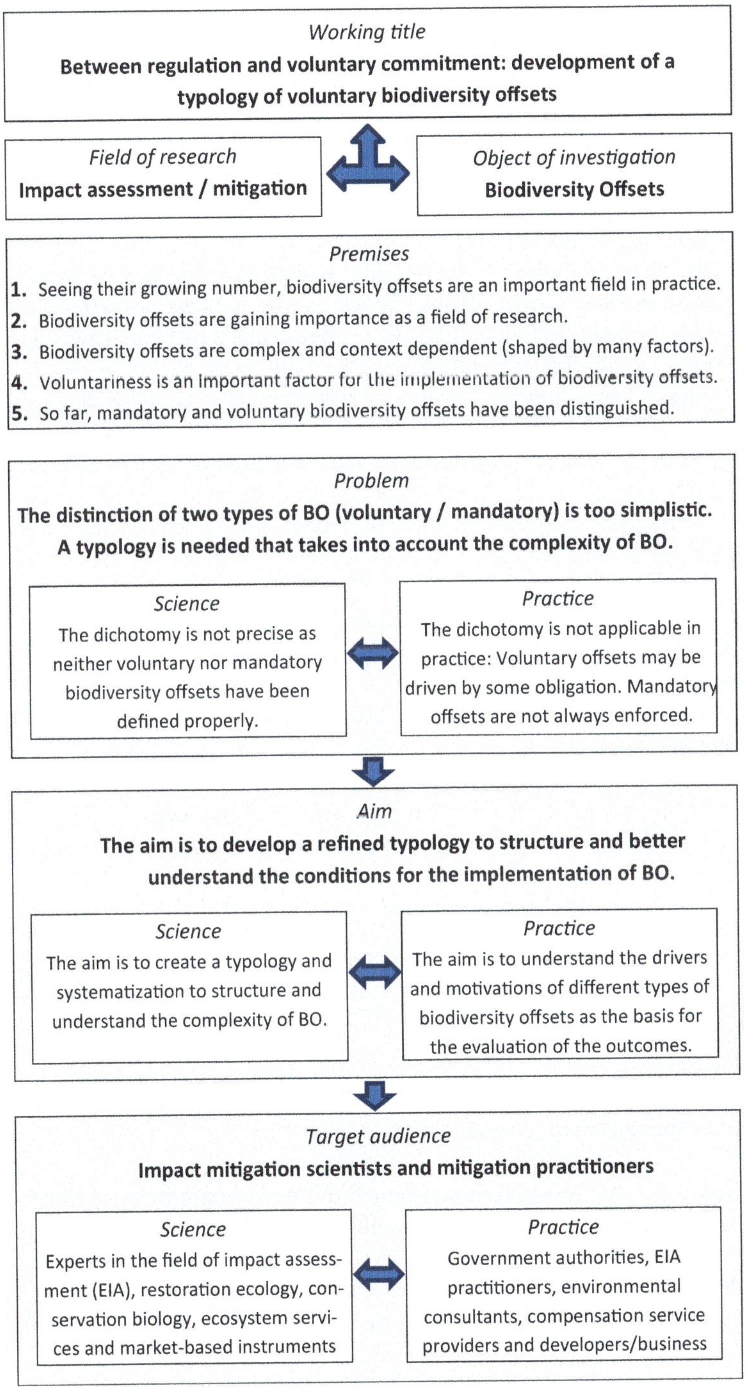

Fig. 1.7 Premises, problem, aim, and target audience. Source: Author

point of view, the aim is to understand the drivers and motivations of different types of biodiversity offsets as the basis for the evaluation of the outcomes.

The target audience (see Fig. 1.7) are on the one hand impact mitigation scientists, including experts in the field of impact assessment (EIA), restoration ecology, conservation biology, ecosystem services, and market-based instruments. On the other hand, mitigation practitioners are addressed, including government authorities, EIA practitioners, environmental consultants, compensation service providers, and developers/business interests.

The present work builds on the hypothesis that differences exist regarding the voluntariness of biodiversity offsets. Types can be built to understand how this influences the outcome of biodiversity offsets. From this the research question arises: Which **types** of **biodiversity offsets** can be built with regard to the degree of **voluntariness**? The following question of how the types influence the outcome of biodiversity offsets is not subject to this study. It could be the starting point of another research with a stronger focus on the evaluation of offsets, their effectiveness, and efficiency.

From the research question, a number of hypotheses are derived that are explored throughout this study:

1. Differences exist regarding the voluntariness of biodiversity offsets.
2. Biodiversity offsets cannot be adequately explained as a dichotomy of mandatory vs. voluntary offsets.
3. The voluntariness of biodiversity offsets can be described as a gradual continuum.
4. A typology of biodiversity offsets (and different types) can be built to analyze and illustrate the space between the two poles of this continuum.
5. These types help to analyze and understand the different outcomes of biodiversity offsets.

Research Design of the Study

In both the working title and the research question three issues are highlighted:

- Biodiversity offsets.
- Voluntariness.
- Methodological considerations: including (1) an internet and expert based research approach and (2) typification.

Consequently, the research design builds on these three pillars as illustrated in Fig. 1.8. These research issues pass through three methodological complexes:

- Establishment of the theoretical and methodological baseline
- Empirical development and analysis of the typology of voluntary biodiversity offsets (theoretically grounded)
- Discussion and conclusions of the typology, outlook

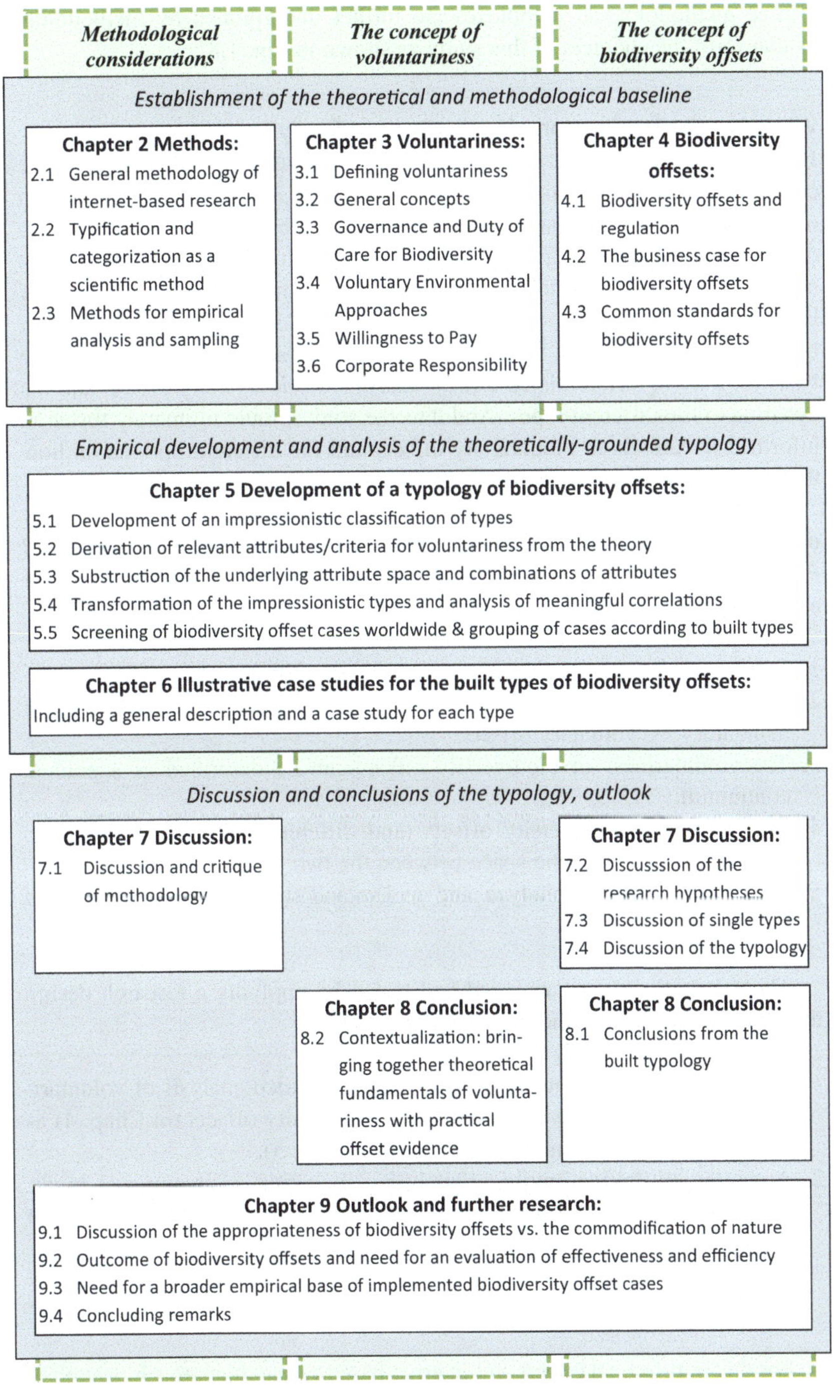

Fig. 1.8 Research design and structure. Source: Author

The three methodological complexes are further underpinned by methodological steps that form the chapters of this study as shown in Fig. 1.8.

What Are the Implications for the Present Study?

Building on the problem setting described, in particular with regard to the complexity and controversy of biodiversity offsets, as well as the discussion on mandatory vs. voluntary biodiversity offsets, the aim of this study is to develop a refined typology with regard to the voluntariness of biodiversity offsets. This shall help to structure and better understand the concept of biodiversity offsets as such and the conditions for the implementation of biodiversity offsets. As a result, the findings of this study shall provide the basis for a more differentiated critical analysis of biodiversity offsets and the outcome of the different types. And thus the study should ultimately foster an informed debate about biodiversity offsets and the different options on how they can be delivered in practice.

The study examines the research question: **Which types of biodiversity offsets can be determined with regard to the degree of voluntariness?** From the research question, a number of hypotheses are derived that are explored throughout this study:

1. Differences exist regarding the voluntariness of biodiversity offsets.
2. Biodiversity offsets cannot be adequately explained as a dichotomy of mandatory vs. voluntary offsets.
3. The voluntariness of biodiversity offsets can be described as a gradual continuum.
4. A typology of biodiversity offsets (and different types) can be built to analyze and illustrate the space between the two poles of this continuum.
5. These types help to analyze and understand the different outcomes of biodiversity offsets.

These hypotheses are analyzed and tested by applying a research design that consists of three methodological complexes:

1. A broader theoretical analysis—including a detailed analysis of voluntariness (in Chap. 3) and voluntariness of biodiversity offsets (in Chap. 4) as well as the development of a typology (in Chap. 5).
2. A shorter illustrative empirical analysis—including a screening of biodiversity offset cases (Chap. 5 and Appendix 1) and short illustrative case studies for the built types (Chap. 6).
3. The integration of both in the discussion and in the conclusion (Chaps. 7 and 8).

For this purpose, an expert and internet based research approach is applied (see Chap. 2 Methods and materials).

References

Aiama D et al (2015) No net loss and net positive impact approaches for biodiversity: exploring the potential application of these approaches in the commercial agriculture and forestry sectors. IUCN

Albrecht J, Schumacher J, Wende W (2014) The German impact-mitigation regulation—a model for the EU's no-net-loss strategy and biodiversity offsets? Environ Policy Law 44(3):317–325

Ambrose RF (2000) Wetland mitigation in the United States: Assessing the success of mitigation policies. In: Wetlands (Australia), vol 19, pp 1–27. Available at http://www.mitigationactionplan.gov/ambrose_wetlandmitigationinus.pdf

Bayon R, Carroll N, Fox J (2012) Conservation and biodiversity banking: a guide to setting up and running biodiversity credit trading systems. Earthscan

BBOP (2007) Biodiversity offsets: a tool for conservation and development. An open discussion. Available at http://www.forest-trends.org/biodiversityoffsetprogram/Presentations/BBOP5/June%2018/BBOP5_K_ten_Kate_summing_up%20_BBOP_LN.pdf

BBOP (2009a) Business, biodiversity offsets and BBOP: an overview. BBOP, Washington, DC. Available at http://www.forest-trends.org/biodiversityoffsetprogram/guidelines/overview.pdf

BBOP (2009b) Compensatory conservation case studies. BBOP, Washington, DC. Available at www.forest-trends.org/biodiversityoffsetprogram/guidelines/non-bbop-case-studies.pdf

BBOP (2010) Principles on biodiversity offsets supported by the BBOP Advisory Committee. Available at http://bbop.forest-trends.org/guidelines/principles.pdf

BBOP (2012a) Standard on biodiversity offsets. BBOP, Washington, DC. Available at http://bbop.forest-trends.org/guidelines/Standard.pdf

BBOP (2012b) Guidance notes to the standard on biodiversity offsets. BBOP, Washington, DC. Available at http://bbop.forest-trends.org/guidelines/Standard_Guidance_Notes.pdf

BBOP (2012c) Glossary. BBOP, Washington, DC. 2nd updated edition. Available at http://bbop.forest-trends.org/guidelines/Updated_Glossary

BBOP (2012d) Biodiversity offset design handbook-updated. BBOP, Washington, DC. Available at http://bbop.forest-trends.org/guidelines/Updated_ODH.pdf

BBOP (2014) To no net loss of biodiversity and beyond. A summary of discussions at the conference. Available at http://bbop.forest-trends.org/documents/files/bbop_nnl_conference_report.pdf

BBOP (2018) Working for biodiversity net gain: an overview of the Business and Biodiversity Offsets Programme (BBOP) 2004–2018. Forest Trends. Washington, DC. Available from https://www.forest-trends.org/bbop_pubs/overview2018

BBOP (n.d.) Biodiversity offsets: the business case. Available at http://bbop.forest-trends.org/pages/business_case

Bekessy SA, Wintle BA, Lindenmayer DB, McCarthy MA, Colyvan M, Burgman MA, Possingham HP (2010) The biodiversity bank cannot be a lending bank. Conserv Lett 3:151–158

Benabou S (2014) Making up for lost nature?: A critical review of the international development of voluntary biodiversity offsets. Environ Soc 5:103–123

Bishop J, Kapila S, Hicks F, Mitchell P (2006) Building biodiversity business: report of a scoping study. Available at http://www.globalnature.org/bausteine.net/f/6644/Building_Biodiversity_Business_-_Draft_Report_%28Oct_2006%29.pdf?fd=2

Brown M, Clarkson BD, Barton BJ, Joshi C (2013) Ecological compensation: an evaluation of regulatory compliance in New Zealand. Impact Assess Proj Apprais 31:34–44

Brownlie S, ten Kate K, Stephens T, Treweek J (2009) Aggregated biodiversity offsets: a preliminary evaluation of their potential and factors affecting success in implementation. Background paper to the BBOP Advisory Committee. Unpublished draft

Bruggeman D, Jones M, Lupi F, Scribner K (2005) Landscape equivalency analysis: methodology for estimating spatially explicit biodiversity credits. Environ Manag 36(4):518–534

Bruggeman DJ, Jones ML, Scribner KT, Lupi F (2009) Relating tradable credits for biodiversity to sustainability criteria in a dynamic landscape. Landsc Ecol 24:775–790

Bull JW (2015) Comparing biodiversity offset methodologies. Decision Point Online, 85/February 2015. Available at http://decision-point.com.au/?article=comparing-biodiversity-offset-methodologies

Bull JW, Brownlie S (2015) The transition from no net loss to a net gain of biodiversity is far from trivial. Oryx 51(1):53–59

Bull JW, Suttle KB, Gordon A, Singh NJ, Milner-Gulland EJ (2013) Biodiversity offsets in theory and practice. Oryx 47(03):369–380

Bull JW, Gordon A, Law EA, Suttle KB, Milner-Gulland EJ (2014) Importance of baseline specification in evaluating conservation interventions and achieving no net loss of biodiversity. Conserv Biol 28(3):799–809

Burgin S (2008) BioBanking: an environmental scientist's view of the role of biodiversity banking offsets in conservation. Biodiv Conserv 17(4):807–816

Burrows L, Butcher B, Treweek J (2011) Offsets for species in the UK planning context: a possible methodology. Practice 71:41–43

Büscher B, Sullivan S, Neves K, Igoe J, Brockington D (2012) Towards a synthesized critique of neoliberal biodiversity conservation. Capital Nat Social 23(2):4–30

Calvet C, Guillaume O, Claude N (2015) Tracking the origins and development of biodiversity offsetting in academic research and its implications for conservation: A review. Biol Conserv 192:492–503

CBD Secretariat (2006) Global biodiversity outlook 2

CBD Secretariat (n.d.) History of the convention. Available at https://www.cbd.int/history/default.shtml

Commission of the European Communities (2007) Green paper on market-based instruments for environment and related policy purposes. COM(2007) 140 final—Brussels, 28.3.2007

Counter balance (2014) Biodiversity offsetting, making dreams come true. Available at http://www.counter-balance.org/biodiversity-offsetting-making-dreams-come-true/

Cuperus R, Canters KJ, de Haes HAU, Friedman DS (1999) Guidelines for ecological compensation associated with highways. Biol Conserv 90(1):41–51

Cuperus R, Bakermans MM, De Haes HAU, Canters KJ (2001) Ecological compensation in Dutch highway planning. Environ Manag 27(1):75–89

Curran MP (2014) Compensating the biodiversity impacts of land use: toward ecologically equal exchange in the North-South context. Diss., Eidgenössische Technische Hochschule ETH Zürich, Nr. 21439

Curran M, Hellweg S, Beck J (2014) Is there any empirical support for biodiversity offset policy? Ecol Appl 24(4):617–632

Darbi M (2010) Biodiversity offsets—a tool for environmental management and biodiversity conservation. In: Sanchez Bengoa D, Powell D (eds) Top biodiversity 2010. Intercollege-Larnaca, Cyprus. Conference proceedings. Intercollege-Larnaca, Larnaca, S289–S301

Darbi M (2014a) Why did the concept of Biodiversity Offsets become so popular? Biodiversity Offsets Blog. Available at http://www.biodiversityoffsets.net/concept-biodiversity-offsets-become-popular/

Darbi M (2014b) Love or leave? The controversy about Biodiversity Offsets. Biodiversity Offsets Blog. Available at http://www.biodiversityoffsets.net/love-leave-controversy-biodiversity-offsets/

Darbi M, Ohlenburg H, Herberg A, Wende W (2010) Impact mitigation and biodiversity offsets—compensation approaches from around the world: a study on the application of Article 14 of the CBD (Convention on Biological Diversity). Naturschutz und Biologische Vielfalt 101. BfN-Schr.-Vertrieb im Landwirtschaftsverl.

Darbi M, Albrecht J, Schumacher J (2016) Die europäische No Net Loss-Initiative und Biodiversity Offsets. Einblicke in die aktuelle Diskussion zur Kompensation von Eingriffen in die Biodiversität aus deutscher Perspektive. In: Jahrbuch des Umwelt- und Technikrechts 2016

Dearden N (2013) Putting a price on nature would be disastrous. The Guardian. Available at http://www.theguardian.com/global-development/poverty-matters/2013/nov/27/price-nature-markets-natural-capital

Defra (2011) Biodiversity 2020: a strategy for England's wildlife and ecosystem services. Available at https://www.gov.uk/government/uploads/system/uploads/attachment_data/file/69446/pb13583-biodiversity-strategy-2020-111111.pdf

Defra (2013) Biodiversity offsetting in England Green paper. September 2013. Available at https://consult.defra.gov.uk/biodiversity/biodiversity_offsetting/supporting_documents/20130903Biodiversity%20offsetting%20green%20paper.pdf

De Oliveira Maciel MA, Filho JS (n.d.) SNUC—Sistema Nacional de Unidades de Conservação. Lei no. 9.985, de 18 de Julho de 2000. Available at http://www.soleis.com.br/L9985.htm

De Sadeleer N (2015) Polluter pays principle. In: Morin J-F, Orsini A (eds) Essential concepts of global environmental governance. Routledge, London, p 155

Doswald N, Barcellos Harris M, Jones M, Pilla E, Mulder I (2012) Biodiversity offsets: voluntary and compliance regimes. A review of existing schemes, initiatives and guidance for financial institutions. UNEP-WCMC, Cambridge, UK. UNEP FI, Geneva, Switzerland

Drechsler M, Hartig F (2011) Conserving biodiversity with tradable permits under changing conservation costs and habitat restoration time lags. Ecol Econ 70(3):533–541

Dunford RW, Ginn TC, Desvousges WH (2004) The use of habitat equivalency analysis in natural resource damage assessments. Ecol Econ 48:49–70

Dyck C, Feeney A, Haque A, Inglis G, Kondylas S (2013) Biodiversity offsets UK. BCM Class Blog. MSc biodiversity conservation & management. Available at https://mscbcmnsep.wordpress.com/2013/12/08/biodiversity-offsets-uk-2/

Edgar PW, Griffiths RA, Foster JP (2005) Evaluation of translocation as a tool for mitigating development threats to great crested newts (Triturus cristatus) in England, 1990–2001. Biol Conserv 122:45–52

Edwards S (2015) A balancing act. IUCN. Available at http://www.iucn.org/?18817. Accessed 24 Feb 2015

EEA (European Environment Agency) (2006) Using the market for cost-effective environmental policy. Market-based instruments in Europe. EEA Report no. 1/2006. Copenhagen

eftec, IEEP et al (2010) The use of market-based instruments for biodiversity protection—the case of habitat banking. Technical report. For European Commission DG Environment

Enetjärn A, Cole S, Kniivilä M, Hårklau SE, Hasselström L, Sigurdson T, Lindberg J (2015) Environmental compensation. Key conditions for increased and cost effective application

Escorcio Bezerra LG (2007) Biodiversity offsets in national (Brazil) and regional (EU) mandatory arrangements: towards an international regime? Available at http://www.forest-trends.org/biodiversityoffsetprogram/library/new/Dissertation%20Biodiversity%20Offsets%20LGB%20IUCN%20BBOP.doc

European Commission (2011) The EU biodiversity strategy to 2020, COM(2011) 244 endg.

European Commission (2012a) Draft mandate for the EU Working Group on No Net Loss of Ecosystems and their Services, unpublished draft

European Commission (2012b) Scoping document for the first meeting of the Working Group on No Net Loss of Ecosystems and their Services, unpublished draft

European Commission (2015a) The state of nature in the European Union, COM (2015) 219 final

European Commission (2015b) The mid-term review of the EU biodiversity strategy to 2020, COM (2015) 478 final

European Commission (2015c) No net loss of biodiversity public consultation. Summary overview of the responses. Available at: http://ec.europa.eu/environment/nature/biodiversity/nnl/pdf/report.pdf

Evans M (2014) Gina vs. the reef? A plea for informed debate on biodiversity offsets. Available at https://mcevansresearch.wordpress.com/2014/08/21/gina-vs-the-reef-a-plea-for-informed-debate-on-biodiversity-offsets/comment-page-1/#comment-1943

Evans M (2015) Offsetting in the context of policy. What happens to the theory when it hits the real world? Decision Point Online, 91/August 2015

Evans MC, Maron M, Gibbons P, Possingham HP (2013) A bird in the hand is worth two in the bush: ecological time preference and biodiversity offsets. In: 15th annual BIOECON conference, Kings College, Cambridge, United Kingdom. Available at bioecon-network.org/pages/15th_2013/EVANS.pdf. Accessed 24 Feb 2016

FERN (2014a) Briefing note 2: What is biodiversity offsetting and why is it problematic? Available at http://www.fern.org/sites/fern.org/files/Biodiversity2_EN.pdf. Accessed 24 Feb 2016

FERN (2014b) Briefing note 3: Biodiversity offsetting in practice. Available at http://www.fern.org/sites/fern.org/files/Biodiversity3_EN.pdf. Accessed 24 Feb 2016

FERN and Friends of the Earth (2014) Case studies of biodiversity offsetting: voices from the ground. Available at http://www.fern.org/sites/fern.org/files/Offset%20stories%20-%20Final.pdf

Forum on the natural commons (n.d.) Available at. http://naturenotforsale.org/

Fox J, Nino-Murcia A (2005) Status of species conservation banking in the United States. Conserv Biol 19:996–1007

Gardner T-A, von Hase A, Brownlie S, Ekstrom J-M-M, Pilgrim J-D, Savy C-D, Stephens R-T-T, Treweek J, Ussher G-T, Ward G, Ten Kate K (2013) Biodiversity offsets and the challenge of achieving no net loss. Conserv Biol 27(6):1254–1264

Garrard G et al (2015) Offset policies don't work. So maybe we should be weighing up the alternatives. Decision Point Online, 91/August 2015

Gibbons P, Lindenmayer DB (2007) Offsets for land clearing: No net loss or the tail wagging the dog? Ecol Manag Restor 8:26–31

Gibbons P, Evans MC, Maron M, Gordon A, Roux D, Hase A, Lindenmayer D, Possingham HP (2015) A loss-gain calculator for biodiversity offsets and the circumstances in which no net loss is feasible. Conserv Lett. https://doi.org/10.1111/conl.12206

Godden D, Vernon D (2003) Theoretical issues in using offsets for managing biodiversity. In: Annual conference, Australian Agricultural and Resource Economics Society, February 2003, Fremantle, Australia

Gordon A, Langford WT, Todd JA, White MD, Mullerworth DW (2011) Assessing the impacts of Biodiversity Offset Policies. Environ Model Softw 144:558–566

Gorrod EJ, Keith DA (2009) Observer variation in field assessments of vegetation condition: implications for biodiversity conservation. Ecol Manag Restor 10:31–40

Herbert M (2015) German impact mitigation regulation—national, European, international. A comparison from the viewpoint of the Federal Agency for Nature Conservation. UVP Rep 29 (3):149–151. Eingriffsregelung; Umweltverträglichkeitsprüfung; Strategische Umweltprüfung; Biologische Vielfalt

Howard K (2007) Voluntary biodiversity offsets: improving the environmental management toolbox. Im Internet: http://www.cortex.org/d-Cortex-%20Biodiversity%20Offsets_01Dec07.pdf

Humphries CJ, Williams PH, Vane-Wright RI (1995) Measuring biodiversity value for conservation. Annu Rev Ecol Systemat 26:93–111

ICH GHK, Bio Intelligence Service (2013) Exploring potential demand for and supply of habitat banking in the EU and appropriate design elements for a habitat banking scheme

ICMM (2005a) Biodiversity offsets: a proposition paper from The International Council on Mining and Metals

ICMM (2005b) Biodiversity offsets—a briefing paper for the mining industry

ICMM IUCN (2012) Independent report on biodiversity offsets. Prepared by The Biodiversity Consultancy. Available at www.icmm.com/biodiversity-offsets

IFC (2012) Performance standard 6. Biodiversity conservation and sustainable management of living natural resource. Available at http://www.ifc.org/wps/wcm/connect/bff0a28049a790d6b835faa8c6a8312a/PS6_English_2012.pdf?MOD=AJPERES. Accessed 24 Feb 2016

Igoe J, Brockington D (2007) Neoliberal Conservation: A Brief Introduction. Conserv Soc 5 (4):432–449
Igoe J, Neves K, Brockington D (2010) A spectacular eco-tour around the historic bloc: theorizing the convergence of biodiversity conservation and capitalist expansion. Antipode 42 (3):486–512. https://doi.org/10.1111/j.1467-8330.2010.00761.x
IUCN (2014) Biodiversity offsets technical study paper. IUCN, Gland, Switzerland, 65pp
Kiesecker JM, McKenney B, Kareiva P (2015) Offsets: factor failure into protected areas. Nature 525:33–33
Köppel J, Peters W, Wende W (2004) Eingriffsregelung, Umweltverträglichkeitsprüfung, FFH-Verträglichkeitsprüfung (No. UTB 2512)
Levrel H, Frascaria-Lacoste N, Hay J, Martin GJ, Pioch S (eds) (2015) Restaurer la nature pour atténuer les impacts du développement—Analyse des mesures compensatoires pour la biodiversité. QUAE, Versailles
Lipton J, Le Jeune K, Calewaert JB, Ozdemiroglu E (2008) Toolkit for performing resource equivalency analysis to assess and scale environmental damage in the European Union. REMEDE (Resource Equivalency Methods for Assessing Environmental Damage in the EU) toolkit. eftec, London, UK
Lopez Arbeláez DM, Quintero Sagre JD (2015) Compensaciones de biodiversidad: experiencias en Latinoamérica y aplicación en el contexto colombiano. Gestión y Ambiente 18(1):159
Madsen B, Moore Brands K, Carroll N (2010) State of biodiversity markets: offset and compensation programs worldwide
Madsen B, Carroll N, Kandy D, Bennett G (2011) Update: state of biodiversity markets. Forest Trends, Washington, DC
Manuel J (2013) Overview of the South African framework for biodiversity offsets. Presentation in a joint workshop of the Business and Biodiversity Offsets Programme and the European Union No Net Loss Working on 22 May 2013 as part of BBOP's Community of Practice and hosted by the European Commission in Brussels, Belgium. Available at http://bbop.forest-trends.org/documents/files/jmanuelppt.pdf
Maron M, Gordon A (2015) Could well-designed offset policies actually increase loss? Decision Point Online, 91/August 2015
Maron M, Dunn PK, McAlpine CA, Apan A (2010) Can offsets really compensate for habitat removal? The case of the endangered red-tailed black-cockatoo. J Appl Ecol 47(2):348–355
Maron M, Hobbs RJ, Moilanen A, Matthews JW, Christie K, Gardner TA, Keith DA, Lindenmayer DB, McAlpine CA (2012) Faustian bargains? Restoration realities in the context of biodiversity offset policies. Biol Conserv 155:141–148
Maron M, Gordon A, Mackey BG, Possingham HP, Watson JE (2015a) Conservation: Stop misuse of biodiversity offsets. Nature 523:401–403
Maron M, Bull JW, Evans MC, Gordon A (2015b) Locking in loss: baselines of decline in Australian biodiversity offset policies. Biol Conserv 192:504–512
Maron M, Gordon A, Mackey BG, Possingham HP, Watson JE (2015c) Interactions between biodiversity offsets and protected area commitments: avoiding perverse outcomes. Conserv Lett. https://doi.org/10.1111/conl.12222
Mathiesen K (2013) Is biodiversity offsetting a 'license to trash nature'? Available at http://www.theguardian.com/environment/2013/nov/12/biodiversity-offsetting-license-trash-nature
McAfee K (2012) The contradictory logic of global ecosystem services markets. Dev Change 43 (1):105–131. https://doi.org/10.1111/j.1467-7660.2011.01745.x
McCarthy MA, Parris KM, Van Der Ree R, Mcdonnell MJ, Burgman MA, Williams NSG et al (2004) The habitat hectares approach to vegetation assessment: An evaluation and suggestions for improvement. Ecol Manag Restor 5:24–27
McKenney B, Kiesecker JM (2010) Policy development for biodiversity offsets: a review of offset frameworks. Environ Manag 45:165–176
McKinney ML (2006) Urbanization as a major cause of biotic homogenization. Biol Conserv 127:247–260

Millennium Ecosystem Assessment (2005) Ecosystems and human well-being. Island Press, Washington, DC, XV, 109 p.

Miller KL, Trezise JA, Kraus S, Dripps K, Evans MC, Gibbons P, Possingham HP, Maron M (2015) The development of the Australian environmental offsets policy: From theory to practice. Environ Conserv 42(04):306–314

Moilanen A, van Teeffelen AJA, Ben-Haim Y, Ferrier S (2009) How much compensation is enough? A framework for incorporating uncertainty and time discounting when calculating offset ratios for impacted habitat. Restor Ecol 17(4):470–478

Monbiot G (2012) The unsung world. Available at http://www.monbiot.com/2012/12/08/the-unsung-world/

Monbiot G (2014) The pricing of everything. Available at http://www.monbiot.com/2014/07/24/the-pricing-of-everything/

Morandeau D, Vilaysack D (2012) Compensating for damage to biodiversity: an international benchmarking study. Evaluation and Integration of Sustainable Development Service

Morris RKA, Alonso I, Jefferson RG, Kirby KJ (2006) The creation of compensatory habitat—can it secure sustainable development? J Nat Conserv 14:106–116

Norton D (2009) Biodiversity offsets: two New Zealand case studies and an assessment framework. Environ Manag 43:698–706

O'Neill J (2014) Markets in biodiversity. Keeping your biodiversity healthy. In: Presentation. 2nd forum on the natural commons. Available at www.fern.org/sites/fern.org/files/O%27Neill%20Nature%20is%20not%20for%20sale.pdf

Parkes D, Newell G, Cheal D (2003) Assessing the quality of native vegetation: The "habitat hectares" approach. Ecol Manag Restor Suppl 4:S29–S38

Peters W, Siewert W, Szaramowicz M (2003) Folgenbewältigung von Ein-griffen im internationalen Vergleich. Endbericht zum F+E Vorhaben. BfN- Skripten 82, 220p

Pilgrim JD, Bennun L (2014) Will biodiversity offsets save or sink protected areas? Conserv Lett 7:423–424

Pilgrim JD, Brownlie S, Ekstrom JMM, Gardner TA, von Hase A, ten Kate K, Savy CE, Stephens RTT, Temple HJ, Treweek JR, Ussher GT, Ward G (2013) A process for assessing offsetability of biodiversity impacts. Conserv Lett 6(5):376–384

Pouzols FM, Burgman MA, Moilanen A (2012) Methods for allocation of habitat management, maintenance, restoration and offsetting, when conservation actions have uncertain consequences. Biol Conserv 153:41–50

PricewaterhouseCoopers LLP (2010) Biodiversity offsets and the mitigation hierarchy: a review of current application in the banking sector. Study completed on behalf of the Business and Biodiversity Offsets Programme and the UNEP Finance Initiative. Available at http://www.unepfi.org/fileadmin/documents/biodiversity_offsets.pdf

Quétier F, Lavorel S (2011) Assessing ecological equivalence in biodiversity offset schemes: Key issues and solutions. Biol Conserv 144(12):2991–2999

Quétier F, Regnery B, Levrel H (2014) No net loss of biodiversity or paper offsets? A critical review of the French no net loss policy. Environmental Science & Policy 38:120–131

Quétier F, Van Teeffelen AJA, Pilgrim JD, von Hase A, ten Kate K (2015a) Biodiversity offsets are one solution to widespread poorly compensated biodiversity loss: a response to Curran et al. Ecol Appl 25:1739–1741

Quétier F, Regnery B, Jacob C, Levrel H (2015b) Les contours flous de la doctrine éviter-réduire-compenser de 2012. In: Levrel H, Frascaria-Lacoste N, Hay J, Martin G, Pioch S (eds) Restaurer la nature pour atténuer les impacts du développement. Analyse des mesures compensatoires pour la biodiversité. Editions Quae, Paris

Quétier F, De Wahter P, Gersberg M, Dessard H, Halleson DN, Ndoutoume EN (2015c) La compensation "volontaire": les standards internationaux face aux réalités de l'Afrique centrale. In: Levrel H, Frascaria-Lacoste N, Hay J, Martin G, Pioch S (eds) Restaurer la nature pour atténuer les impacts du développement. Analyse des mesures compensatoires pour la biodiversité. Editions Quae, Paris

Quigley JT, Harper DJ (2006) Effectiveness of fish habitat compensation in Canada in achieving no net loss. Environ Manag 37(3):351–366

Rainey HJ et al (2015) A review of corporate goals of no net loss and net positive impact on biodiversity. Oryx 49:232–238

Rhineland-Palatinate (Germany), Hufen F, Jutzi S, Westenberger N (2011) Landesrecht Rheinland-Pfalz: Textsammlung. Nomos: Landesverordnung über die Bestimmung von Eingriffen in Natur und Landschaft vom 19. Dezember 2006

Robertson MM (2004) The neoliberalization of ecosystem services: wetland mitigation banking and problems in environmental governance. Geoforum 35(3):361–373

Rundcrantz K, Skärbäck E (2003) Environmental compensation in planning: a review of five different countries with major emphasis on the German system. Eur Environ 13(4):204–226

Salzman J, Ruhl JB (2000) Currencies and the commodification of environmental law. Stanford Law Rev 53:607–694

Salzmann J, Ruhl JB (2002) Paying to protect watershed services: wetland banking in the United States. In Pagiola S, Bishop J, Landel-Mills N (eds) Selling forest environmental services: market-based mechanisms for conservation. Routledge, London

Scharks T, Masuda YJ (2016) Don't discount economic valuation for conservation. Conserv Lett 9 (1):3–4

Spash CL (2015) Bulldozing biodiversity: the economics of offsets and trading-in nature. Biol Conserv 192:541–551

Sullivan S (2013) After the green rush? Biodiversity offsets, uranium power and the 'calculus of casualties' in greening growth. Hum Geogr 6(1):80–101

Temple H, Edmonds B, Butcher B, Treweek J (2010) Biodiversity offsets: testing a possible method for measuring losses and gains at Bardon Hill Quarry, UK. Practice 70:11–14

ten Kate K (2005) Biodiversity offsets: good for business and biodiversity? Available at https://www.cbd.int/financial/offsets/g-offsetsbusiness.pdf

ten Kate K, Bishop J, Bayon R (2004) Biodiversity offsets: Views, experience, and the business case. IUCN, Gland, Switzerland and Cambridge, UK and Insight Investment, London, UK

The Natural Capital Initiative (2010) Towards no net loss, and beyond. Addressing practical challenges for biodiversity offsetting in the UK. A one day inter-disciplinary workshop, organised by the Natural Capital Initiative. Summary Report For Policy Makers. Workshop held on Tuesday 22nd June, 2010 at Charles Darwin House, London

The Wildlife Trusts (n.d.) Thoughts on biodiversity offsetting. Available at http://www.wildlifetrusts.org/biodiversityoffsetting. Accessed 24 Feb 2015

Treweek J et al (2009) Scoping study for the design and use of biodiversity offsets in an english context. Final report to Defra. Available at http://www.forest-trends.org/publication_details.php?publicationID=2400

Treweek J, Butcher B, Temple H (2010) Biodiversity Offsets: possible methods for measuring biodiversity losses and gains for use in the UK. Practice 69:29–32

Trezise J (2015) Of apples, oranges and offsets. And the importance of scientific advocacy in shaping long-term policy. Decision Point Online, 91/August 2015

Tucker G, Allen B, Conway M, Dickie I, Hart K, Rayment M, Schulp C, van Teeffelen A (2013) Policy options for an EU No Net Loss Initiative. Report to the European Commission. Institute for European Environmental Policy, London

United Nations (1992) Convention on biological diversity

Van Teeffelen A (2018) The Netherlands. In: Wende W, Tucker GM, Quétier F, Rayment M, Darbi M (eds) Biodiversity offsets: European perspectives on no net loss of biodiversity and ecosystem services. Springer, Cham

Verpoest B (2014) Thousands urge European Commission to ditch biodiversity offsetting. Available at http://www.fern.org/sites/fern.org/files/Thousands%20demand%20European%20Commission%20to%20drop%20biodiversity%20offsetting%20legislation_FinalFERN_0.pdf. Accessed 22 Oct 2014

Villarroya A, Barros AC, Kiesecker J (2014) Policy development for environmental licensing and biodiversity offsets in Latin America. PLoS One 9:e107144

Walker S, Brower AL, Stephens RTT, Lee WG (2009) Why bartering biodiversity fails. Conserv Lett 2:149–157

Walters S (2014) Alvecote wood. Biodiversity offsetting in relation to ancient woodland. Presentation. 2nd Forum on the natural commons. Available at http://www.fern.org/sites/fern.org/files/Walters%20Alvecote%20Wood.pdf. Accessed 24 Feb 2015

Wende W (2018) Other EU member states. In: Wende W, Tucker GM, Quétier F, Rayment M, Darbi M (eds) Biodiversity offsets: European perspectives on no net loss of biodiversity and ecosystem services. Springer, Cham

Wende W, Tucker GM, Quétier F, Rayment M, Darbi M (eds) (2018a) Biodiversity offsets: European perspectives on no net loss of biodiversity and ecosystem services. Springer, Cham

Wende W, Tucker GM, Quétier F, Rayment M, Darbi M (2018b) Introduction: biodiversity offsets—the European perspective on no net loss of biodiversity and ecosystem services. In: Wende W, Tucker GM, Quétier F, Rayment M, Darbi M (eds) Biodiversity offsets: European perspectives on no net loss of biodiversity and ecosystem services. Springer, Cham

Wende W, Bezombes L, Reinert M (2018c) Introduction of a European strategy on no net loss of biodiversity. In: Wende W, Tucker GM, Quétier F, Rayment M, Darbi M (eds) Biodiversity offsets: European perspectives on no net loss of biodiversity and ecosystem services. Springer, Cham

Wende W, Albrecht J, Darbi M, Herbert M, May A, Schumacher J, Szaramowicz M (2018d) Germany. In: Wende W, Tucker GM, Quétier F, Rayment M, Darbi M (eds) Biodiversity offsets: European perspectives on no net loss of biodiversity and ecosystem services. Springer, Cham

Wissel S, Wätzold F (2010) A conceptual analysis of the application of tradable permits to biodiversity conservation. Conserv Biol 24(2):404–411

Developing a Typology of Biodiversity Offsets Using an Internet and Expert Based Approach: Methods and Materials

2

Abstract

Despite their increasing recognition in selected public discourses, biodiversity offsets remain a highly specific subject in expert circles in various locations worldwide. However, these circles are increasingly connected and exchanging, in particular using information and communication technologies online. Therefore, an expert and Internet based research approach was chosen for this study. It follows a netnographic approach in terms of an "offering strategy" to start an interaction and to engage others in a research process. For this purpose, the "Biodiversity Offsets Blog" was created. It served as a content management system to store, structure, and comment information, and it provided the base for a personal learning network, together with exploratory qualified Internet research using the Google search engine as well as chosen social network sites (LinkedIn, ResearchGate, and Twitter) and online curation services (Scoop.it).

In research practice, the process of typification often follows simplified methodologies or is implicitly carried out (intuitive types). By contrast, this study explicitly reflects the process of how types are built. Building on the method of empirically grounded typification, this study consists of an extensive theoretical analysis of voluntariness and biodiversity offsets. This is empirically underlined by two consecutive steps: (1) a screening of potential biodiversity offsets cases and (after grouping these cases to the built types) and (2) illustrative case studies for each of the built types. Due to the exploratory nature of the study, both the screening and the choice of illustrative case studies built on maximum variation, i.e., a spatially and functionally diverse set of cases.

Keywords

Internet-based research · Biodiversity Offsets Blog · Science 2.0 · Cyberscience · Research 2.0 · Netnography · RUDE procedure · Typification · Empirically grounded typification · Attribute space · Typological classification

M. Darbi, *Biodiversity Offsets Between Regulation and Voluntary Commitment*,
https://doi.org/10.1007/978-3-030-25594-7_2

2.1 General Methodology of an Internet-Based Research

The present study is located in the field of qualitative social research (Flick et al. 2000). The general methodological approach applied can be described as iterative and thus very much corresponds to the natural evolution of most research projects. Furthermore, it is largely based on Internet and expert based research and follows a participatory observational ("netnographic") approach (cf. Kozinets 2010).

What this means can be best illustrated using an example of how this study differs from "traditional" methodological approaches in qualitative social research: i.e., in the scope of this research, no interviews (neither personal, nor oral or written) are conducted despite the expert-based approach. Instead, making use of the tools of the "web 2.0" (Nentwich and König 2012, Nielsen 2012), an active network of experts is established (mostly via the "Biodiversity Offsets Blog"). Web 2.0 implies understanding and using the Internet as an (actively shaped) interactive platform as opposed to the "web 1.0" that is understood as a (passively used) data source (Nentwich and König 2012).

Both the general line of argumentation and the specific case studies are elaborated subsequently drawing on the experiences and findings from this network, e.g., via forum discussions, including comments and links to external sources, and personal communication (most notably LinkedIn messages and emails). This kind of research is often referred to as "science 2.0" or "cyberscience" (Nentwich and König 2012).

Figure 2.1 shows the interrelation of web 2.0, cyberscience, and science 2.0. In this context, cyberscience refers to "scientific activities taking place in the information and communication technologies, a space in which scientists increasingly circulate while remaining at their desk" (Nentwich 1999, translated in Nentwich and König 2012). Nentwich and König (2012) clarify the often misleading relation of science 1.0 and science 2.0 as follows:

> Science 2.0 [. . .], however, is in our view sometimes used merely to provide a contrast with 'Science 1.0', i.e. traditional science before the advent of the Internet, and so corresponds more closely to cyberscience (1.0). It is true, though, that the expression 'Science 2.0' also appears in connection with Web 2.0 sites such as ResearchGate, which explicitly applies the model of the new Web 2.0 social networks to the setting up of a scientific community.

Building on this, the approach of this thesis is more precisely described as "cyberscience 2.0" as shown in Fig. 2.1. Nevertheless, the definition of the term

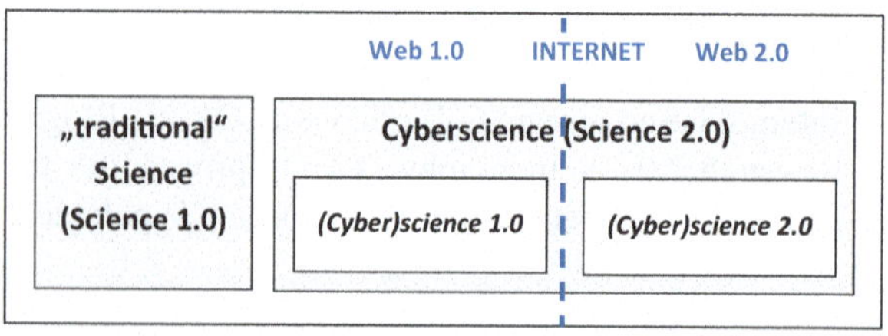

Fig. 2.1 Cyberscience in the age of the Internet. Source: Author

"science 2.0" has evolved more toward an interactive web 2.0-based concept as is shown in the Wikipedia article excerpt (Science 2.0 n.d.):

> Science 2.0 is a suggested new approach to science that uses information-sharing and collaboration made possible by network technologies. The roadmap [...] rests on four concepts; modern collaboration, communication, publication and participation. It is similar to the open research and open science movements and is inspired by Web 2.0 technologies [...] Science 2.0 uses collaborative tools like wikis, blogs and video journals to share findings, raw data and "nascent theories" online. Science 2.0 benefits from openness and sharing, regarding papers and research ideas and partial solutions [...] Emerging methods of scientific collaboration include models where scientists share data and methods before publication, and collaborate online in the process. Proponents of so-called "Open science" or "Science 2.0" stress the benefits of such a model, although detractors suggest that the more established method of conducting research privately, and publishing papers in peer-reviewed journals, has a fairer system of establishing credit and paying peer-reviewers.

In the following subchapters, an overview is given on the possibilities and particularities of the web 2.0/cyberscience and the resulting implications for research (Sect. 2.1.1). Building on this, the netnographic approach (Kozinets 2010) is introduced (Sect. 2.1.2) and the research process applied to this thesis is described (Sect. 2.1.3). This is then completed by the choice of tools (Sect. 2.1.4) and the presentation of the key tool that has been chosen in the scope of this research, the Biodiversity Offsets Blog.

Why Was This Research Approach Chosen for the Present Study?

Despite their increasing recognition in selected public discourses, biodiversity offsets remain a highly specific subject in expert circles—located at the interface between nature conservation and development/business. Notwithstanding the technical concentration of biodiversity offsetting to expert circles, it is spatially dispersed—a global phenomenon. This is the result of increasingly globalized production and supply chains. Multinational enterprises producing goods (e.g., for European consumers) harvest their products in other parts of the world and cause huge environmental impacts (e.g., mining in Madagascar, palm oil production in Indonesia).

Thus, experts on biodiversity offsets are spatially disperse on various locations worldwide, but they are increasingly connected and exchanging, in particular using information and communication technologies online. Many experts are very active in discussions in networks and blogs, use webinars and teleconferences (via Skype and other services). They exchange and share information via twitter and social network sites and collaborate over the distances of continents. Given these preconditions, an expert and internet based approach is particularly suitable.

Table 2.1 Global Internet usage

Global region	Estimated population (2008)	Internet users (2008)	Internet penetration (% of population)	Internet use (% of total global use)	Growth in Internet usage (2000–2008)
Africa	955,206,348	51,065,630	5.3	3.5	1031.2
Asia	3,776,181,949	578,538,257	15.3	39.5	406.1
Europe	800,401,065	384,633,765	48.1	26.3	266.0
Middle East	197,090,443	41,939,200	21.3	2.9	1176.8
North America	337,167,248	248,241,969	73.6	17.0	129.6
Latin America/ Caribbean	576,091,673	139,009,209	24.1	9.5	669.3
Oceania/Australia/ NZ	33,981,562	20,204,331	59.5	1.4	165.1
Global total	**6,676,120,288**	**1,463,632,361**	**21.9**	**100.0**	**305.5**

Source: Kozinets (2010)

2.1.1 Introduction to Web 2.0 and the Role of Cyberscience for Academic Research

In the face of the digital era, social interaction increasingly occurs through the internet and related information and communications technologies (ICT). As can be seen from Table 2.1, the global Internet usage has more than tripled between 2000 and 2008 and continues to grow (Kozinets 2010).

But not only is the Internet the new global arena for communication and interaction, it has also led to new ways of communicating. In particular, the trend toward the Web 2.0 has marked a fundamental change in different aspects (Ullrich and Schiek 2015):

- Never before so many communication forms have been available and used simultaneously (email, SMS, forums, chats, social networking, weblogs, microblogging, etc.).
- Never before so many people have used forms of written communication so often and intensively.
- Never before communication has been so widely documented.

Consequently, types of dissemination of communication have changed from personal communication (face-to-face) over forms of unilateral information (one-to-many) to multilateral interaction (many-to-many) (see Table 2.2).

There is a general trend toward the use of Web 2.0 in social interaction. Thus, the question can be raised which implications this has for science and research:

> What role does the digital social culture triggered by Web 2.0 play in the academic world at present and what are the potentials of platforms such as Twitter, Facebook, and Wikipedia? What impact will the emerging socio-technical practices have? (Nentwich and König 2012).

Table 2.2 Changing types of dissemination of communication

	Face-to-face	One-to-many	Many-to-many
Attention	To be there and awake	Turned on and slot chosen	Turned on, accessed/opened
Attitude	Yes or no response	No switch/switch off	Like/Dislike functions
Scope	Iteration, ratifications, and validations	Simulation and substitution	Keying by (time-shifted) Comments
Audience	Exclusive circle	Delegation of the all-included public	Multilateral staggered inclusion of potentially unlimited connected posts
Online	Pre Internet	Web 1.0	Web 2.0

Source: Translated from Corsten and Herma (2015)

Nielsen (2012) speaks of the "era of networked science" and new ways of knowledge production, using wikis, blogs, and other forms of online collaboration. The meaning of this paradigm shift can be best explained using one groundbreaking example (however, it is clear that those spectacular cases are not the norm)—the Polymath project.

In 2009 Tim Gowers, a mathematician at Cambridge University, decided to try something new to solve an important and difficult mathematical problem. Instead of working in a closed research team, he posted an open invitation asking other people to help out to his blog. Little more than a month later, Gowers successfully closed the Polymath project, claiming that it had solved not just the original problem, but an even greater problem that included the original as a special case (Nielsen 2012).

But how was this possible? The answer lies possibly in two aspects:

- Bringing together the right people at the right time and place (that this is easily feasible is a major achievement of online culture).
- Allowing for a hands-on trial-and-error process (as opposed to the traditional publishing system that is time consuming and less favorable to draft ideas and nonproven endeavors).

In fact, more than two dozen people, including genuine experts, wrote 800 mathematical comments and developed, refined, and discarded ideas at incredible speed (Nielsen 2012):

> You see top mathematicians making mistakes, going down wrong paths, getting their hands dirty following up the most mundane of details, relentlessly pursuing a solution. And through all the false starts and wrong turns, you see a gradual dawning of insight.

Thus, the success of the Polymath project marked a new era of online collaboration in research.

Nentwich and König (2012) conclude that the transition from traditional science to the so-called "cyberscience" can potentially "bring about changes in all dimensions of scientific activity, including organizational space." The Web 2.0 has induced changes of the scholarly communication system.

The particularities of this Web 2.0-enabled research and communication are displayed in Table 2.3. It is characterized not only by a high degree of multimodality or multimedia capability using text, visuals, audio, video, and links (Schirmer et al. 2015), but also by a high degree of interactivity. Furthermore, the dynamic and experimental character is a novel characteristic that rises from applying a science 2.0 approach. Links and cross references are also frequently encountered. They provide the opportunity to navigate away from the original site with just a few clicks and explore new content. This, however, bears the risk of getting distracted or lost in irrelevant details and veering away from the focus of the research.

The notion of cyberscience is not as new as one might expect, it already appeared in the literature almost 20 years ago (Wouters 1996; Thagard 1997, cited in Nentwich and König 2012). Nevertheless, the question regarding the role of cyberscience for academia does not (yet) provides a generalizable answer.

While Nentwich and König (2012) state that "since the turn of the millennium, cyberscience matured and is now ubiquitous" and "[p]ractically all researchers of most fields are cyberscientists as they spend a considerable and increasing amount of time not only in front of a computer screen, but also communicating with their objects of research, their peers, and the extra-academic world," they see the active exploration of the Web 2.0 and its tools restricted to a minority of academics. Similarly, Nielsen (2012) fears a "tragedy of lost opportunity":

> We're at a unique moment in history: [. . .] We have an opportunity to change the way knowledge is constructed. But the scientific community, which ought to be in the vanguard, is instead bringing up the rear, with most scientists clinging to their existing way of working, and failing to support those who seek a better way. (Nielsen 2012)

Concluding, Nentwich and König (2012) note an increase in and necessity for cyberscientific research to keep pace with the increasingly digital world outside academia. However, the questions need to be dealt with "whether the new tools are functional, whether they satisfy genuine needs of the highly differentiated and professionalized academic world [. . .] At the present state of development, the problem of multiple channels and information overload, or of lacking participation in evolving processes of bottom-up quality control, for instance, show that there is still need for further development and adaptation, both technically and regarding the social embedding of these technologies." (Nentwich and König 2012)

What Are the Implications for the Present Study?

The Internet (and in particular the Web 2.0) enables a huge (and growing) variety of tools that can be used both for personal and professional communication and exchange. This bears also a large potential for research to keep pace

(continued)

with societal trends. It is more than any other medium bringing together the academic and the public sphere.

Biodiversity offsets are at the intersection of the technical and scientific discussion on the one hand and the political and societal debate about their implementation on the other hand. Much of the latter is being exchanged online (e.g., Monbiot 2014; FERN n.d.; BirdLife International 2008; Harris 2014). As this thesis focuses on the different form of voluntary implementation of biodiversity offsets, it is thus crucial to include these various online sources. However, as has been shown the particularities of online communication need to be taken into consideration, most notably with regard to its multimediality (comprising a wider range of sources, e.g., videos, podcasts, articles, and short messages), interrelatedness (linked content) and interactivity (multilateral interaction between various people from various contexts worldwide), dynamic and temporal aspects (content is mostly not static, but changing). The often experimental and discursive character of online communication makes it inevitably necessary to carefully evaluate the reliability, validity, and context of online retrieved or generated information.

Table 2.3 Particularities of the Web 2.0 compared to offline research, problems, and opportunities

Particularities of the Web 2.0 and science 2.0	
Criteria	Description
Multimediality and multimodality	*"The academic communication channels became increasingly diversified, more and more based on electronic means, and ubiquitous. The number of ways to reach a typical researcher is greater than ever in the past."* (Nentwich and König 2012) *"[T]hose new tools that are generally associated with the term Web 2.0 have already started to play a significant role in science and research: academics blog about their research, communicate via microblogging services, contribute to collaborative wiki resources, activate profiles in social network sites, and some even populate virtual worlds."* (Nentwich and König 2012)
Interactivity	*"Since the distinct feature of Web 2.0 is the users' possibility to participate, interactivity has to be seen as a crucial category for assessing its impact."* (Nentwich and König 2012)
Open access	*"Since 2003, the open access movement has gained in strength and challenges the commercial scholarly publishing system. Today, an estimated 20 to 25 percent (Björk et al. 2009; Björk et al. 2010; Gargouri et al. 2010) of all research literature in journals is available online and open access."* (Nentwich and König 2012)

(continued)

Table 2.3 (continued)

Particularities of the Web 2.0 and science 2.0	
Criteria	Description
Collective intelligence—wisdom of crowds (and collective stupidity as well)	*"Many books and magazine articles have been written about collective intelligence. Perhaps the best-known example of this work is James Surowiecki's 2004 book The Wisdom of Crowds , which explains how large groups of people can sometimes perform surprisingly well at problem solving. Surowiecki opens his book with a striking story about the scientist Francis Galton. In 1906, Galton was attending an English country fair, and among the fair's attractions was a weight-judging contest, where people competed to guess the weight of an ox. Galton expected that most of the competitors would be far off in their estimates, and was surprised to learn that the average of all the competitors' guesses (1,197 pounds) was just one pound short of the correct weight of 1,198 pounds. In other words, collectively, if one averaged the guesses, the crowd at the fair guessed the weight almost perfectly."* (Nielsen 2012)
Bringing together the academic and the public sphere	*"[T]he overall tendency of blurring the boundaries between academia and the public is not entirely new, but the result of a longer development that has started before the advent of Web 2.0. Internet-based technologies are not its direct cause, but their popularity may partly be explained by their fit to this overall tendency, which, in turn, they speed up to some degree."* (Nentwich and König 2012)
Relativization of time and space	*"Cyberscience 2.0 potentially completes this process of relativization of time and space for academia." (Nentwich* 2003, *cited in Nentwich and König* 2012*)*
Dynamic character	*"[T]he Internet in general and Web 2.0 in particular are moving targets, and so are the potential roles and impacts they may play in academia."* (Nentwich and König 2012)
Experimental character	*"Science 2.0 generally refers to new practices of scientists who post raw experimental results, nascent theories, claims of discovery and draft papers on the web for others to see and to comment on."* (Waldrop 2008, cited in Nentwich and König 2012)
Linked content	*"On the side of users, it can also happen that they follow a link and move ever further away from the original text—and no more possibly come back. This could, of course, happen even with a (scientific) book; however, the distraction on the Internet can be reached technically easily, so that the probability of this is likely to be relatively large."* (Wenninger 2015)

Source: Nentwich and König (2012), Nielsen (2012) and Wenninger (2015)

2.1.2 The Netnographic Approach

The research procedure draws on the netnographic approach developed by Kozinets (2010) as a specific online form of participant observation grounded in qualitative social sciences. Netnography emphasizes the role of online communities and computer-mediated communications.

> Netnography is "a type of online, or Internet, ethnography; netnography provides guidelines for the adaptation of participant-observation procedures—planning for fieldwork, making a cultural entrée, gathering cultural data, ensuring a high-quality ethnographic interpretation, and ensuring strict adherence to ethical standards—to the contingencies of online community and culture that manifest through computer-mediated communications." (Kozinets 2010)

Kozinets (2010) suggests a simplified flow of a netnographic research project, including the following five steps:

1. Definition of research questions, social sites, or topics to investigate.
2. Community identification and selection.
3. Community participant-observation (engagement, immersion) and data collection (ensure ethical procedures).
4. Data analysis and iterative interpretation of findings.
5. Write, present, and report research findings and/or theoretical and/or policy implication.

The most important step in the beginning is locating an appropriate online field, choosing from varieties of online social experience. Kozinets has formulated guidelines for site choice and entrée that build on a number of criteria:

- Relevant
- Active
- Interactive
- Substantial
- Heterogenous
- Data-rich

Decisions that need to be made by a netnographer include determining appropriate forms of social interaction, research questions and topics, use of research engines, etc. Different potential forms of netnographic participation in an online community can be used. This includes a "learning" component on the one hand, i.e., identifying similar others and absorbing culture and rules to gain a sense of membership of this particular community, and a "doing" component on the other hand, i.e., actively contributing to this community by asking questions, commenting (possibly), and ultimately taking a leadership role.

Kozinets suggests that an "offering strategy" is usually beneficial to start an interaction and to engage others in a research process (e.g., via a research web page or a blog).

What Are the Implications for the Present Study?
This study follows the netnographic approach in many aspects, e.g., in terms of appropriate forms of social interaction (learning and doing components) as well as site choice and entrée. Most notably, the "offering strategy" encouraged by Kozinets (2010) has been chosen by creating a blog dedicated to the "Information and Exchange on Biodiversity Offsets and the Mitigation Hierarchy." However, this thesis differs from the pure form of netnography in that it is not focused on an online ethnography, i.e., it does not study a particular online community as such, but rather uses it to elaborate on the topic of biodiversity offsets that is transported by this online community.

2.1.3 Research Process of Internet-Based (Web 1.0 and Web 2.0) Research: The RUDE Procedure

According to Müller et al. (2013), online research is a process that builds on four steps: (1) Find, (2) Evaluate, (3) Keep/document, and (4) Stay up-to-date. Building on these four steps, the **R** (Research)—**U** (Update)—**D** (Documentation)—**E** (Evaluation)—**procedure** has been developed for the present study as presented in Fig. 2.2 and described below:

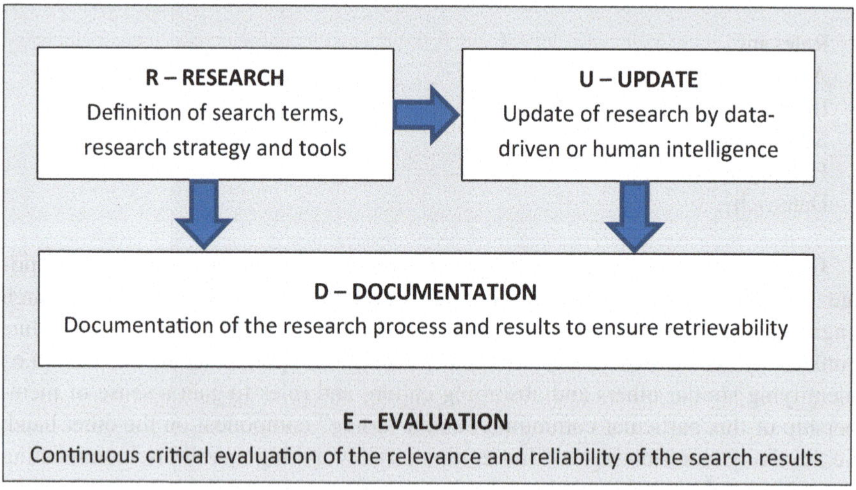

Fig. 2.2 The RUDE procedure: steps in the research process of Internet-based web 1.0 and 2.0 research. Source: Author, building on Müller et al. (2013)

- **R—Research:** This step is core to any project, including the definition of search terms, research strategy, and tools. A necessary prerequisite is to precisely define the research question and goal and finding appropriate criteria. Thereby, tags and keywords can be used. With regard to research strategies three approaches can be distinguished: (1) From the specific case to general, (2) From general to the specific case, and (3) "snowballing" or "pearl growing."
- **U—Update:** As research is understood as a process, it is important to keep the research findings up to date, in particular in a quickly changing online environment. This can either be done technology-based (through algorithms), i.e., via RSS feeds etc. or human-based through social networking, i.e., via a personal learning network (Twitter, Facebook, LinkedIn, etc.). Despite the necessity of the update step, the question when to close research is crucial to prevent it from becoming a "never ending story."
- **D—Documentation:** For any scientifically sound project, documentation is key to ensure retrievability and traceability. Online a number of tools facilitate are reliable and accurate documentation, for both the researchers and the third parties, e.g., via bookmarks and curation services (Google scholar, My Library, Citation, Scoop.it) or the use of a blog as a Content Management System (CMS).
- **E—Evaluation:** As any research, research in the Web 2.0 requires not only documentation, but also critical reflection of the research process. Evaluation is not particularly online, but related to the person of the researcher, her experiences, and knowledge. Evaluation refers to a process in the head of the researcher that is done permanently in the background. This continuous process is embedded into all steps of research process and needs to focus on relevance and reliability of websites and other sources as well as the search results.

What Are the Implications for the Present Study?

The four research steps developed by Müller et al. (2013) have been adapted to the R (Research)—U (Update)—D (Documentation)—E (Evaluation)—procedure.

1. Research: This step includes a combination of the three mentioned approaches with emphasis on snowballing (this is particularly suitable and efficient as the present study aims at an exploratory rather than a representative typology).
2. Update: This step is human-based through social networking, i.e., via a personal learning network (online).
3. Documentation: In addition to downloading and storing digital copies of relevant sources, the newly created Biodiversity Offsets Blog is used as a Content Management System.

(continued)

4. Evaluation: In order to ensure quality and relevance, continuous evaluation is crucial and encloses the three other steps of research, in particular with regard to the evaluation of all sources for their reliability.

Details regarding the specific tools chosen to underpin these steps are presented in Sect. 2.1.4.

2.1.4 Variety and Choice of Tools for Research in a Web 2.0 Environment

Online communication builds on a number of innate features, characterized by specific structures and processes. These have different functions, including:

1. Visualizations using pictures, audio, and video make information more vivid and clear and foster a quick reception, but also capture the attention of the recipient/user.
2. Connection of contents through links enables the user to quickly explore a particular field of interest by navigating from one link to another (both within one site and to external sites).
3. Keywords, tags, and tag clouds describe and categorize the content, the foster filtering for relevant information.
4. Information is increasingly spread as short extracts—namely sharing "tweets" (short messages that others can forward, comment, etc.) using the microblogging platform "twitter." Online resources may respond to this by explicitly highlighting key messages.

Building on these features, a large number of different tools or forms of communication exist in Web 2.0 environment. The most commonly used tools are presented in Table 2.4.

It can be observed that many of these tools are no longer restricted to a specific function, but offer a set of different functions. Thus, the user and/or researcher can often choose between different tools to perform the same task. For example, the web search (using keywords or tags) is not restricted to web search engines. Social network sites and bookmarking, blog, and microblogging portals, forums, wikis, and audio and video platforms can now be increasingly used for research purposes. Most of them even contains an advanced search function (e.g., YouTube, Twitter, and Wikipedia).

However, even if different tools can be used comparably well for the same task, the results will be different. Thus, the choice of tools is crucial and it is often necessary to combine multiple tools or instruments. It is important to note that this combination of methods reflects a thoughtful (yet also pragmatic) choice based on

Table 2.4 Tools in Web 2.0 research

Web search engines	• Distinction between general web search engines (e.g., Google and Yahoo) and special search engines • Potential of the Google portfolio (Google Websearch, Google Scholar, Google Books) for academia: complexity of algorithms (but also lacking transparency)
Social Network Sites (SNS)	• Different forms, e.g., professional (Xing, LinkedIn), academic (ResearchGate, Academia.edu) and broad (Facebook, Google+) • Consist of user profiles and built on common interests • Include different communication forms (e.g., public, private, and semi-public, e.g., in thematic groups)
(Social) bookmarking	• Development and maintenance of a Social Learning Network through sharing of and reacting to information provided by others (e.g., using LinkedIn, Google+, Facebook) • Saving and organizing links and websites through bookmarks (e.g., in browser), social bookmarks (e.g., Google+, Facebook, and LinkedIn), curation services (e.g., Scoop.it), and citation managers (e.g., Citavi, Endnote, and Zotero)
Email	• Nowadays by far the most common form of direct communication between researchers
Blogs	• Blogs cover a wide range of general blogs and blog portals as well as a multitude of highly specific and individual blogs, including science blogs • Software platforms (e.g., Wordpress, Blogger, or Joomla) provide an intuitive user-friendly interface that enables to set up a blog easily
Microblogging	• Microblogging has evolved as a specific form of blogging, mostly using Twitter • Can be used for scientific marketing and communication
Forums	• Wide range includes bulletin boards or forums, chat rooms, playspaces, virtual worlds • Differentiation regarding open (public) and closed group discussions, types of contributions (e.g., oral chats/written forums), synchronous/asynchronous communication
Wikis	• Crowd-sourced open access online encyclopedias (e.g., Wikipedia, Wikimedia, Wikimindmap, numerous specific wikis with different target groups)
Audio(-visual) sites	• Videos can be embedded on websites or uploaded on video platforms (e.g., YouTube) • Presentation platforms (e.g., Slideshare and vimeo) • Podcasts and vodcasts

Source: Author, building on and Nentwich and König (2012) and Schirmer et al. (2015)

the actual research needs rather than arbitrariness. Thus, the choice of tools depends on the focus of the research, the target audience, temporal aspects (when and over what period), and available capacity. Table 2.5 lists the abovementioned tools with regard to their suitability for the steps of the R (Research)—U (Update)—D (Documentation)—E (Evaluation)—procedure: research, update, and documentation.

Table 2.5 The cyberscience toolbox and its usages for the steps of the RUDE procedure

	Suitable for		
Cyberscience tools	Research	Update	Documentation
Search engines	a	b	c
Google web search	a	b	c
Google scholar	a	b	b
Google books	a	b	b
Google translator	b	c	c
Other general search engines	a	b	c
Specific search engines	a	b	c
Social network sites	b	a	b
Facebook	b	a	b
LinkedIn	b	a	b
ResearchGate	a	a	b
Academia.edu	b	a	b
Google+			
Weblogs			a
Maintaining an own blog	b	b	a
Blogportale	a	b	c
Wordpress, Blogger, etc.	b	b	c
Science blogs	b	b	c
Microblogging	b	a	b
Twitter	b	b	c
Share and update (Facebook, LinkedIn, Google+, etc.)	b	a	c
Bookmarks	c	b	a
Browser bookmarks	c	b	a
Social bookmarks	c	b	a
Scoop.it	b	a	a
Wikis	a	b	b
Wikipedia/media	a	b	b
Wikimindmap	b	c	c
Audio/visual sites	a	c	c
Slideshare	a	c	c
YouTube	a	c	c
Instagram	b	c	c
Flickr	b	c	c
Pinterest	b	c	c
Bulletin boards, groups, forums	a	b	c
Facebook groups	b	b	c
LinkedIn groups	b	b	b
ResearchGate groups	b	b	b
Open forums (general/specific)	b	b	b
Semi-public forums	b	b	b

(continued)

Table 2.5 (continued)

Cyberscience tools	Suitable for Research	Update	Documentation
Chat rooms	b	c	c
(Mailing) Lists	c	a	c
Email	b	b	b
RSS	c	a	b

Source: Author
[a]Particularly suited
[b]Suited with limitations
[c]Not particularly suited

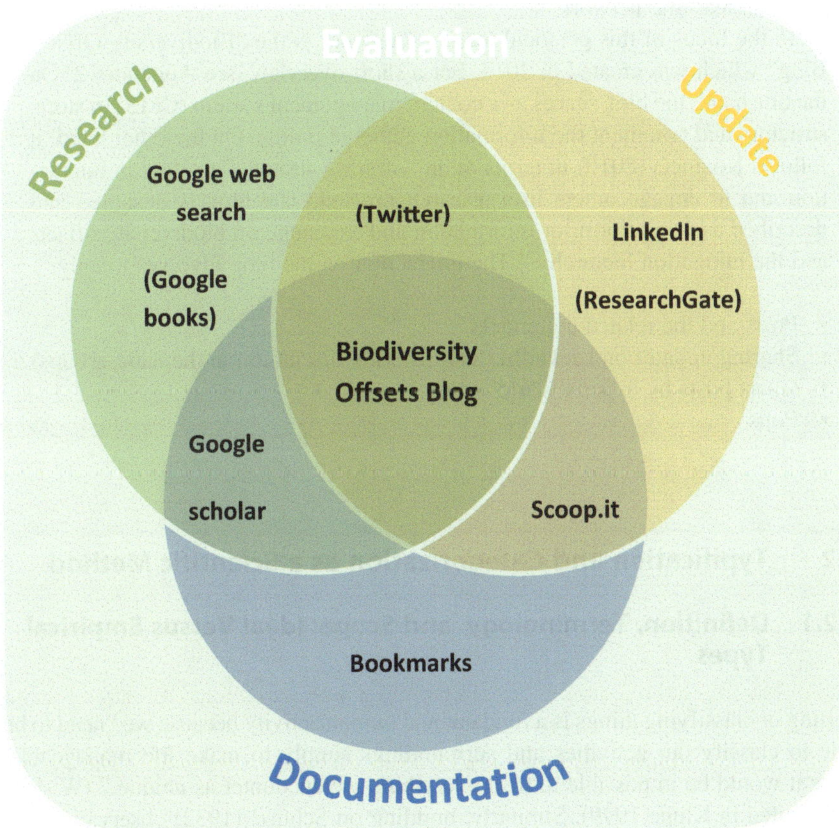

Fig. 2.3 Tools of Internet-based research used in the scope of this study. Source: Author

What Are the Implications for the Present Study?

The Internet (and in particular the Web 2.0) enables a huge (and growing) variety of tools that can be used both for personal and professional communication. When using such tools, the most obvious and important task is to reduce this multitude and to choose the most appropriate (or efficient) for the specific research purpose.

Figure 2.3 displays the tools that have been chosen for this study from the different Web 1.0 and Web 2.0 tools. Exploratory research is based on qualified Internet research using the Google search engine (accompanied by Google books and Google scholar) as well as chosen social network sites, most notably the professional network LinkedIn (and to some extent ResearchGate and Twitter) and the online curation service Scoop.it. These tools serve as a personal learning network.

In the focus of this personal learning network is the "Biodiversity Offsets Blog" which was created in 2014. For a short overview, see Appendix 2. On the one hand, the blog serves as a content management system (CMS) to store, structure and comment the information gathered online. On the other hand, it follows Kozinets (2010) in terms of an "offering strategy" to start an interaction and to engage others in a research process. The blog is designed and described as "a platform for information and exchange on biodiversity offsets and the mitigation hierarchy." The interaction on the blog involves mainly:

- Posts and the related comments
- Sharing updates on LinkedIn (and fostering discussion in thematic groups)
- Guest posts by experts worldwide
- Polls

2.2 Typification and Categorization as a Scientific Method

2.2.1 Definition, Terminology, and Scope: Ideal Versus Empirical Types

Sorting or classifying things is a fundamental human activity because we "need to be able to classify our activities and surroundings simply to make life manageable, since it would be impossible to treat everything we encounter as unique." (Wishart 1987, cited in Kluge 1999). Similarly, building on Schütz' (1932) observations on "the phenomenology of the social world," Srubar (1979) claims that the reality of the world presents itself as a meaningful context structured by types. He claims that typification is necessary for the acting person to orientate herself in the world she is part of, to assume meaningful actions of others and act accordingly.

> The typifying detection of an action sequence consists of [...] a synthesis of recognition, which the interpreting person performs in self-interpretation of his experiences. Subject of this recognition are objectively visible "in-order-to" and "because" motives of the scenario. Namely, similarly recurring storylines, that realize by setting similar means similar action targets, are associated with similar "in-order-to" motives or [...] "because" correlations of the experiences of each actor. These are assumed to be constant and invariant to all modifications which they may experience in the lively consciousness of the acting person. (translated from Schütz 1932)

Thus, since their introduction in the 1940s (cf. Schütz 1932; Hempel and Oppenheim 1936; Lazarsfeld 1937), types and typification play an important role in qualitative social research (Kluge 1999). They help to capture and understand (to the most possible degree) complex social realities and meaningful context relations (Kluge 1999) and subsequently to build or explain theoretical systems or models. Types define multidimensional evaluation groups and thus have a structuring function that enables understanding of complex interrelations (Kluge 1999).

> It is important to note [...] that types function as theory. This capability is built into them, since as composites they are given a structure with functional consequences, and hence types are systems. (McKinney 1969, cited in Kluge 1999)

For the purpose of typification, two levels have to be distinguished (Kluge 1999):

1. The level of the typology (general) is focused on both the similarities between the types in order to relate them to each other and the differences in terms of scope or spectrum of the studied subject or field, i.e., if the heterogeneity between the types is high, they can be more clearly distinguished from another.
2. The level of the type (individual) is focused on the common attributes or similarities of one individual type, i.e., each type should ideally be as homogenous as possible.

Typification is not only characterized by a multitude of steps, processes, and tools, but it also grounds on a variety of conceptions of definitions of types (Kluge 1999). In the literature, a number of different types are distinguished, i.e., ideal and normal types, constructed and empirical/existential types, extreme or representative types, etc. (Kluge 1999).

Max Weber (1968, cited in Bohnsack 2007) has contrasted the "normal type" based on representativeness with the "ideal type" that aims at the representation of deeper meaningful structures. Bohnsack (2007) notes a broad consensus in qualitative research that the formation of ideal types provides the methodological key for generalization; however, he claims that we have to go beyond the conventional understanding of types as coined by Max Weber and Alfred Schütz.

Similarly, Kluge (1999) argues that types can neither be reduced to ideal (or "constructed") types nor empirical or real types, but must include a combination of both. Building on the work of Lazarsfeld (1937) and Barton (1955), Kluge has explored the methodology of empirically grounded typification ("empirisch begründete Typenbildung"). She explains why the distinction between pure theoretical and pure empirical types becomes blurred and how types can be built in practice.

What Are the Implications for the Present Study?
In research practice, the process of typification often follows simplified methodologies or is only implicitly carried out (building of intuitive types). By contrast, this study explicitly reflects the process of how types are built.

Building on the method of empirically grounded typification after Kluge (1999), typification in the scope of this study is understood as a process that necessarily includes both empirical evidence and previous theoretical analysis.

2.2.2 Theoretical Concepts of Empirically Grounded Typification: The Attribute Space After Lazarsfeld and Barton

Hempel and Oppenheim (1936) define types as multidimensional structuring or classification terms. Consequently, types and typologies are the results of a particular combination of (selected) criteria or attributes. Thus, Bailey (1973, cited in Kluge 1999) notes that the "[c]onstruction of any typology, whether heuristic, empirical, or classical, requires prior selection of the component dimensions."

Lazarsfeld (1937) and Barton (1955) highlight the multidimensionality of typologies and refine the concept of the "attribute space" (as introduced by Hempel and Oppenheim 1936).

> Whenever a writer uses such a typological classification, he should substruct to it a corresponding attribute space and the reduction connected therewith, in order to be aware of what is logically implied in his enumeration of types. [...] the writer would see whether he has overlooked certain cases, he could make sure that some of his types are not overlapping, and he would probably make the classification more valuable for actual empirical research. (Lazarsfeld 1937)

The attribute space is used as a tool to make the (sometimes implicit) underlying combination of attributes of typologies explicit and visible. After carefully identifying and choosing the attributes most relevant (to the subject studied and the research purpose envisaged), these are combined in a matrix with a minimum of two (usually two to four) dimensions. Figure 2.4 shows an example of a three-dimensional attribute space (three attributes with two values each, i.e., a theoretical maximum of eight possible types).

Lazarsfeld and Barton further introduce typological procedures that can be applied to analyze this attribute space, namely

- Reduction, i.e., reducing or simplifying the dimensions for functional or pragmatic reasons (e.g., two or several cases from the matrix above may be grouped to one type).
- Substruction, i.e., finding the underlying attribute space from a given typology.
- Transformation, i.e., complementing a typology and the attribute space through arising evidence, or integrating different theoretically possible typologies from one given situation (through variation in choice and number of underlying attributes).

	Criterion A			
	A1		A2	
	Criterion B		Criterion B	
Criterion C	B1	B2	B1	B2
C1	Case 1	Case 2	Case 3	Case 4
C2	Case 5	Case 6	Case 7	Case 8

Fig. 2.4 Example of a simple three-dimensional attribute space. Source: Author, after Kluge (1999) and Barton (1955)

Furthermore, Lazarsfeld and Barton argue that it is usually beneficial if the researcher first intuitively builds an interim typology on a completely impressionistic basis (Lazarsfeld 1937, Kluge 1999). Such an interim typology helps to reduce the complexity. Through the process of substruction the underlying attribute space, i.e., combination of attributes, is then specified. The attribute space displays all possible attribute combinations and can thus help to identify gaps in the impressionistic classification. The typology is analyzed and transformed or rectified.

> Often, especially if many attributes are at stake, it might be much better for the student to become deeply acquainted with his material and then bring order into it by first blocking out a few main types on a completely impressionistic basis. Only thereafter would he reconsider the matter and substruct to his own typological intuitions an adequate attribute space and bring into relief the reduction which he has used implicitly. The best results probably will be gained in just this combination of a first general survey and a subsequent systematic analysis. (Lazarsfeld 1937)

Finally, it must be noted that typification (in particular as part of participatory observational research) is highly normative and as such influenced by the individual perspective of the researcher. Consequently, Lazarsfeld (1937) notes that there may be more than one attribute space for a given number of types and vice versa:

> The problem comes up whether to every given system of types there corresponds only one attribute space [. . .] The answer is probably 'no'.

What Are the Implications for the Present Study?
Building on the methodological considerations of Kluge (1999), Lazarsfeld (1937), and Barton (1955), typification in the scope of this study follows the argumentation that it is usually beneficial to first build an interim typology on a completely impressionistic basis in order to better understand the research field and to reduce complexity. In addition, the concept of the attribute space is used

(continued)

to construct the typology (as shown in Fig. 2.4). Summarizing these premises, the development of a typology of biodiversity offsets in the scope of this study consists of four consecutive steps:

1. Development of an impressionistic classification of types.
2. Derivation (and reduction) of relevant attributes/criteria for voluntariness from theory.
3. Substruction of the underlying attribute space and combinations of attributes.
4. Transformation (rectification) of the impressionistic types and analysis of meaningful correlations (i.e., integration of the impressionistic classification of types and the theoretically grounded attribute space).

2.3 Methods for Empirical Analysis and Sampling

2.3.1 Methods and Materials of Data Collection

In order to develop a typology and group single biodiversity offset cases to distinct types, a two-stage empirical research is conducted:

1. Screening of potential biodiversity offset cases is carried out to identify a variety of biodiversity offset cases worldwide (see Sect. 5.5 and Appendix 1).
2. Case studies are examined to analyze and illustrate each of the types that have been built.

The successful implementation of biodiversity offsets is a long-term process, which is time consuming and faces a number of obstacles and controversies. Therefore, to date, the recent rise in the interest in biodiversity offsets worldwide has not yet resulted in a correspondingly high number of implemented cases.[1] Thus, the screening of potential offset cases only delivers a significant number of cases within a worldwide geographical scope (cf. Darbi et al. 2010).

Furthermore, seeing the complexity and highly specific nature of offsets as such, documentation of offset cases is (publicly) available to a limited extent and is mainly restricted to expert circles.

Therefore, methods for data collection have to take into consideration three conditions:

[1]This observation refers to offsets at global scale. Notwithstanding, there are single offset schemes with a long tradition and numerous projects implemented, e.g., US Wetland Mitigation and German Impact Mitigation.

1. The broad geographical scope.
2. The limited number of implemented and documented offset cases.
3. The concentration of knowledge in expert circles.

Furthermore, given these conditions, data collection methods must consider the following organizational demands:

- The requirement to reach destinations worldwide (travel is too time and cost intensive, inefficient).
- The need for and access to highly specialized/trained experts.

As has been outlined (see Sect. 2.1) an Internet and expert based participatory observational research approach has been chosen to respond to these demands, in particular with regard to the following characteristics:

- It is desk based, using tools of online communication and research.
- It is of exploratory and not representative nature (see Sect. 2.3.2 purposive sampling).
- It targets experts in their specific field of interest and action (e.g., members of online communities, networks and group discussions, webinar attendees, authors, and commenters on articles and opinion pieces).

The focus of this study is on written materials (marginally including videos, audio materials, and pictures) that are extracted from various sources, including scientific and gray literature, websites, blogs, social networks, the biodiversity offsets blog, emails, LinkedIn group discussions, and personal messages.

What Are the Implications for the Present Study?

As has been outlined, the empirical analysis rests on two consecutive steps: (1) a screening of potential biodiversity offsets cases and (after grouping these cases to the built types) (2) and illustrative case studies for each of the built types.

Building on the specifics of the research field and subject, and consequently the chosen internet and expert based participatory observational research strategy, an iterative multilevel approach to data collection is applied in the scope of this study:

1. Gain overview: Preliminary (orienting) Internet search (using Google, Wikipedia, "snowballing") with the goal to define the scope and terminology.
2. Collect and weigh data: Identify and approach experts and key players by (a) using Google scholar to identify relevant literature and authors, search

(continued)

for specific publications and authors, and b) using the Google search engine and social networks.

3. Analyze and verify data: Read and evaluate collected data, verify, and deepen information using experts contact (e.g., via email, LinkedIn messages, and discussions).
4. Arrange and present results: Create order and logical connection, develop synthesis, and conclusions, present results as the written text.

It is important to note that the mentioned (new) online tools (see Sect. 2.1.4) are combined with conventional methods, in particular document review.[2]

Materials used are mainly (but not restricted to) written sources (all sorts of texts from articles to forum discussions and transcripts of audio material) that are available online.

2.3.2 Methods for Sampling as Basis for Choice of Case Studies

Sampling is used to choose a subset from the whole of entities studied, e.g., a subset of the population, to represent the whole population, in order to identify and analyze "processes that are meaningful beyond the particular cases, individuals or sites studied" (Nonprobability sampling n.d.). Broadly, two forms of sampling can be distinguished: whereas probability sampling, or random sampling, is a sampling technique in which the probability of getting any particular sample may be calculated, nonprobability sampling does not meet this criterion and is usually applied for in-depth qualitative research targeting complex (social) phenomena. This enables the "discovery" and identification of patterns and causal mechanisms (Nonprobability sampling n.d.). While nonprobability sampling leads to a sample which is not representative of the whole (e.g., population), it is nonetheless possible to draw generalizations (e.g., propose a new theory, propose policy) from studies based on nonprobability samples, using the notion of "theoretical saturation" and "analytical generalization" instead of on statistical generalization (Yin 2014, cited in Nonprobability sampling n.d.).

The most commonly used types of nonprobability sampling include:

- Convenience, haphazard, or accidental sampling—The sample is chosen based on the relative ease of access, e.g., friends or co-workers.
- Snowball sampling—The first respondent refers to someone, who then also refers to someone, and so on.

[2]According to Silverman (2006, cited in Earl et al. 2010), document review involves systematic categorization of information into either predetermined or emergent themes.

Table 2.6 Types of purposeful sampling

Maximum variation sampling	(heterogeneous sampling), is used to capture a wide range of perspectives related to the study subject.
Homogeneous sampling	aims to achieve a sample whose units (e.g., people and cases) share the same or very similar characteristics or traits.
Typical case sampling	is used to study the normality/typicality of the units (e.g., people, cases, events, settings/contexts, places/sites).
Extreme (or deviant) case sampling	is used to focus on cases that are special or unusual (i.e., highlight notable outcomes, failures, or successes).
Critical case sampling	is particularly useful in exploratory qualitative research, research with limited resources, or where a single case can be decisive in explaining the phenomenon studied.
Total population sampling	includes an entire population (i.e., the total population) that have a particular set of characteristics.
Expert sampling	is used when research needs to gain knowledge from individuals who have particular expertise.

Source: Purposive sampling (2012)

- Purposive sampling or judgmental sampling—The researcher chooses the sample and selects the units (e.g., people, cases/organizations, events, pieces of data) based on his or her judgment who or what would be appropriate to be included for the study.

From the above named, purposive sampling represents not one, but a group of different nonprobability sampling techniques as shown in Table 2.6.

What Are the Implications for the Present Study?

The major part of this study consists of an extensive theoretical analysis of voluntariness and biodiversity offsets. This is empirically underlined by practical evidence through illustrative case studies. The empirical part builds on a screening of biodiversity offset cases. In line with the research design (expert-based participant–observational study grounded in qualitative social research), this screening is carried out using purposive sampling, a form of nonprobability sampling, i.e., non-exhaustive and non-representative.

From the cases that are grouped according to the built types characteristic examples are chosen (one for each type) and analyzed with regard to quality criteria (for voluntariness and biodiversity offsets, respectively). They serve to illustrate the general description of the built types.

Due to the exploratory nature of the study, both the screening and the choice of illustrative case studies built on maximum variation (maximum variation sampling), i.e., a spatially and functionally diverse set of cases.

References

Barton AH (1955) The concept of property-space in social research. In: Lazarsfeld PF, Rosenberg M (eds) The language of social research. Free, New York, pp 40–53

BirdLife International (2008) Finding ways to offset private sector impacts on biodiversity. Presented as part of the BirdLife State of the world's birds website. Available at http://www.birdlife.org/datazone/sowb/casestudy/206. Checked 24/02/2016

Bohnsack R (2007) Typenbildung, Generalisierung und komparative Analyse: Grundprinzipien der dokumentarischen Methode. In: Bohnsack R, Nentwig-Gesemann I, Nohl A-M (eds) Die dokumentarische Methode und ihre Forschungspraxis. Leske + Budrich, Opladen, pp 225–253

Corsten M, Herma H (2015) Internetbasierte Daten als sprachsoziologische Rätsel. In: Schirmer D, Sander N, Wenninger A (eds) Die qualitative Analyse internetbasierter Daten. Springer, Wiesbaden, pp 199–226

Darbi M, Ohlenburg H, Herberg A, Wende W (2010) Impact mitigation and biodiversity offsets-compensation approaches from around the world: a study on the application of Article 14 of the CBD (Convention on Biological Diversity). BfN-Schr.-Vertrieb im Landwirtschaftsverl.

Earl G, Curtis A, Allan C (2010) Towards a duty of care for biodiversity. Environ Manag (4):682–696

FERN (n.d.) Biodiversity offsetting. Available at http://www.fern.org/campaign/biodiversity-offsetting

Flick U, Kardorff EV, Steinke I (2000) Qualitative forschung. Ein Handbuch, vol 337. Rowohlt, Reinbek

Harris R (2014) Is biodiversity offsetting just a tool for the elite to extract more wealth from natural resources? LinkedIn discussion. Available at https://www.linkedin.com/groups/3667510/3667510-5829721851814776836

Hempel CG, Oppenheim P (1936) Der Typusbegriff im Lichte der neuen Logik. Wissenschaftstheoretische Untersuchungen zur Konstitutionsforschung u. Psychologie. Sijthoff, Leiden

Kluge S (1999) Empirisch begründete Typenbildung. Zur Konstruktion von Typen und Typologien in der qualitativen Sozialforschung. Leske + Budrich, Opladen

Kozinets RV (2010) Netnography. Wiley, London

Lazarsfeld PF (1937) Some remarks on the typological procedures in social research. Zeitschrift für Sozialforschung 6(1):119–139

Monbiot G (2014) The pricing of everything. Available at http://www.monbiot.com/2014/07/24/the-pricing-of-everything/

Müller R, Plieninger J, Rapp C (2013) Recherche 2.0: Finden und Weiterverarbeiten in Studium und Beruf. Springer, Wiesbaden

Nentwich M (2003) Cyberscience: research in the age of the internet. Austrian Academy of Sciences Press, Vienna, pp 479–489

Nentwich M, König R (2012) Cyberscience 2.0: research in the age of digital social networks, vol 11. Campus, Frankfurt

Nielsen M (2012) Reinventing discovery: the new era of networked science. Princeton University Press, Princeton, NJ

Nonprobability sampling (n.d.) Wikipedia. Available at https://en.wikipedia.org/wiki/Nonprobability_sampling

Purposive sampling (2012) Laerd dissertation. Available at http://dissertation.laerd.com/purposive-sampling.php

Schirmer D, Sander N, Wenninger A (eds) (2015) Die Qualitative Analyse internetbasierter Daten: Methodische Herausforderungen und Potenziale von Online-Medien. Springer, Wiesbaden

Schütz A (1932) Der sinnhafte Aufbau der sozialen Welt. Eine Einleitung in die verstehende Soziologie. Springer, Wien

Science 2.0 (n.d.) Wikipedia. Available at https://en.wikipedia.org/wiki/Science_2.0

Srubar I (1979) Die Theorie der Typenbildung bei Alfred Schütz. Ihre Bedeutung und ihre Grenzen

Ullrich CG, Schiek D (2015) Forumsdiskussionen im Internet als reaktives Instrument der Datenerhebung. Ein Werkstattbericht. In: Schirmer D, Sander N, Wenninger A (eds) Die qualitative Analyse internetbasierter Daten. VS Verlag für Sozialwissenschaften, Wiesbaden, pp 133–159

Wenninger A (2015) Hermeneutische Analysen neuer Kommunikationsformen im Internet. In: Schirmer D, Sander N, Wenninger A (eds) Die qualitative Analyse internetbasierter Daten. Soziologische Entdeckungen. Springer, Wiesbaden

3 Applying the Concept of Voluntariness to Explain Behavior Towards Environmental Conservation

Abstract

An in-depth view and critical analysis of voluntary biodiversity require the operationalization and qualification of what voluntariness means under different contexts. In practice, voluntariness has a long tradition, i.e., it is anchored in societal and ethical discourses, with regard to solidarity, the provision of common goods and voluntary commitment. However, voluntariness is a relatively new phenomenon in a scientific context.

Therefore, this study aims to define and understand voluntariness and its prerequisites. This includes defining the term voluntariness and when an action is considered voluntary. Furthermore, voluntariness is a normative concept. This implies asking why humans act voluntarily, and specifically why businesses and others implement biodiversity offsets voluntarily. While the ultimate goal of voluntary commitment is the contribution to public goods, it may also create added value. Voluntary biodiversity can thus be driven by both altruistic and selfish motivations, and these motivations need to be examined carefully in every individual case.

Building on this, different theoretical concepts were explored with regard to their suitability to operationalize and explain voluntariness of biodiversity offsets and ultimately derive criteria that can be used as attributes for building a typology of biodiversity offsets. On the one hand, this includes more general theories, i.e., altruism, economic theories and willingness to pay. On the other hand, the theoretical concept of voluntariness is transferred and examined as to its application in the scope of nature conservation and care for the environment, i.e., with regard to voluntary environmental approaches and corporate (environmental) responsibility.

M. Darbi, *Biodiversity Offsets Between Regulation and Voluntary Commitment*,
https://doi.org/10.1007/978-3-030-25594-7_3

Keywords

Voluntary commitment · Voluntary environmental approaches · Corporate environmental responsibility · Vis compulsive · Egoism–altruism paradigm · Reciprocal altruism

Voluntary biodiversity offsets have been mentioned in a few reports and articles (Howard 2007; Darbi et al. 2010; Doswald et al. 2012; Benabou 2014). However, what is most lacking in previous works is a fundamental definition and analysis of the term "voluntary" or "voluntariness." It is often construed in relation to its antonym "mandatoriness," which is mostly associated with some form of legal regulation or requirements. However, "voluntariness" is a normative term that has different connotations depending on the context in which it is applied. Thus, using a simplified definition of voluntariness and voluntary biodiversity offsets means missing a lot of its potential to explain and understand different forms of biodiversity offsets and the motivations and reasons for their emergence under various circumstances.

In practice voluntariness has a long tradition, i.e., it is anchored in societal and ethical discourses, with regard to solidarity and the provision of common goods, voluntary commitment, and work. Consequently, Badelt (2004) concludes that voluntariness is one of the most important phenomena of our civil society. In contrast, it is a relatively new phenomenon in scientific context (Hasse 2008) and a very underresearched field (Amman 2008).

Building on this scientific discussion, in the following chapter different theoretical concepts are explored with regard to their suitability to operationalize and explain voluntariness of biodiversity offsets and ultimately derive criteria that can be used as attributes for building a typology of biodiversity offsets.

The first subchapter aims at defining and understanding the notion of voluntariness and its prerequisites. This includes both a descriptive and a normative approach, which, however, cannot be sharply separated. The former focuses on the questions what the term voluntariness literally means and when an action may be considered voluntary. The latter looks at the role of voluntariness and asks why voluntariness matters and consequently, why humans act voluntarily, i.e., why businesses and others implement biodiversity offsets voluntarily.

The following subchapters aim to theoretically analyze and underpin these motivations. Therefore, different theoretical concepts are analyzed with regard to their suitability to explain the different degrees of voluntariness in biodiversity offsets. On the one hand, this includes more general theories, i.e., altruism, economic theories, and willingness to pay. On the other hand, the theoretical concept of voluntariness is transferred and examined as to its application in the scope of nature conservation and care for the environment, i.e., with regard to voluntary environmental approaches and corporate (environmental) responsibility. While willingness-to-pay focuses on a cost–benefit relation, i.e., the value of benefit, the concept of altruism focuses on the moral obligation to conserve biological diversity, i.e., the ethical value. By contrast, the concept of Environmental or Corporate Responsibility combines both moral and use value and therefore is particularly suited to examine the motivation for the implementation of voluntary biodiversity offsets.

Why Is It Necessary to Establish a Theoretically Grounded Description of Voluntariness for the Present Study?
An in-depth view on and critical analysis of (voluntary) biodiversity offsets (in particular with regard to implementation and responsibility) requires the operationalization and qualification of what voluntariness means under different contexts. Furthermore, the colloquial distinction of voluntary action from legal requirements, i.e., any action that is not mandated by legal regulations is in principle voluntary (and voluntary biodiversity offsets are all compensation measures that are not required by law), is too simplified. Voluntariness and regulation rather are intertwined concepts in various aspects. Therefore, criteria and thresholds need to be developed and derived from the literature to differentiate voluntary biodiversity offsets from mandatory biodiversity offsets and to identify, distinguish, and characterize different forms of (voluntary) biodiversity offsets.

3.1 Defining and Understanding the Notion of Voluntariness and Its Prerequisites: Definition, Terminology, and Scope

3.1.1 Descriptive Approach

As is shown in Fig. 3.1, voluntariness can refer to: (1) The nature of a decision or motive and to (2) The type of an action in terms of contributing to the public welfare/another's good. The latter can also be referred to as "voluntary commitment." Voluntary commitment is an umbrella term for different forms of voluntary, charitable action (Priller 2008).

Colloquially, voluntariness refers to the absence of coercion, or more precisely "compelled or coerced bodily movements and manipulated, programmed, or otherwise externally induced or controlled behaviors" (Flathman 1992; cf. Badelt 2004). Similarly, Aristotle (fourth century BC), whose reasoning has greatly influenced the discussion on voluntariness, states that something is non-voluntary if it is done by coercion or ignorance. Something is considered to be coerced if it is externally imposed and cannot be influenced by the acting person. Aristotle argues that any action that is characterized by some kind of willful decision can be categorized voluntary (even if the possible courses of action are significantly or completely reduced) (Nikomachische Ethik, Gutmann 2000).

Coercion traditionally (since Aristotle) distinguishes "vis absoluta" and "vis compulsiva" (see Fig. 3.2). The first refers to the immediate coercion on an individual, making him/her a "tool," so that this cannot be judged as an act of the coerced person anymore. Therefore, it is not suitable as a theoretical concept to analyze the

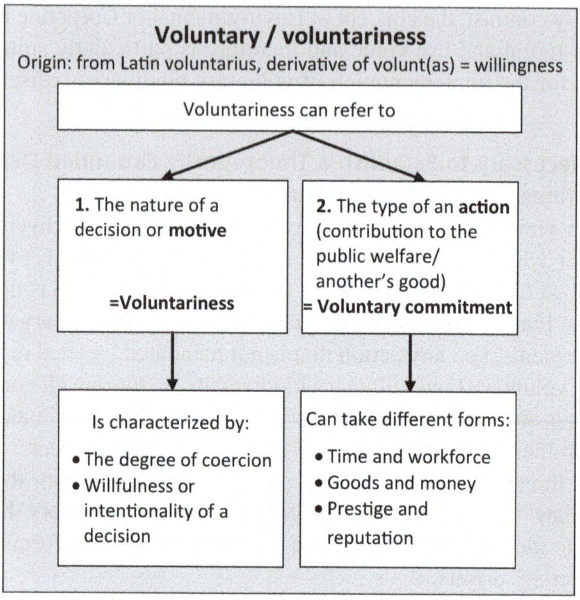

Fig. 3.1 Definition of voluntariness. Source: Author, building on Amman (2004)

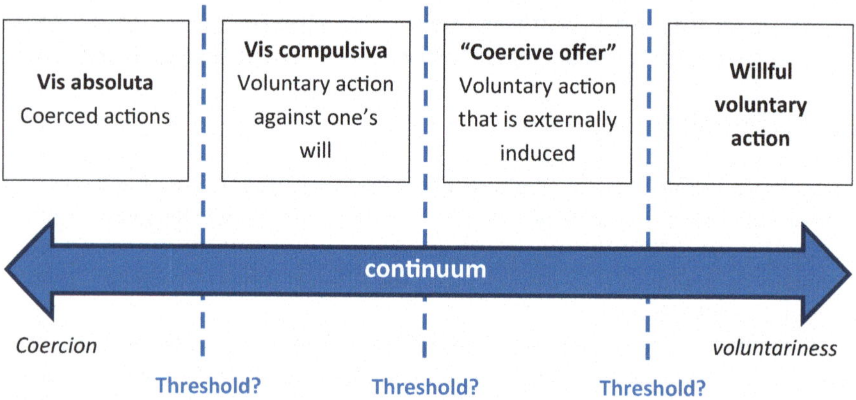

Fig. 3.2 Voluntariness as a gradual continuum. Source: Author

voluntariness and motivations of individual action. The second, "vis compulsiva" is coercion through influence on the will or more precisely the action goals of the acting person. By changing the desirability of his/her options through imposing different consequences, alternative behavior is forced. While the acting individual will change his/her action goals because his/her intrinsic preference has become unattractive by the imposed sanctions, he/she still decides voluntarily and following his/her own preferences, even though he/she acts under pressure. Vis compulsiva is the situation most likely found in practice (Gutmann 2000).

Following Gutmann (2000), voluntariness in the narrow sense of the word focuses on the volitional yield of the term, i.e., spontaneity and non-coercion. Nevertheless, acting free or voluntarily does not mean to act without any influence. By contrast, following a wide interpretation of voluntariness, Thomas Hobbes (1962, cited in Gutmann 2000) argues that any action based on motivations is in principle free. The kind of motivation itself is irrelevant for the voluntariness. As such, complying with the law for fear of legal consequences has to be considered a voluntary action.

The dichotomy between the narrow and the wider definition of voluntariness is described as the coactus volui paradox, i.e., whether or how the action of a person, who acts under the pressure of expected sanctions and against his/her own will, can still be considered somewhat voluntary, because he/she decides voluntarily between different options, thus based on his/her free will. Building on the coactus volui paradox, the question arises whether the "free will" or "free choice" is the most relevant criterion for voluntariness and whether actions that are the result of social pressure can be deemed voluntary (Jakobs 2008; Flathman 1992).

It may help to note that human action is the result of two consecutive filter processes: first, the theoretically possibilities are reduced to the viable courses of action; second, these are further reduced according to subjective preferences (this is where voluntary and non-voluntary actions can be distinguished) (Gutmann 2000). According to Benn and Weinstein (1971, cited in Gutmann 2000) freedom exists where the options are not restricted, i.e., only those actions are restricting the freedom or voluntariness of an individual that reduces the number of options or choices (option theory of freedom, Gutmann 2000).

In this context, another fundamental question arises as to whether positive incentives or offers can reduce the attractivity or possibility of choices and thus limit the voluntariness. However, Nozick (1969, cited in Gutmann 2000) argues that "offers of inducements, incentives, rewards, bribes, consideration, remuneration, recompense, payment do not normally constitute threats, and the person who accepts them is not normally coerced." Thus, he sees threats and offers as two distinct categories:

> When someone does something because of offers it is his own choice, whereas when he does something because of threats it is not his own choice but someone else's.

The discussion about voluntary action must consider which reason or rationality and which conception of individual autonomy is the prerequisite for such action. Autonomy and voluntariness are interrelated but not identical concept, i.e., voluntariness is a basic part of the individual autonomy (Gutmann 2000). Personal or individual autonomy means the normative responsibility and actual ability of a person to shape one's life and decide upon its course of action following one's own maxims (Gutmann 2000).

Autonomy must not necessarily be defined in absolute, ideal terms, but it can also be described as a threshold concept, i.e., persons do not necessarily need to fulfil an ideal concept of autonomous life, rather it is sufficient to be acceptably autonomous to determine one's own life (Gutmann 2000). Similarly, an action can be considered voluntary if certain conditions are met. Standard conditions for the voluntariness of an action can be formulated. In this regard an action is not or not entirely voluntary,

1. If it is automatic or without conscience
2. If it is unintentionally (by mistake) or
3. If it is under coercion by other individuals (Gutmann 2000)

While the first two criteria are relatively easy to distinguish, the question whether an action is coerced or externally imposed cannot be answered simply, as has been laid out above (i.e., with regard to vis compulsiva and coercive offer). This can theoretically result in two extremes, i.e., almost every action is voluntary versus almost none is voluntary (cf. vis compulsiva). This is because any threshold is highly normative and can differ from one case to another. Consequently, voluntariness can neither be derived from an ideal concept, nor from a threshold concept of autonomy (see Fig. 3.2). As Gutmann (2000) observes, autonomy can be best explained as a gradual phenomenon, i.e., a person can be more or less autonomous. In the same line of argumentation, Wolff (1740, cited in Gutmann 2000) qualifies the normative judgment of the voluntariness of actions not as a binary yes/no, but rather a "quantitative continuum."

From the legal perspective, a higher degree of regulation corresponds to a likely higher degree of coercion, i.e., deregulation results in less coercion (Gutmann 2000).

Amman (2004) describes voluntariness as a "may" rule ("Kann-Norm"), i.e., non-compliance with this rule will not result in negative sanctions, whereas complying with this rule can result in positive sanctions. This, however, does mean no relief from normal enforcement related to complying with all other laws and regulations. Thus, voluntariness is "enforcement neutral" (US Environmental Protection Agency 1992).

According to Field (2001) variations of the scope or distance of voluntary actions arise "according to historical or case specific factors." Siegenthaler (2008) argues that routines of modern state-owned solidary communities have made voluntary action at the national level partly dispensable. Instead, the most important public goods are those that are established and cared for at the global level, e.g., fighting climate change. However, voluntary action to reach global goals is more complex and likely to fail, i.e., combatting climate change will either be a global success or a global failure and can only be achieved if its significance is acknowledged worldwide. Thus, an extension of the scope of voluntary action can be observed.

What Are the Implications for the Present Study?

The terminology in the scope of this study builds on the motivational aspect of voluntariness, i.e., referring to the nature of a decision or motive. However, in practice there is a strong link between motivations and actions in that the

(continued)

motivations condition actions and motivations are uncovered by the resulting actions. Acknowledging this overlap between both terms—voluntary commitment and voluntariness—this study applies an inclusive approach of both terms and points to the differences where applicable.

Building on Aristotle's broad understanding of voluntariness, something is considered to be coerced if it is externally imposed and cannot be influenced by the acting person, i.e., any action that is characterized by some kind of willful decision can be categorized voluntary. From this stems a broad definition of the scope of voluntary biodiversity offsets, which, however, is not suitable to explain or compare the variety of biodiversity offset evidence.

Furthermore, "vis absoluta" (the immediate coercion on an individual) and "vis compulsiva" (coercion through influence on the will or more precisely the action goals of the acting person) can be distinguished.

Under vis compulsiva, the desirability of options of the acting person is changed through imposing different consequences, thus forcing alternative behavior. Thus can be concluded that acting free or voluntarily does not mean to act without any influence, i.e., actions that are the result of social pressure or even under the pressure of expected sanctions, can still be considered somewhat voluntary, if the acting person decides voluntarily between different options, thus based on free will (coactus volui paradox).

Seen that vis compulsiva is the most likely to find a situation in practice (both in general and for the specific case of biodiversity offsets), a typology of voluntary biodiversity offsets needs to specify and compare these different influences (threats/pressures and offers/incentives).

Moreover, building on the findings of Gutmann (2000) and Wolff (1740) voluntariness in practice can be explained as a gradual phenomenon (i.e., an action can be more or less voluntary) rather than an ideal or threshold concept (i.e., a binary yes/no threshold). Applying this to the case of biodiversity offsets requires to explore whether criteria exist which enable to compare the degree of voluntariness (in the sense of "case a is more voluntary than case b").

Derived potential criteria for voluntariness[1]: Influenceability (7), degree of obligation (9), choice, eligibility or alternatives (12), charitable/directed towards public good or welfare (17), scope or distance (22), free choice/ one's own initiative (23).

3.1.2 Normative Approach

The term voluntariness is a legal and moral philosophical problem, i.e., it cannot be understood without legal and socio-philosophical concepts, e.g., justice (Gutmann

[1]see Sect. 5.2 and refer to corresponding numbers in Table 5.1.

2000). Thus, according to Raz (1986, cited in Gutmann 2000) there is no value-free, purely descriptive definition of voluntariness—at least none that is appropriate for moral, legal, or political phenomenas. Coercion and voluntariness are "essentially contested concepts" (Gallie 1956, cited in Gutmann 2000), i.e., complex normative terms that are highly context-dependent and subject to normative interpretation.

Voluntariness means identification with the values and concerns that are shared in a certain society and shared by the voluntary commitment of individuals (Boothe 2004), i.e., it is shaped by the societal framework, with its cultural, social, and political institutions, in which it arises (Priller 2008; Flathman 1992). Putnam (2000) notes:

> Social connections encourage giving for many reasons. Joiners may be generous souls by nature, but involvement in social networks is a stronger predictor of volunteering and philanthropy than altruistic attitudes per se.

Therefore, an analysis of voluntary commitment needs to be seen in its societal and organizational, i.e., social and cultural contexts. While Amman (2001) supposes that voluntariness has an anthropologic core, that is independent of the organization and structure of the society within which it arises, i.e., it is an intrinsic human characteristic to care for weaker parts or members of society (altruism, see Sect. 3.2.1), observations by Priller (2008) suggest that the influence of country-specific historic and cultural contexts shapes the societal structures and thus the voluntary commitment. As can be noted, voluntary commitment is stronger in some countries and weaker in others. The specific causes include national traditions and values, currently legal, economic, and general societal conditions (Priller 2008). In light of this context dependency, Flathman (1992) argues that "genuinely voluntary conduct is possible, but only for persons who are "situated" in a society or culture with these characteristics."

What can be concluded from this discussion is that the source of voluntary commitment may lie either in social norms and ties or in the individual (or most likely in a combination of both), but the societal context shapes or conditions how this voluntary commitment is translated into action (Fig. 3.3).

Many consider voluntary commitment to have an important function to complement governmental activities (e.g., Priller 2008; Hasse 2008; Carigiet 2004), i.e., it is neither in competition with nor an alternative to the welfare state. By contrast, some not only consider its function complementary, but rather raise the question whether the same results could not be reached with less effort and thus more efficiently (e.g., Khanna 2001; Alberini and Segerson 2002; Segerson and Li Dawson 2001). This, however, does not imply questioning the goals and responsibilities of government (Hasse 2008). Instead, Carigiet (2004) highlights the role of the state as enabling, motivating, and activating institution. To conclude, Priller (2008) suggests an integrative understanding of government-mandated action and voluntary commitment. He claims that government, business, and society will increasingly need to work together in the future.

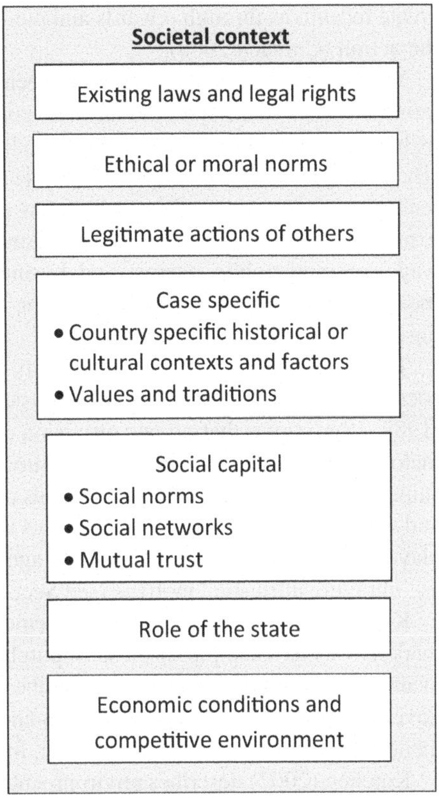

Fig. 3.3 Influence of the societal context on voluntary commitment. Source: Author (building on Priller 2008; Flathman 1992; Hasse 2008; Carigiet 2004; Putnam 2000)

According to Cranach (2008) voluntary actions get their direction by a goal and their energy by motivation. The goal of voluntary commitment is oriented toward public matters and thus can be distinguished from actions merely or mostly oriented toward private benefit (Priller 2008). Similarly, Amman (2004) defines voluntariness as the contribution of an individual to the public welfare or public goods[2] that the individual has chosen freely.

Seen that the individual rationality conditions voluntary commitment, from an individual, psychological perspective, the question of the motivation is central, i.e., why humans act voluntarily. Motivations can be distinguished according to two criteria. They can, on the one hand, be directed toward selfish or selfless (altruistic, see Sect. 3.2.1) goals and on the other hand be intrinsically or extrinsically motivated. Intrinsic motivations include freedom of utilitarian purpose, self-determination, being happily absorbed in an action or for a certain goal and stimulating social integration (Cranach 2008; Badelt 2004). Extrinsic motivations

[2]Public goods are those that create generally and publicly accessible use and benefits (Siegenthaler 2008).

create incentives through rewards and punishments that are not directly connected to the action (Cranach 2008).

With regard to the connection between altruistic and selfish motivations, voluntariness can (in addition to the ultimate goal of being oriented toward public matters as has been mentioned above) create added value for the voluntarily acting person (Badelt 2004). Consequently, Olk (2004) suggests that the motivations and goals in real-world situations are the result of new compositions and combinations of the two extremes common public good (selflessness) and self-interest. In the best possible way, common public interest and business sense can collude promisingly. This means that an action can be serving the public good without being altruistic and unselfish (Siegenthaler 2008).

Thus, voluntary engagement not only represents an instrument to fulfil certain societal tasks, but carries a high intrinsic value and societal relevance. Andreoni (1988, 1990) states that private provision of public goods can be motivated by many factors, apart from being purely altruistic, i.e., striving for prestige, respect, friendship, and other social and personal goals on the one hand and social pressure, guilt, and sympathy on the other. He also notes that "simply a desire for a warm glow may play important roles in the decision of agents" (Andreoni 1990). He describes these as "impurely altruistic" motives (see Sect. 3.2.1).

Kotchen (2006) cites markets for environmentally friendly goods, i.e., "green markets," as an example of these impurely altruistic motivations and the increased availability of impure public goods[3] in the economy. Green markets are promoted by governments, NGOs, and industries to encourage the private provision of environmental public goods (such as a healthy, intact environment).

Kotchen (2007) describes environmental offsets as a particular case of voluntary provision of public goods. He suggests that "provision may occur, in part, to offset other activities that have an adverse effect on a public good." He observes a growing number of examples motivated in this way in the environmental context (citing carbon offsets as the most prominent example). Extrinsic motivations associated with social approval or signaling could be important (Kotchen 2007) and businesses, NGOs, and others can use it to legitimate themselves and their actions, e.g., in the scope of their public relations (Hasse 2008). Kotchen (2007) concludes that in larger economies "environmental offsets will always persist, as compensating for consumption of the polluting good becomes even more important."

[3]According to Kotchen (2006), individuals typically have three relevant choices: a conventional (pure private) good, a direct donation to an environmental (pure public) good, and a green (impure public) version of the good, whereas the latter is the result of a joint production of a private good and an environmental public good and thus combines characteristics of the other two.

What Are the Implications for the Present Study?

Voluntariness is a normative concept and highly context dependent. This means that developing a typology of voluntary biodiversity offsets in the scope of this research needs to build on a number of normative decisions to be made by the researcher (e.g., choice of criteria). Consequently, the suggested typology cannot claim to be universal (there may be others similar to the developed one or completely different from it), rather it is a result of the focus on a specific research subject (and the context into which it is embedded) from a particular research perspective.

Given the context dependency of voluntary commitment, Priller (2008) suggests that the influence of country-specific historic and cultural contexts shapes the societal structures and thus the voluntary commitment. Therefore, an analysis of voluntary biodiversity offsets needs to include their societal and organizational context with regard to the legal/political (including role of government), economic/competitive, and social/cultural environment.

This study follows Cranach (2008) in that voluntary actions get their direction by a goal and their energy by a motivation. The goal of voluntary commitment is oriented toward public welfare, i.e., the contribution of an individual to public goods.

In addition to the ultimate goal of being oriented toward public matters (i.e., the contribution of an individual to public goods), voluntary commitment can create added value for the voluntarily acting person. It can thus be driven by both altruistic and selfish motivations. This means that common public interest and business sense can collude promisingly. Examples for this are green markets that are increasingly promoted for the private provision of environmental public goods. Environmental or biodiversity offsets are a particular case of voluntary provision of public goods, that may be driven by a variety of motivations, e.g., motives associated with social approval or signaling could be important. Thus, the question of the motivation of voluntary biodiversity offsets needs to be examined carefully in every individual case.

Derived potential criteria for voluntariness[4]: Motivation (3), Ultimate goal (4), Recompense/benefit (5), Public relations benefits (6), Role of government (11), Context (14), Social and individual commitment/societal context (16), Charitable/directed towards public good or welfare (17), and Cost efficiency (20).

---

[4]see Sect. 5.2 and refer to corresponding numbers in Table 5.1.

3.2 General Theoretical Concepts to Analyze or Explain Voluntariness

3.2.1 The Egoism–Altruism Paradigm

When discussing the motivations of voluntary action and private provision of public goods, this ultimately leads to the contrasting juxtaposition of "egoism" and "altruism" as inherently human characteristics. Thus, the question can be asked, whether voluntary action is altruistically or egoistically motivated. While Darwinist explanations of evolution and biological correlations focus on egoistic motivations in the sense of the struggle for existence, i.e., "humans are driven by material self-interest" (Field 2001), newer observations and findings conclude that humans are just as much disposed to cooperation, support, and solidarity as competition and fight. The dedication for the good of others thereby also adds to one's own well-being (Cranach 2008).

In this regard, Batson (2014) argues that according to advocates of universal egoism "everything we do, no matter how noble and beneficial to others, is really directed toward the ultimate goal of self-benefit." This includes obvious external material rewards or public recognition on the one hand and internal self-rewards, i.e., congratulating ourselves for being kind and caring or escaping shame and guilt. Other self-referenced motivations are formulated as "a taste for having other people better off" (Margolis 1982, cited in Liebe et al. 2011), the preference "to do good" (Liebe et al. 2011), "a warm glow of giving," or "purchase of moral satisfaction" (Kahneman and Knetsch 1992; Andreoni 1990). Liebe et al. (2011) conclude that individuals "may derive utility from altruistic behavior per se, independent of the fact that others will be better off."

In contrast, advocates of altruism believe that humans are also capable of pursuing the ultimate goal of benefiting someone else. Field (2001, "Alturistically inclined") has pointed out the thesis that humans are born with a predisposition to act altruistically and serve the public good.

As is obvious from the above considerations, altruism and egoism are closely related concepts that have much in common, i.e., one cannot be explained or understood without the other (similar to "voluntariness" and "coercion"). This contrasting juxtaposition (and the term "altruism" as such) was coined by nineteenth century philosopher Auguste Comte.

> Comte considered altruism and egoism to be two distinct motives within the individual. He did not deny the existence of self-serving motives, even for helping: the impulse to seek self-benefit and self-gratification he called egoism. But Comte believed that some social behavior was an expression of an unselfish desire to 'live for others.' (Batson 2014)

He described both egoism and altruism as motivational states that differ ultimately regarding the goal of their motivation, i.e., the distinction between egoistic and altruistic motivation is qualitative, not quantitative. Following up on this, Batson (2014) suggests defining altruism as "a motivational state with the ultimate goal of increasing another's welfare." Consequently, egoism is defined as "a motivational state with the ultimate goal of increasing one's own welfare." These definitions include three aspects (Batson 2014):

1. The focus on "a motivational state."
2. The orientation toward "the ultimate goal."
3. The aim of "increasing another's welfare" vs. "increasing one's own welfare."

According to Lewin (1935, cited in Batson 2014), motivation is energy and a force within the individual. This energy translates into some goal-directed motivation, which is characterized by five key features (Batson 2014):

- A person perceives "a negative discrepancy between [one's own or] another person's current state and potential state on one or more dimensions of well-being" (a need), thus aiming for some desired change (goal).
- A force of some magnitude exists (motivation), driving the person toward this goal (i.e., reduction of the need).
- If a barrier prevents direct access to the goal, alternative routes are sought.
- The force disappears on the attainment of the goal.
- Reaching the goal is likely to be pleasurable for the person.

As is obvious from these key features, "[f]or a goal to be an ultimate goal, it must be an end in itself and not just an intermediate means for reaching some other goal. If a goal is an intermediate means for reaching some other goal, and a barrier arises, then alternative routes to the ultimate goal that bypass the intermediate goal will be sought. And if the ultimate goal is reached without the intermediate goal being reached, then the motivational force will disappear and pleasure will be felt." (Batson 2014).

As per definition, altruism is directed toward the benefit of others. However, this is not restricted to human beings but also to environmental properties and conditions (Liebe et al. 2011).

> Altruistic motivation can lead to perceived obligations to contribute to the preservation of environmental goods. For example, people might feel that they are obliged to do something about deforestation of rain forests. It follows that a financial contribution to a specific environmental good may give personal satisfaction and thus yields individual utility when there are feelings of moral obligation. (Liebe et al. 2011)

Pure altruism means to sacrifice something (time, goods, prestige) to the benefit of someone other than oneself, without expecting any personal (direct or indirect) benefits in return. This theoretical model of altruism is, however, very idealized and seldom holds to be true in practice. Batson (2014) explains that an altruistic motivation does not imply self-denial, i.e., an action can be considered altruistically motivated even if it involves some form of self-benefit or individual advantages, as long as the ultimate goal is oriented toward another person's benefits (Batson 2014).

On the contrary, such advantages can provide incentives for individual and collective actors to undertake voluntary actions (this is especially true for the motivation of corporations to demonstrate Corporate Responsibility, see Sect. 3.6). The most commonly cited advantages are (Hasse 2008):

- Prestige and (societal) recognition and reputation
- Increase in societal status
- Access to social networks (social capital)
- Acquisition of competences and qualifications

Thus, altruistic behavior is usually rewarded in terms of societal status and recognition at the least. It is even more common that voluntary and altruistic action entails some form of reciprocity. Robert Trivers (1971, cited in Field 2001) has established the theory of reciprocal altruism to explain "impure altruistic" behavior. Putnam (2000) explains reciprocal altruism and the inherent altruism–egoism relation as a temporal phenomenon:

> The touchstone of social capital is the principle of generalized reciprocity—I'll do this for you now, without expecting anything immediately in return and perhaps without even knowing you, confident that down the road you or someone else will return the favor. (Putnam 2000)

Voluntary contribution to improving environmental quality can be explained by different theoretical concepts, such as fairness, inequity aversion, and reciprocity (Liebe et al. 2011). In fact, as has been outlined above, reciprocal altruism is most commonly encountered in practice.

Despite the existing multitude of motivations, including selfish and selfless, intrinsic and extrinsic as well as hidden motives or goals, it is obvious to note, that the concept of altruism is crucial to include in any normative evaluation of voluntariness. Jakobs (2008) argues that voluntary action is something that is done for predominantly altruistic reasons, without external obligation or coercion. Similarly, Hasse (2008) states that one common attribute across the broad spectrum of voluntary actions is that the aim is not primarily oriented toward one's own interests, but toward a "noble cause," i.e., the good of others (especially socially or otherwise disadvantaged) or widely accepted abstract principles (e.g., of environmental protection and sustainability).

What Are the Implications for the Present Study?

The motivations of voluntary action and private provision of public goods, can ultimately be reduced to the contrasting juxtaposition of "egoism" and "altruism" as inherently human characteristics. Altruism and egoism are close, contrasting motivational states within the individual that differ only in their goal. Single motives are either egoistic or altruistic, but (1) true motives may

(continued)

be hidden (even to the acting person) and (2) different motives can coexist. Furthermore, whether a motive forces an action, depends on the context (Batson 2014).

Thus, the question can be asked, whether voluntary action is altruistically or egoistically motivated. Building on this, two general reasons can be distinguished for the implementation of voluntary biodiversity offsets, which build on the value assigned to biological diversity by humans:

1. Value of benefit: Implementing biodiversity offsets creates added value (egoism).
2. Ethical value: Moral motivation for the implementation of biodiversity offsets (altruism).

Voluntary action is commonly understood to be done not for one's own interests (above all), but for predominantly altruistic reasons, i.e., directed toward the benefit of others, including humans but also the preservation of the environment. Thus, it is obvious to note, that the concept of altruism is crucial to include in any normative evaluation of voluntariness. The theoretical model of altruism is, however, very idealized and in practice voluntary action often entails some form of reciprocity. Reciprocal altruism implies no self-denial of the acting person, i.e., an action can be considered altruistically motivated even if it involves some form of self-benefit or individual advantages, as long as the ultimate goal is oriented toward another person's benefits. For the case of voluntary biodiversity offsets, this means that the potential advantages that a business may gain from implementing an offset (e.g., societal recognition and reputation, access to social networks) can provide incentives for individual and collective actors to undertake voluntary actions. It is thus important (but also difficult) to explore the existing multitude of motivations, including selfish and selfless, intrinsic and extrinsic, as well as hidden motives or goals.

Derived potential criteria for voluntariness[5]: Motivation (3), Ultimate goal (4), Recompense/benefit (5), and Charitable/directed towards public good or welfare (17).

3.2.2 Economic Theories

Historically, economic approaches have focused on the rationality and selfishness of human actions. According to Field (2001) economists "have been perfectly happy appealing to Darwin as justification for the assumption that humans are driven by material self-interest advanced by a general-purpose reasoning capability (humans

[5]see Sect. 5.2 and refer to corresponding numbers in Table 5.1.

are self-interested and pursue their goals rationally)". In this line of reasoning, rational choice theory can be considered a useful tool to analyze voluntariness and voluntary action, i.e., that actions of a person are usually reasonably or rationally determined and the result of a cost–benefit analysis (Kausch 2008). Similarly, Olk (2004) points to the homo oeconomicus as the central character of rational choice theory. If homo oeconomicus acts voluntarily, this is considered a wise form of pursuing individual interests, i.e., it is a more reasonable and promising form of egoism. Thus, voluntariness is seen as a social exchange process, because homo oeconomicus expects some physical or intangible trade-offs (cf. the concept of reciprocal altruism above). Badelt (2004) adds that this exchange can be implicit or explicit.

Rational choice theory analyses the question how rational individuals have to act to realize their subjective preferences in terms of priority relations (Gutmann 2000). The individual evaluation of the consequences of an action and the respective building of preferences are more dependent on the frame, in which a problem occurs, i.e., the same problem in different frames/situations will most likely result in different individual preferences (Gutmann 2000). Social arrangements (including a legal system) can foster or hinder self-constraints under the form of a commitment or negotiated agreement. This means an influence on the own preferences, either by intentionally trying to change his preferences or by trying to resist changes (Gutmann 2000). For the individual, however, it may be considered rational to restrict or sacrifice possible courses of action for the own long-term autonomy.

One frequently used approach in economics to analyze and predict the actions and motives of humans is game theory. While in evolutionary biology (where game theory is also used) there is no consideration of motives and no consideration of conscious thought or choice, or intentionality (and everything is measured in terms of increased "fitness" of oneself or others to spread the organism's genes in future generations), in game theoretic models in economics, the emphasis is on rational choice, i.e., driven by rational decisions based on motives, and "the payoffs are usually in the form of material benefits" (which may but will not necessarily increase fitness) (Field 2001).

Following the works of von Neumann and Morgenstern on game theory, the individual acts as homo oeconomicus by striving to reach the maximum use (Gutmann 2000). Nevertheless, Gutmann (2000) also points to the limits of homo oeconomicus: homo oeconomicus is neither an unsocial egoist nor an altruistic or aesthetic goals be explained. Field (2001) notes:

> In economic models motives may matter in determining whether an action will be characterized as altruistic. And assumptions that agents act in ways contrary to their individual interest will often be attacked ab initio on the grounds that such behavior is not consistent with rationality.

In fact, decision rationality is restricted in practice, i.e., in actual decision situations persons do not tend to maximize their use value, but instead aim at an acceptable solution with least effort. Thus, in reality, individuals do not act fully rational, rather they tend to rely also on experience, emotions, and intuition (Gutmann 2000).

Altruistic or nonprofit behavior so far was mostly explained by sociology as a natural consequence of socialization processes. In recent times, this traditional distinction has become blurred by the introduction of the new field of "neuro-economics." As opposed to the rationality of the "homo oeconomicus" people often do not act selfishly and resign selfish goals for the sake of other goals considered to be more important (Siegenthaler 2008).

This can be analyzed using a game theoretic model. In real-world settings, so called Prisoners' Dilemma situations often arise. The Prisoners' Dilemma is the best-known game of strategy in social science. It helps us understand what governs the balance between cooperation and competition in business, in politics, and in social settings.

> In the traditional version of the game, the police have arrested two suspects and are interrogating them in separate rooms. Each can either confess, thereby implicating the other, or keep silent. No matter what the other suspect does, each can improve his own position by confessing. If the other confesses, then one had better do the same to avoid the especially harsh sentence that awaits a recalcitrant holdout. If the other keeps silent, then one can obtain the favorable treatment accorded a state's witness by confessing. Thus, confession is the dominant strategy (see game theory) for each. But when both confess, the outcome is worse for both than when both keep silent. (Dixit and Nalebuff 2008)

The implications of the Prisoners' Dilemma in social interaction suggest the pursuit of personal benefit to the disadvantage of others. This leads to a social dilemma, because either one party or all parties are worse off in the end. An example of this phenomenon of the exploitation of others is the provision of public goods. As Liebe et al. (2011) note, the essential non-excludability of public goods, i.e., no one can be excluded from using it, independently of whether he has contributed to its provision, leads to a social dilemma.

> Social dilemmas are situations in which individual rationality leads to collective irrationality. [. . .] As individuals we are each better off when we make use of a public resource, such as public television, without making any contribution, but if everyone acted on this conclusion, the public resource would not be provided and we would all be hurt. (Kollock 1998, cited in Liebe et al. 2011)

This is the free rider dilemma, i.e., individual rationality provides an incentive to rely on the contributions of others and use the public good as "free rider."

Despite this free rider problem, Field (2001) finds that "in complex social settings where people have established reputations they care about and where they may anticipate repeat engagements with their counterparties, most people most of the time play cooperate when finding themselves in PD-like situations." This means that they do not try to be better off at the expense of others. Cooperation thus seems to be a highly dominant action strategy not only for pure altruists who solve the single play Prisoner's Dilemma without difficulty (Field 2001).

> Most people in most situations tend to cooperate without even thinking about whether this is the "smart" thing to do. In many instances such behavior turns out, ex post, to benefit the actor. Indeed, the very fact that such behavior later appears to have been "smart" or "intelligent" sometimes feeds skepticism that there is anything altruistic about the behavioral motivators that underlie it. (Field 2001)

While helping someone can also benefit the helping individual without direct reward, many people seem to cooperate if and only if others cooperate in return. This so-called "conditional cooperators" act is dependent on the (perceived or anticipated) contribution of others (Liebe et al. 2011).

Thus, conclusions can be drawn that people are more cooperative:

- With individuals they will most likely interact with in the future.
- If their helping behavior is communicated, publicly announced, or observed by others.
- If they see and experience cooperative behavior by others.

Reciprocal altruism (as has been outlined in Sect. 3.2.1) is a game theoretic example of cooperation.

> Each individual act in a system of reciprocity is usually characterized by a combination of what one might call short-term altruism and long-term self-interest: I help you out now in the (possibly vague, uncertain, and uncalculating) expectation that you will help me out in the future. Reciprocity is made up of a series of acts each of which is short-run altruistic (benefiting others at a cost to the altruist), but which together typically make every participant better off. (Taylor 1982, cited in Putnam 2000)

What Are the Implications for the Present Study?

The economic model of utility maximization is based on the rational choice of "homo oeconomicus." In this sense, acting voluntarily ultimately means pursuing individual interests, i.e., driven by rational decisions based on motives, thus expecting payoffs in the form of material benefits or intangible trade-offs. Consequently, in rational choice theory, voluntary commitment can be interpreted as a cost–benefit analysis and social exchange process.

Social arrangements (including a legal system) can foster or hinder self-constraint under the form of a commitment or negotiated agreement. This can also be explained using game theory (one frequently used approach in economics to analyze and predict the actions and motives of humans). Accordingly, most people most of the time choose cooperation and voluntary action, even though this might not be directed toward their own short-term interest. However, voluntariness is largely based on generalized reciprocity, i.e., conditional cooperation if others are perceived or anticipated to cooperate in return.

(continued)

Applied to the present study this means that both the (social) context (in particular in terms of interrelation and cooperation with others, e.g., competitors, regulators, local communities) and the expected benefits (which may be closely tied to the social context, e.g., in terms of reputation) influence or drive the implementation of voluntary biodiversity offsets.

Derived potential criteria for voluntariness[6]: Intentionality (1), Motivation (3), Recompense/benefit (5), Influenceability (7), Context (14), Competitive environment (15), Social and individual commitment/societal context (16), Charitable/directed towards public good or welfare (17), Influence.

Table 3.1 Policy instruments to address biodiversity loss

Policy instruments to address biodiversity loss		
1	Conservation reserves	Direct land acquisition by governments and non-government organizations
2	Regulation	"Command-and-control," self-regulation, cross-compliance measures
3	Voluntary approaches	Engagement of landholders or community groups for natural resource management activities by incentives
4	Education and awareness	Information and technical support for building social and human capital in natural resource management activities
5	Economic approaches	Property-based instruments (e.g., leasehold agreements and covenants/easements that attach to title), revolving funds for land purchase and resale, tax rebates, market-based instruments (auctions, offset schemes with tradable rights)

Source: Author (building on Earl et al. 2010)

3.3 Governance and Duty of Care for Biodiversity

Earl et al. (2010) assert that it is globally accepted that "no single policy instrument will bring about optimal outcomes for natural resource management." They conclude it is best to use multiple approaches or a mix of instruments, including regulation (see Table 3.1).

This observation is accompanied by a trend "from government to governance" based approaches. As Paavola et al. (2009) note, definitions and understandings of environmental governance range from "new ways of achieving social objectives" (with the state being restricted to a participatory role) to "all attempts to address environmental dilemmas or to resolve environmental conflicts" by suitable institutional arrangements (Paavola et al. 2009). The goal of governance regimes is "to deliver public policy objectives in an effective, efficient, equitable, transparent and accountable way." The role of the state has been said to have shifted "from provider

[6]see Sect. 5.2 and refer to corresponding numbers in Table 5.1.

Table 3.2 Key characteristics of duty of care and stewardship

Duty of care	Stewardship
Moral basis	Moral basis
Externally imposed	Internal, personal
Community standard	Individual standard
Obligatory	Voluntary
Articulated, codified	May show in behavior
Enshrined in law	Not enshrined in law
Specific	All encompassing

Source: After Earl et al. (2010)

and controller to facilitator and enabler," fostering the capacities of non-state actors to realize public policy objectives. In this regard, the use of new environmental policy instruments (NEPIs) is being increasingly supported, i.e., market- and information-based instruments and voluntary agreements.

Earl et al. (2010) distinguish duty of care and stewardship, whereby the duty of care is defined as the legal concept, a responsibility imposed by society, to avoid causing any damage; and stewardship is defined as the responsible management of something not for oneself, but for others, including biodiversity and a healthy environment. They differentiate the duty of care as related to the present and stewardship as related to the future. Table 3.2 opposes key characteristics of duty of care and stewardship.

> Specifically 'duty of care' has a moral basis but is usually enshrined in legislation. It is externally imposed and obligatory, but it reflects a standard that is widely held by society, in effect a social norm, or one that governments seek to embed in society. It is usually articulated and/or supported by codes or guidelines, and is specific to particular entities. 'Duty of care' also appears to be about the present. 'Stewardship' also has a moral basis, but is clearly a personal ethic, internally driven and reflecting an individual, voluntary standard. It may show in behavior, but other concerns may limit its expression. 'Stewardship' is never translated into legislation. In contrast to 'duty of care', it tends to be more worldly and holistic, and often has a spiritual foundation. An emergent and related characteristic of 'stewardship' is its focus on 'eternity' rather than the immediate. (Earl et al. 2010)

What Are the Implications for the Present Study?

Earl et al. (2010) claim that increasingly a mix of instruments is needed for best outcomes for a global natural resource management. This includes regulation, market-based instruments, and voluntary approaches. The use of multiple approaches also means a shift "from government to governance," i.e., a change in the role of the state "from provider and controller to facilitator and enabler." Thus, new institutional arrangements encourage the inclusion of the capacities of non-state actors to realize public policy objectives. This may refer to a justification of both voluntary environmental approaches in general and voluntary biodiversity offsets in particular. However, only a limited number of criteria can be derived from this.

(continued)

Derived potential criteria for voluntariness[7]: (Degree of) obligation (9), Role of government (11), Context (14), Competitive environment (15), Social and individual commitment/societal context (16), Taking over responsibility (24), Ethics (25).

3.4 Voluntary Environmental Approaches

Development and Terminology of Voluntary Environmental Approaches

In the 1960s, 1970s, and beginning of the 1980s regulation was considered crucial in environmental policy to ensure adequate environmental protection and control of environmental externalities. "Command-and-control" regulations (compulsory standards) that impose requirements or restrictions (e.g., the imposition of liability for damages, direct regulation of processes or product quality) have been widely applied (Blowfield and Murray 2014). This "government-push" approach has encouraged end-of-pipe pollution controls by prescribing emissions thresholds and specific pollution control and abatement technologies. On the one hand, this approach has proven to be effective, i.e., improved the environment, on the other hand, economists have criticized traditional command-and-control regulation not only for its lack of flexibility, but also for the high costs associated to it (i.e., lacking efficiency) (Khanna 2001).

As a result, a broadening of responsibility for environmental protection beyond governmental action can be observed. This has led to two developments: first, economic incentives have become more common and market-based instruments have been increasingly used for environmental regulation by the late 1980s (market-based pressures that create demand for self-regulation by firms) (Lyon and Maxwell 2002). This include taxes, subsidies, emissions charges, and tradable permits (e.g., emissions trading programs), certification, or labeling.

Second, Maxwell and Lyon (2001) note that fundamental changes in the 1990s (mounting and increasingly complex legislation, technological innovation and scientific discoveries, regulatory budget cutbacks, increasing use and effectiveness of citizen lawsuits), especially in Europe, have led to the exploration of voluntary environmental approaches as a promising tool in future environmental policy and as alternative to traditional regulation (Delmas and Terlaak 2001; Bosch et al. 1995; Carraro and Siniscalco 1996; Segerson and Miceli 1998; Wu and Babcock 1999).

In this regard, in 1996 the European Commission has issued a Communication on Voluntary Agreements (COM(96 561, OJ (1996] L333/69), signaling a "cautiously positive view" as regards their use at national level. Consequently, the Fifth Environmental Action Program of the EU suggests a new line of cooperation with business (van Calster and Deketelaere 2001):

[7]see Sect. 5.2 and refer to corresponding numbers in Table 5.1.

> Whereas previous environmental measures tended to be prescriptive in character with an emphasis on the "thou shalt not" approach, the new strategy leans more towards a "let's work together" approach. This reflects the growing realization in industry and in the business world that not only is industry a significant part of the (environmental) problem but it must also be part of the solution.

Voluntary programs or approaches are referred to using a large variety of terms that are mostly used synonymously: voluntary agreements, voluntary initiatives, business-led environmental strategies, corporate environmentalism, self-regulation, negotiated agreements, environmental covenants, voluntary codes etc. (Lyon and Maxwell 2002).

> The terms 'voluntary agreement' or 'voluntary approach' can be defined in a number of different ways, with the differences driven mainly by the scope of which types of actions are included. The narrowest definition is from Rauscher (1997), who includes in this category only those abatement actions taken completely out of self-willingness or altruism. An alternative definition defines a voluntary agreement as 'an agreement between government and industry to facilitate voluntary action with a desirable social outcome, which is encouraged by the government, to be undertaken by the participant based on the participant's self-interest' (Storey et al. 1999). (Segerson and Li 1999)

Thus, depending on the definition, voluntary environmental approaches provide an umbrella under which a variety of different tools fit (environmental programs, agreements etc.).

Types of Voluntary Environmental Approaches

A number of authors have tried to distinguish different forms or types of voluntary approaches (Khanna 2001; Alberini and Segerson 2002; Carraro and Lévêque 1999; Segerson and Li Dawson 2001; Segerson and Li 1999; Lyon and Maxwell 2002). Although the terminology differs slightly, the different classifications basically refer to the same three types as shown in Table 3.3: (1) Unilateral Business-led initiatives/ commitments by the industry, (2) Bilateral (negotiated) environmental agreements between industry and government, and (3) Voluntary government programs/ schemes.

These three types ultimately differ regarding the role or degree of involvement of government: passive (1), collaborating (2), and active (3). Thus, regulators can actively implement innovative voluntary programs (and determine both the rewards and obligations from participation, as well as the eligibility criteria), that encourage firms to "overcomply" and go beyond defined minimum standards (Arora and Cason 1995).

By contrast, business-led initiatives are characterized as "unilateral action by a single polluter or a group of polluters, which occurs without any regulatory involvement" (Alberini and Segerson 2002). Khanna (2001) further specifies unilateral business-led initiatives and divides them into three sub-groups:

> Firms can: (a) develop their own plans or management systems to improve their own environmental performance (b) participate in codes of conduct or guidelines developed by trade associations, and (c) meet the environmental performance standards for registering with a certifying organization, such as the International Organization for Standardization.

Table 3.3 Three types of voluntary approaches for environmental protection

(1)	Unilateral Business-led initiatives/ commitments by the industry	The initiative is taken by the firms or the industry as a whole. The government does not play an active role in determining the industry response
(2)	Bilateral (negotiated) environmental agreements between industry and government	Environmental agreements are formally negotiated between industry and public authorities. This entails an exchange of voluntary environmental protection measures against some concession granted by the government (e.g., the government could guarantee that it would not impose a certain regulation or control on the industry or firm)
(3)	Voluntary schemes by the government	Governments might also unilaterally develop a voluntary scheme and seek for industry participation, e.g., by incentives such as payments or other benefits in exchange for voluntary actions to protect environmental quality

Source: After Segerson (1999)

Motivations for Voluntary Environmental Approaches

Besides the question of whether voluntary environmental approaches can deliver the same or sometimes even better outcomes than environmental policy (efficiency), another fundamental question arises as to why business and industry would participate in voluntary initiatives. Segerson (1999) and Alberini and Segerson (2002) suggest that the decision whether to adopt a voluntary approach or not is the result of a cost–benefit analysis (cf. rational choice theory and utility maximization as outlined in Sect. 3.2.2), which depends on several interacting factors (cf. Table 3.4):

1. The expected change in net revenues (through shifts in demand).
2. The likelihood that mandatory controls will be imposed if the firm chooses not to adopt a voluntary approach.
3. The cost difference between meeting the standard voluntarily and being forced to meet it.
4. The legal rules related to payments for damages from contamination.
5. The availability of any direct financial inducements from government.

As Alberini and Segerson (2002) point out, there is potentially a trade-off associated with the participation of polluters in voluntary programs. They describe the conditions for participation in a voluntary program as a bargaining process between regulator and polluter(s). This interpretation of the participation in voluntary environmental programs as a tradeoff, with related bargaining processes and

Table 3.4 Motivations and incentives for voluntary approaches

Motivation (pressure/incentive) and source/initiator	What are the specific motivations? How do they work?
Environmental stewardship *Source/initiator*: Business internal	• *Intrinsic motivation*: Personal satisfaction or utility gained from undertaking environmental protection measures (under individual control)
Market-based incentives *Source/initiator*: Consumers and competitors	• *Consumer demand and affordability*: Consumers with "green preferences" can create demand for "green products" (generating price premiums for green products and eco-labeling), in particular in developed countries, where higher income levels have led to an increased demand for cleaner environmental records as compared to their counterparts in less developed countries • *Competitive advantage*: Firms can use environmental quality and performance (and the related "green" advertising) as criteria to stand out in competition (through changing in production processes or product characteristics, i.e., by differentiating their products and services) thus weighing higher costs against increased market shares and firm profits • *Competitive pressure*: Firms' proactive efforts may provide direct regulatory relief and shape future environmental standards (thus, exerting pressure on competitors and creating a barrier to entry by new firms) • *Industry concentration and pressure*: Voluntary initiatives are more likely in concentrated industries with fewer firms
Government-created incentives *Source/initiator*: Government	• Government-created incentives can either target polluters individually or a whole industry or group of polluters collectively. Furthermore, "carrot" and "stick" approaches can be distinguished • *"Carrot" approach*: Regulatory agencies create positive incentives under the form of technical assistance, financial subsidies or cost-sharing schemes, thus increasing the likelihood that private benefits exceed private costs, i.e., firms' costs of learning and abatement can be lower than under existing mandatory achievements • *"Stick" approach*: Regulatory agencies create negative incentives by inducing an explicit or implicit threat to impose a policy, thus for polluters participation in a voluntary abatement program may be the "lesser of two evils" (fear of liability for environmental contamination leads to preempting regulatory threats)
Incentives through capital markets *Source/initiator*: Financial institutes and capital markets	• *Financial performance*: Environmental performance has an indirect effect on the firm's financial performance through capital markets. Polluters can participate in voluntary programs to improve their financial performance or image • *Sanctions by investors and capital markets*: Suppliers of capital can also exert pressures on firms. Investors react negatively to information revelation, e.g., regarding a higher than expected level of toxic emissions. Thus, the environmental performance of a firm can affect access to

(continued)

Table 3.4 (continued)

Motivation (pressure/incentive) and source/initiator	What are the specific motivations? How do they work?
	capital, i.e., prevent or increase their ability to obtain loans and sell equities and enable access to certain markets
Community pressure *Source/initiator*: Communities and consumers, NGOs	• *Improved stakeholder relations/stakeholder pressure*: Public opinion and improved relations with customers and the stakeholders (including community, environmental, and industry group pressure) drive firms' actions, e.g., campaigns and consumer boycotts of firms that violate accepted clean environmental practices • *Public environmental reporting*: Publicly available environmental records (e.g., public reporting of releases) influence the public image of a company (public recognition through press releases, newsletters, and awards) • *Proximity to consumers*: Firms that produce products that are more visible to consumers are more likely to participate in voluntary programs
Firm-specific incentives *Source/initiator*: Business internal	• Technical and organizational feasibility is influenced by several firm-specific characteristics • *Past environmental performance*: Firms that had poorer environmental performance, and thus were likely to experience public pressure or larger costs of compliance/ penalties, are more likely to undertake voluntary initiatives • *Size*: Larger firms are more likely to participate due to lower marginal costs of abatement (economies of scale, more personnel) and greater exposure to liabilities • *Financial health*: More profitable and less risky firms seem to be more likely to participate in voluntary initiatives (because this requires long term commitment) • *Research and Development*: Firms in industries with greater expenditures in Research and Development are more likely to participate • *Voluntary program interaction*: If a company participates in one voluntary environmental program it is also more likely to participate in another voluntary initiative

Source: Author, after Alberini and Segerson (2002), Maxwell et al. (2000), Segerson and Li (1999), Brau and Carraro (1999), Arora and Gangopadhyay (1995), Segerson and Li Dawson (2001), Segerson and Miceli (1998), Khanna (2001), Maxwell and Lyon (2002), Arora and Cason (1996)

consideration of individual payoffs, leads to conclude that voluntary participation or commitment is more likely to be based on the rational choice of business decisions rather than on altruistic motivations (Alberini and Segerson 2002). Consequently, "quite simply, voluntary environmental protection appears to make good business sense" (Maxwell and Lyon 2001). This means "that large firms undertake voluntary corporate environmental actions for solid economic reasons" (Maxwell and Lyon 2001), thus weighing benefits and constraints of voluntary approaches against each other. The benefit that a firm has from applying a voluntary approach "will reflect not only net revenue from the sale of its product, including safety or increased shelf-life, but also any public relations benefit (e.g., increased current or future demand because of increased "good will" toward the firm)". Similarly, Caswell and Henson (1997, cited in Alberini and Segerson 2002) argue, "that the loss of reputation and market sales are likely to be of more importance to firms than the direct damage costs imposed through liability."

Motivations for the participation can differ according to various incentives and their respective sources or initiators as shown in Table 3.4. In this line of reasoning Alberini and Segerson (2002) note:

> there are a variety of motives for participation in voluntary environmental approaches. In assessing a given voluntary program or approach, a key component is to determine whether there are sufficient incentives for participation and what the source of those incentives might be. In particular, it is necessary to assess whether markets alone will provide sufficient incentives for voluntary abatement, or whether instead some form of government inducement is necessary.

As is shown in Table 3.4, government incentives can be classified into "carrot" and "stick" approaches. According to Segerson and Miceli (1998), the latter is considered less voluntary, because the firm is essentially choosing the lesser of two evils.

Outcome of Voluntary Environmental Approaches

With regard to their outcomes, voluntary environmental approaches have a number of potentials and risks as shown in Table 3.5.

According to Delmas and Terlaak (2001), voluntary environmental agreements have a positive environmental impact. Similarly, Voigt (2000) examines the use of

Table 3.5 PROs and CONs of voluntary approaches

PROs (potentials)	CONs (risks)
• There is a potential for cost savings from increased flexibility relative to inefficient regulations and freedom to find solutions adapted to the specific situation (cost effectiveness and efficiency) • Building on a common understanding of environmental problems and mutual responsibilities, industry is encouraged to act proactively and in cooperation with government (thus reducing conflicts between regulators and industry) • Shorter and more cooperative processes between polluters and regulators can improve information flows and reduce implementation lags (faster and more efficient achievement of targets)	• The lack of binding enforcement bears the risk that voluntary measures do not ensure adequate environmental protection • Polluters will not be forced to go beyond what they might have done anyway • The focus might be shifted from the worst polluters to those willing to abate voluntarily

Source: After Alberini and Segerson (2002), Segerson and Li (1999), Segerson and Miceli (1998), van Calster and Deketelaere (2001)

voluntary pollution constraints[8] as an example for unilateral business initiatives in environmental policy. Despite the criticism and controversy of this instrument, he concludes that voluntary pollution constraints can result in behavioral changes toward the voluntary provision of public goods and environmental protection. Lyon and Maxwell (2002) also point out that voluntary approaches can generally be considered welfare-enhancing, but note that in certain situations welfare reductions might occur, characterized by one or more of the following conditions:

1. Overarching corporate lobbying.
2. Biased political objectives of regulators.
3. Lacking regulatory incentives due to fear of public criticism.

Furthermore, Khanna (2001) notes that "[m]ost of them focus only on the means (proactive efforts) for pollution control rather than the ends (actual performance improvement)." Thus, Segerson and Miceli (1998) conclude that the environmental impact of voluntary agreements can be either positive or negative, depending on several factors, i.e. the allocation of bargaining power between the regulator and the firm, the magnitude of the background threat (to impose a regulation) and the social cost of funds. Or as Khanna (2001) puts it:

> Voluntary initiatives do not require or guarantee an improvement in environmental performance because they provided flexibility in the extent of improvement in environmental performance and lacked any sanctions for non-improvement [. . .] firms may undertake these initiatives to disguise poor performance and adopt the outward form of these initiatives but not undertake the effort required to really improve environmental performance.

Evaluation of Voluntary Environmental Approaches

Arora and Cason (1996) conclude that it is necessary to carefully evaluate the success of voluntary environmental programs. This needs to take into consideration that any evaluation is normative (Alberini and Segerson 2002). This is especially true for voluntary approaches, because voluntariness is an inherently normative concept.

Furthermore, any evaluation of the outcome of voluntary initiatives requires a benchmark or baseline against which the success or failure of such a voluntary approach is assessed. Any evaluation of the outcome of voluntary initiatives, therefore, has to compare scenarios with and without a certain voluntary initiative and include other factors that could also have influenced performance (Khanna 2001). This is especially difficult as the counterfactual (baseline or "business as usual" scenario) is often implicit and simply assumed rather than carefully constructed.

Keeping in mind that this remains hypothetical to a great degree, the effects need to be evaluated against the likely alternatives and their estimated effects (i.e.,

---

[8] He defines voluntary pollution constraints as declarations of a group of polluters to achieve a certain environmental policy goal in a fixed period of time.

changes in environmental quality with or without a specific voluntary approach) (Lyon and Maxwell 2002).

Building on Bohm and Russell (1985) and OECD (1999), Alberini and Segerson (2002) cite as criteria for the normative evaluation of environmental policy instruments:

- Environmental effectiveness
- Economic efficiency
- Administrative costs
- Effect on competition
- Dynamic effects and innovation incentives
- Soft effects and diffusion of information
- Political acceptability

Evaluation mostly focuses on environmental effectiveness, i.e., the effectiveness in meeting environmental quality goals and the level of environmental protection that is reached. The effectiveness, in turn, can be evaluated taking into account three indicators (Alberini and Segerson 2002):

- The number of participants (polluters).
- The amount of pollution abatement per participant.
- The impact on the number of polluting firms, i.e., impacts on competition.

In contrast, evaluations of the efficiency of voluntary approaches can alternatively focus on minimizing costs or maximizing environmental outcome. Cost efficiency examines whether a specific environmental goal can be best (i.e., with the least costs) reached with a specific voluntary approach. Efficiency with regard to increasing environmental quality examines whether the outcome is greater with a specific voluntary approach than without (Alberini and Segerson 2002).

Segerson and Li (1999) conclude that voluntary approaches can be effective, but there is no guarantee for their success. Success of any voluntary approach hinges on the participation by polluting firms (Segerson and Li Dawson 2001), i.e., individual firms need to have a strong incentive to participate in or to apply a certain voluntary approach. Building on this prerequisite, the Union of Industrial and Employers' Confederations of Europe (UNICE) defined conditions for the success of voluntary agreements (van Calster and Deketelaere 2001):

- The goals of the exercise have to be clearly defined.
- The negotiating process has to be transparent.
- Top management of the sectors or enterprises concerned has to be involved.
- Free riders have to be discouraged.
- Compliance with the agreement has to be ensured by accompanying measures (e.g., tax advantages, subsidies for research and development, stricter legislation for those who do not participate in the agreement).

Nevertheless, voluntary environmental approaches may fail not only because they are poorly conceived—even if they are considered successful in terms of participation, they could result in limited improvements in environmental quality, i.e., success of a voluntary approach cannot be measured in terms of participation only (Arora and Cason 1996; Alberini and Segerson 2002).

What Are the Implications for the Present Study?

Traditionally environmental protection is targeted by command-and-control regulation and end-of-pipe pollution control. This has changed since the 1990s when a broadening of responsibility for environmental protection beyond governmental action can be observed: voluntary environmental approaches mark a new line of cooperation between business and government (or other entities in society).

Similarly to voluntary biodiversity offsets (understood here in the wider sense of any offset beyond regulatory compliance), voluntary environmental approaches have been and are examined as a promising tool in future environmental policy and as an alternative to traditional regulation. In fact, voluntary biodiversity offsets can be understood as a specific case of voluntary environmental approaches. Thus, from the discussion and literature on voluntary environmental approaches insights can be gained for biodiversity offsets. This refers in particular to the question what the role of government is, why a business would participate (motivation) and whether better outcomes could be reached with a voluntary approach (efficiency).

With regard to the degree of government involvement, it can be observed that often government plays a crucial role (e.g., in bilateral negotiated agreements between government and industry or by setting up voluntary schemes). Is thus assumed that the same holds to be true for biodiversity offsets beyond regulation and that government is one of the different drivers (initiators or facilitators respectively) for voluntary biodiversity offsets. Other drivers include altruistic motivation based on environmental stewardship, market-based incentives (drawing on competitive advantage and competitive pressure), incentives through capital markets (most notably sanctions by investors) and community pressure.

The literature on voluntary environmental approaches not only delivers insights on the initiators or sources of incentives for business to adopt them, but also points to the fact that there need to be sufficient incentives for participation, in terms of need to make good business sense. Large firms adopt voluntary environmental approaches for solid economic reasons, i.e., businesses may gain market-driven and social benefits from adopting voluntary environmental approaches. This can also be observed with regard to voluntary biodiversity offsets (the "business case," "social license to operate").

(continued)

In terms of efficiency and evaluation of the outcome of voluntary environmental approaches, it can be noted that this needs to be seen in comparison to the counterfactual, i.e., comparing scenarios with and without a certain voluntary initiative. This is an important consideration for the use of biodiversity offsets, too.

Summarizing, a lot can be learned from the literature on voluntary environmental approaches and a large set of potential criteria can be extracted to be applied and tested at the specific case of biodiversity offsets.

Derived potential criteria for voluntariness[9]: Intentionality (1), Additionality (2), Motivation (3), Recompense/benefit (5), Public relations benefit (6), Influenceability (7), (Degree of) obligation (9), Community pressure/public response (10), Role of government (11), Competitive environment (15), Social and individual commitment/societal context, Charitable, directed toward public good/welfare (17), Outcome/effectiveness/success (18), Flexibility (19), Lower costs/cost efficiency (20), Ethics (25).

3.5 Willingness to Pay

It is increasingly being asked whether and how much both individuals and organizational/business entities are willing to pay for the use of certain public goods to ensure their sustainable provision in perpetuity, e.g., access to clean water. As this usually builds on goods and services that are provided by the natural environment this is referred to as "payments for ecosystem services" (PES). PES are often based on analyses of Willingness to Pay or Willingness to Accept. Willingness to pay is an important economic concept, that is also used in public policy and environmental economics to explain how individuals use and contribute to (and thus value) public environmental goods.

Knetsch (1990) examines the relation between willingness to pay and compensation in environmental loss assessments. Building on his findings, losses are generally valued more than gains, i.e., "the minimum compensation people demanded to give up a good or an asset was reported to be several times larger than the maximum amount they were willing to pay to keep or acquire a commensurate entitlement." Thus, compensation for environmental loss seems to be more efficient in terms of increased environmental protection than willingness to pay for environmental gains. This makes clear that Willingness to pay corresponds very much to PES, but not so much to compensation and biodiversity offsets (PES and biodiversity offsets can in practice lead to similar outcomes for the natural environment; however, the two concepts are not congruent but rather complementary). As a result, concepts other

[9]see Sect. 5.2 and refer to corresponding numbers in Table 5.1.

than willingness to pay are better suited to examine motivations, thresholds, and quality (i.e., environmental protection outcomes) of biodiversity offsets.

What Are the Implications for the Present Study?
While willingness to pay, like other voluntary and market-based approaches, is targeted at contributing to public goods by paying for their use, this is more suitable as a concept to analyze and explain payments for ecosystem services (PES) schemes than biodiversity offsets. While biodiversity offsets and PES can in practice lead to similar outcomes, they build on complementary notions: compensation for environmental loss as opposed to the willingness to pay for environmental gains. The focus of this study being on biodiversity offsets, willingness to pay is not further explored.

3.6 Corporate (Environmental or Social) Responsibility

Development and Terminology of Corporate Responsibility
Already in the mid-1960s and 1970s, a basic shift from personal responsibility to corporate responsibility has significantly changed the field of business ethics (Reed 1999). More precisely, Blowfield and Murray (2014) trace back a shift in business behavior toward the environment starting in the 1980s:

> The environment is one area where companies in the past relied on government to say what was permissible, but which in the face of global challenges from climate change to water availability to deforestation they are, at least to some degree, willing to go beyond legal compliance. This is part of a trend commencing in the 1980s when the effectiveness of command-and-control regulatory solutions started to be questioned, and both companies and regulators began to accept that preventing pollution could be a more effective way forward than simply punishing it [...] In the 1990s, environmental management standards, such as BS 7750, the EU Eco-Management and Audit System (EMAS), and the ISO 14000 series provided new ways for companies to understand and manage their environmental impacts. Companies began to realize that, in certain situations, improved environmental performance could have a positive impact on the financial bottom line [...] what came to be called the 'greening revolution.'

At the beginning of the twenty-first century, the speed and scope of market globalization have led to increasing concerns about global business ethics and corporate (environmental) responsibility (Maak 2008; Egri and Ralston 2008). For business, this includes new management challenges arising from the need to comply with various legal obligations from different national or regional contexts and to respond to various societal expectations (OECD 2001a). In addition, the corporate scandals in recent years have significantly added to opening a broad public, political, and academic debate on the impacts—both positive and negative—and role of business in society, i.e., on its legitimacy, obligations, and responsibilities (Maak 2008). The basic notion of corporate responsibility, i.e., that corporations are moral

Table 3.6 Dimensions of corporate (social) responsibility

Dimensions	The dimension refers to
Environmental dimension	The natural environment
Social dimension	The relationship between business and society
Economic dimension	Socioeconomic or financial aspects, CSR as part of business operations
Stakeholder dimension	Stakeholders or stakeholder groups
Voluntariness dimension	Actions not prescribed by law
Ethical/moral dimension	Business as a moral actor
Self-reflection dimension	How business admits and deals with its impacts and activities

Source: Author, building on Dahlsrud (2008), Carroll (1983), Kilcullen and Kooistra (1999), Business for Social Responsibility (2003), Ethical Performance (2003)

actors with obligations and responsibilities toward other constituencies than their shareholders, is "winning the ground" (Halme et al. 2009; Reed 1999). Consequently, Halme et al. (2009) raise questions "about the capacity of corporate responsibility to solve some pressing problems in the neediest parts of the globe" and Gordon (2001) notes that "now is a particularly promising time for global instruments to have a major impact on international business behavior and to play a prominent role in the public debate about the respective roles of companies, governments and individuals in ensuring [...] improved economic, social and environmental welfare."

Corporations are responding to these concerns and challenges with managerial innovations, presenting themselves as "good corporate citizens," i.e., "explaining why and how they care about a sustainable future" (Maak 2008).

The concept and definition of corporate responsibility have evolved over several decades (Mohan 2003, cited in Bakker et al. 2005). Corporate responsibility has a number of synonyms and sister concepts, including corporate social responsibility, corporate sustainability, business in society, or corporate citizenship (Halme et al. 2009). The terms of corporate responsibility and corporate social responsibility are mostly used interchangeably. Despite the fact that the former by definition encompasses all different aspects of responsibility (cf. Table 3.6) and therefore is broader than corporate social responsibility, the latter is more frequently used—both in practice and in theory. This can be interpreted as a result of historic development (earlier definitions have a stronger focus on social aspects, i.e., ethical expectations, societal good, and philanthropy).

According to Halme et al. (2009), generally speaking, corporate responsibility reflects the idea that business is responsible for some of the wider societal good. As the country of origin matters, more specific definitions diverge according to the different geographical, national, and cultural contexts. However, not only spatial, but also temporal aspects have influenced the multitude of different existing definitions. Some of these are exemplarily outlined in Appendix 3 in Chap. 10. As can be seen, numerous authors from science, politics, business, and society have attempted to determine a clear and unbiased definition of corporate responsibility over the course of almost half a century. Nevertheless, as Dahlsrud (2008) finds the existing definitions are to a large degree congruent and that "the confusion is not so much about how CSR is defined as about how CSR is socially constructed in a specific context." In this regard, a certain number of keywords can be extracted (underlined in Appendix 3 in Chap. 10) that in different combinations figure in most of the definitions. Dahlsrud (2008) has classified and grouped these keywords into five dimensions as shown in Table 3.6.

Consequently, the scope of corporate codes of conduct usually encompasses a detailed set of commitments that are relevant for the economic, social, and environmental welfare (the so called "triple bottom line") of the societies within which they operate. They are oriented toward different stakeholders and targeted to go beyond legal compliance (Heikkurinen 2012; Dahlsrud 2008). Building on the variety of definitions (e.g., Carroll 1983; Kilcullen and Kooistra 1999; Business for Social Responsibility 2003; Ethical Performance 2003), two further dimensions can be added: the ethical and moral dimension (judging business as a moral actor) and the self-reflection dimension, i.e., how business admits and deals with its impacts and activities (see Table 3.6).

Corporate responsibility has become an "ubiquitous phenomenon" in both business practice and research (Maak 2008). It can be best understood as an umbrella term to describe "much of what is done in terms of ethics-related activities in firms around the globe" (Maak 2008) or more generally for the various relations between business and society—in terms of definition, management, and action.

Building his observation that "CSR, business ethics, stakeholder theory, corporate citizenship, etc. are all important research domains in their own right," Maak (2008) claims there is a need for a "framework that connects these domains in plausible and workable ways." Consequently, he suggests a shift in terminology from the corporate responsibility toward what he calls "corporate integrity," with the latter being characterized by the "7Cs" of integrity: commitment, conduct, content, context, consistency, coherence, and continuity (see Fig. 3.4).

> to master integrity requires integrative efforts to ensure alignment of intention and purpose (commitment), conduct, responsibilities (content), relationships both distant and close (context), words and deeds (consistency), principles and action as well as internal and external conditions (coherence), on an ongoing, life-long basis (continuity). (Maak 2008)

Despite this enthusiasm for and plausibility of the concept of corporate integrity, both corporate responsibility and corporate social responsibility remain the most frequently used terms in scientific discussion (e.g., Dahlsrud 2008; Castelló and Lozano 2011; Heikkurinen 2012; Blowfield and Murray 2014) and dominating concepts in practice.

Fig. 3.4 The 7Cs of corporate integrity. Source: Author, based on Maak (2008)

Context of Corporate Responsibility and Relation of Business and Government Corporate responsibility is determined by the interplay between contextual factors, i.e., the institutional, political, legal, economic, and cultural setting within which business is practiced and systems of governance internal to the company (Halme et al. 2009). In the face of globalization, context dependency makes corporate responsibility more and more complex (OECD 2001a).

> Large, highly visible corporations now coexist alongside smaller companies that also have international reach. The boarders of the firm have become blurred, as companies have deepened and extended relationships in supply chains as well as other business partnerships [...] business entities themselves are often quite fuzzy. (OECD 2001b)

Halme et al. (2009) argue that if the significance of context is great, "then there will be a move away from standardized CR practices toward corporate frameworks which provide room for local branches or operations to follow a path in line with the framework on the one hand and local conditions on the other."

Corporate responsibility requires balancing and managing the relation between states, markets, and civil society (OECD 2001a). This refers particularly to the role of government and the degree of regulatory influence, which are among the core notions of corporate responsibility, i.e., government influence on (voluntary) actions and initiatives is actually greater than one might expect.

On the one hand, this includes compliance with national laws and regulations. However, "[i]n terms of creation of standards, government regulation can set only minimum or average standards" (OECD 2000). On the other hand, "on a voluntary basis, the level of ambition can be higher; the companies can be encouraged [by government] to go beyond the set standards, even if they do not figure in legal codes" (OECD 2000).

Moreover, businesses can use different corporate responsibility strategies in their relation with government and regulation (Maxwell and Lyon 2001):

1. Preempting tougher government regulations.
2. Weakening forthcoming regulations (in situations where pre-emption is impossible).
3. Reducing the extent of monitoring by regulatory agencies.
4. Signaling regulators to persuade them to raise rivals' costs.

Building on the characteristic of "going beyond compliance" (with or without government involvement), Blowfield and Murray (2014) conclude that "corporate responsibility as a field is about voluntary not mandatory actions. The distinction between voluntary and compulsory behavior may be useful for demarcating a particular subset of corporate responsibility theory and practice."

The voluntary nature of corporate responsibility is in close relation to voluntary environmental approaches (see Sect. 3.3). In fact, the two are to a large degree congruent as voluntary initiatives form an important part of corporate responsibility.

> Voluntary initiatives in the area of corporate responsibility have been among the major trends in international business in recent years. Business surveys show that most large OECD-based multinational enterprises have participated in this trend in one way or another. (OECD 2001a)

As a study by the OECD (2001a) finds, "some initiatives are more voluntary than others", i.e., some firms are under stronger pressures to adopt voluntary initiatives than others (e.g., than those with low public visibility). These pressures can arise from different sources as shown in Table 3.7: government pressure, government failure, market pressure, market failure, and pressure from consumers and civil society.

In response to these pressures a "new kind of interaction between government and business is emerging in which both partners see the need for co-operative rather than adversarial relationships" (OECD 2001a) and a trend towards private–public partnerships can be noted. Accordingly, the best way to proceed could lie in complementary and mutually reinforcing efforts by governments, companies, and other stakeholders in a spirit of shared responsibility and partnership (OECD 2000). In fact, there is no dichotomy between voluntary approaches, such as corporate codes of conduct, and regulatory action taken by governments (OECD 2000).

Table 3.7 Pressures for businesses to adopt voluntary initiatives

Source	Pressure	Failure
Market	Pressures for social and environmental regulation increasingly arise out of the market, in terms of self-regulation	Corporate responsibility of businesses builds on the premise to rectify market failures and to take ownership of the externalities generated in business operations
Government	As has been noted government can actively shape voluntary initiatives, e.g., through institutional and regulatory arrangements	In countries where there are gaps in legal enforcement, social provision, and governance, companies are under heightened expectations to fill those gaps
Civil society	To protect brand or reputation capital, companies have to respond to pressure from consumers, NGOs, and other stakeholders to adhere to more demanding behavioral norms	

Source: Author, building on OECD (2000), Halme et al. (2009), Arrow (1973, cited in Orlitzky et al. 2011), Crouch (2006, cited in Orlitzky et al. 2011)

> [E]merging regulatory practice within the OECD tends to treat public regulation and private initiatives as complementary, rather than competing, approaches. Judiciously used, private initiatives help to lower the costs of regulation and to enhance its effectiveness. The difficulty for both public and private actors is determining the appropriate mix of private and public action. (OECD 2001a)

This complementary nature holds especially true at global scale, i.e., any global system of norms usually includes combinations of public and private action (OECD 2001a).

> The future of corporate social responsibility is not in replacing government responsibility. In fact, it will only fully realise its potential when it can operate on a more level playing field in the context of effective global rules. (Baker 2001)

In this line of argumentation, Fougère and Solitander (2009) believe "that a certain amount of (international) regulation will be needed to mitigate the overwhelming potential for social and environmental nuisance of the global free market." Nevertheless, much compliance heavily relies on (informal) social norms and conventions, that are not written down as formal law (OECD 2001a, c), i.e., individuals believe in the law and they may be under social pressure to comply with it. This can be interpreted as "softer" forms of social control of business (OECD 2001a). In this line of reasoning, the concept of "voluntary compliance" has been coined, i.e., compliance based on personal belief or social pressure, rather than on enforcement practices (OECD 2001a). In this sense, the distinction between voluntary initiatives or voluntary compliance and binding systems of law and regulation becomes blurred in that voluntary initiatives and formal law and regulation draw on the same sets of social capital (or consensus on appropriate behavior).

> This blurring has become more pronounced as a result of recent enforcement trends [. . .] Newer enforcement techniques try to promote voluntary compliance in the corporate sector, but also to draw on the sector's compliance expertise and to integrate this private expertise into public enforcement strategy. (OECD 2001a)

Stakeholder Orientation of Corporate Responsibility and Corporate Reputation

As Castelló and Lozano (2011) note, corporate legitimacy or credibility, i.e., the need to gain acceptance in society, has become one of the most critical issues for corporations, in particular, if they are operating globally. The different forms of corporate legitimacy and associated rhetoric point to the importance of having a clear stakeholder focus. Several scholars thus call for a conceptualization of corporate responsibility from a stakeholder perspective (Wood et al. 2006; Waddoch 2002, cited in Hillenbrand and Money 2007). Thus, corporations, or organizations more generally, strive to comply with norms, standards, values, and beliefs in order to meet stakeholder expectations, "by introducing new initiatives and a new rhetoric in their communications with stakeholders" (Castelló and Lozano 2011).

In light of the question how to define the optimal performance level above regulatory requirements (or when no regulations exist), Dahlsrud (2008) notes "the only conclusion to be made from the definitions is that the optimal performance is dependent on the stakeholders of the business." This is also crucial seen the often conflicting concerns of the stakeholders and the need to balance between them.

Furthermore, involving stakeholders in defining an organization's responsibilities and building a corporate responsibility strategy can add not only to corporate legitimacy, but also enhance corporate reputation. According to Hillenbrand and Money (2007), there is considerable similarity between the concepts of corporate responsibility and corporate reputation in that both are "expressed through [. . .] and understood in terms of similar and overlapping corporate behaviors and [. . .] stakeholder perceptions." They conclude that "rather than viewing reputation and responsibility as two separate concepts, they may more usefully be thought of as two sides of the same coin."

Consequently, activities under the scope of corporate responsibility, such as embracing CSR standards, philanthropic giving and building relationships and trust with stakeholders, will potentially have an impact on corporate reputation. For example, some see activities in the scope of corporate responsibility as building their personal reputations as good global citizens or as a signal of the quality of their products and services (Harjoto and Jo 2011). Especially top management "tends to over-invest in CSR activities to build their own personal reputation as good global citizens (over-investment hypothesis)" and to "generate support from social and environmental activists in order to reduce the probability of CEO turnover in a future period (strategic-choice hypothesis)" (Barnea and Rubin 2010, cited in Harjoto and Jo 2011).

The Business Case and Motivations for Corporate Responsibility

As has been noted, one crucial point of the discussion on corporate responsibility relates to the question how corporate behavior is embedded into the societal context and ultimately, what drives or motivates it.

Heikkurinen (2012) finds motivations for corporate responsibility spanning from "hard-core profit making" and "oppression of others" to moral reasons and even spirituality. Blowfield and Murray (2014) group this whole range of motivations into two broad categories:

1. Values motivation: Companies, like people, have values that guide their interaction with other society members.
2. Materiality motivation: To succeed companies need to manage their relationship with wider society.

Drawing on this distinction, "companies face three competing cases that will ultimately determine their corporate responsibility strategy: the moral case (obligations that the company has to society); the rational case (taking proactive steps that will minimize the restrictions society imposes on business); the economic case (adding financial value to the company by preserving its legitimacy with its stakeholders)" (Blowfield and Murray 2014).

Several authors suggest that corporate responsibility is mostly based on the materiality motivation (i.e., as a combination of the rational case and the economic case) and that companies and organizations define corporate responsibility in terms of commercial benefits (e.g., Heikkurinen 2012; Blowfield and Murray 2014; Fougère and Solitander 2009). Milton Friedman (1962, cited in OECD 2001a) claims that "there is one and only one social responsibility of business—to use its resources and engage in activities designed to increase its profits" and Fougère and Solitander (2009) note that a corporation "can only engage in 'ethical duties' when this engagement provides a bottomline payoff." Consequently, Heikkurinen (2012) relates the commitment toward corporate responsibility to a cost–benefit analysis. Thus, corporate responsibility in and of corporations is seen as a means to gain a competitive advantage on the "market for virtue" (Maak 2008).

Harjoto and Jo (2011) find examples for the instrumental (economic) value in that firms use corporate responsibility activities to signal their product quality (in particular in highly competitive markets) and to reduce conflicts of interest between managers and (investing and non-investing) stakeholders. In addition to external benefits (e.g., in terms of reputational advantage), businesses may gain internal benefits, i.e., develop new resources and capabilities which are related namely to know-how and corporate culture (Heikkurinen 2012). Apart from these, there are potentially numerous benefits of corporate responsibility activities for the organizations adopting them (OECD 2001a; Spurgeon 2008; Harjoto and Jo 2011):

1. Positively influencing operating performance and firm value
2. Reducing costs within the firm

3. Managing risk (of costly criminal prosecutions, litigation, and damage to reputation) and uncertainty
4. Managing relations with shareholders and with actors in the societies in which they operate (conflict resolution, obtaining a license to operate)
5. Environmental product differentiation (product signaling)
6. Managing competitors
7. Redefining markets
8. Preempting government regulation (In some instances, corporate responsibility activities have been used to deflect momentum for government regulation)

These likely advantages highlight that there may be a business case for corporate responsibility. Going even further, Fougère and Solitander (2009) consider the existence of a business case a necessary precondition for business to engage in corporate responsibility:

> [A]ll corporate promises regarding environmental and social issues can only be delivered if there is a clear 'business case', i.e. if it will create value for the shareholders. This represents a fundamental limit on what CR, consisting of voluntary corporate action as it does, can deliver. Corporations may not be 'irresponsible' but they should be expected to be by definition 'aresponsible' towards society and nature [i.e. they need to focus on the economic bottomline].

Strategic Corporate Responsibility

"[T]he business-environment relationship became, for some companies, less a costly problem than a strategic opportunity (Blowfield and Murray 2014)", i.e., it is becoming more and more common to address their corporate responsibility strategically. Thus, many firms are employing corporate responsibility as a part of a differentiation strategy at the product, business and corporate levels (Orlitzky et al. 2011). These strategically oriented corporate responsibility activities have been framed as "strategic corporate responsibility."

Spurgeon (2008) outlines that there are "various alternative corporate environmental strategies with certain types of strategy favoring different types of organization." He builds on the model developed by Orsato (2006) whereby a company tends to adopt either one or another of four different approaches (see Fig. 3.5).

Orsato's model is composed as a two-by-two matrix, relating two competitive foci (organizational processes and products and services), displayed on the x-axis, to the two main competitive advantages (low cost and differentiation), displayed on the y-axis. The result is four different corporate environmental strategies (Orsato 2006):

- Strategy 1: Eco-efficiency relates to minimizing costs through improved organizational processes associated with natural resource use and waste.
- Strategy 2: Beyond compliance, leadership relates to continually seeking first-mover advantage, and publicizing it.
- Strategy 3: Eco-branding is based on obtaining a premium for differentiating products and services on environmental grounds.

Fig. 3.5 Environmental strategies based on competitive advantage and focus. Source: After Orsato (2006)

- Strategy 4: Environmental cost leadership is based on companies focusing on radical product innovation that provide first-mover advantage.

In terms of designing and implementing corporate strategically, not only functional, but also temporal aspects play an important role. Consequently, the evolution of corporate responsibility at individual level (i.e., for a single business, corporation, or organization) can be broken down to seven key development steps in the specific case of corporate action on biodiversity:

1. Make the business case for biodiversity.
2. Identify a senior level biodiversity champion.
3. Carry out a biodiversity assessment.
4. Secure board-level endorsement.
5. Develop a corporate biodiversity strategy.
6. Develop a corporate biodiversity action plan.
7. Implement the corporate biodiversity action plan.

Scaling this up to the broader societal context (i.e., at aggregate level), Zadek (2004, cited in Spurgeon 2008) identifies four stages of issue maturity:

1. Latent stage: A new issue is largely ignored or dismissed.
2. Emerging stage: Societal awareness grows and leading businesses experiment with approaches to dealing with it.
3. Consolidating stage: Sector wide voluntary initiatives get underway.
4. Institutionalization stage: Legislation is introduced and approaches become embedded within business excellence models.

Outcome of Corporate Responsibility

Corporate activities are subject to a "heightened public awareness and scrutiny by critical stakeholders as to how corporations run their businesses and what constitutes legitimate business behavior" (Maak 2008). The question is raised what the outcomes and benefits to society are. Thus, corporate responsibility needs to be judged with respect to its impact in terms of effectiveness or meaningfulness. Is it true that "what CR is claimed to be about by its business proponents tends to be very different from what a great deal of evidence shows it to be" (Fougère and Solitander 2009)?

To date, there is little research about the outcomes and consequences of corporate responsibility, especially for society (Halme et al. 2009). In fact, there is considerable disagreement about its meaningfulness. While some view it as "first tentative steps towards global convergence of business practices" (OECD 2001a), there are also skeptical voices: some consider corporate social responsibility a "tortured concept" (Maak 2008) and others see them mostly as a "public relations effort by the business community" (OECD 2001a). Critics believe that "what CR tends to do is to disguise its fundamentally reactive, crisis-addressing—or at best risk-mitigating—function into a proactive care for the social and the environment" (Fougère and Solitander 2009).

With regard to the question, what makes corporate responsibility initiatives meaningful in terms of changing business behavior (or whether they can be meaningful at all), some claim that corporate responsibility can only be credible if it is subject to formal deterrence, i.e., external verification or legitimization through monitoring and enforcement, binding laws, trade sanctions, etc. (OECD 2001a). They argue that a problem of accountability is inherent to the voluntary nature of corporate responsibility, i.e., business itself defining its goals and responsibilities (Fougère and Solitander 2009). Others hold against this the difficulty of external verification and the importance of voluntary compliance (OECD 2001a).

Thus, the question whether corporate responsibility and its related activities are effective "cannot be answered by looking only at what businesses do. If they are effective, it is because societies manage to formulate and channel reasonable pressures for appropriate business conduct" (OECD 2001a). Consequently, corporate responsibility is only as strong or effective as the factors shaping it. In addition, CR is not only a corporate sector trend, but many constituencies such as the World Bank, OECD; EU and UN put trust in and encourage it (Baughn et al. 2007, cited in Halme et al. 2009). Given the involvement of major global institutions, one would assume that this raises expectations that corporate responsibility and its related activities are beneficial for society and the environment (Halme et al. 2009).

It should be noted, that above all it is certainly an achievement of corporate responsibility to provide a platform for discussion of social and environmental challenges at a global scale (Fougère and Solitander 2009). Furthermore, voluntary, decentralized approaches under the scope of corporate responsibility are more flexible to adapt to regional and sectoral circumstances and to acquire necessary knowledge and experience (OECD 2001a). So despite public concerns about the economic, social, and environmental impacts of business activities, most notably those of multinational enterprises (MNEs) on the one hand and the instrumentalization of sustainable development in corporate responsibility practice, on the other hand, the importance of corporate standards and responsibilities must be highlighted.

However, there are two main challenges for business to ensure the effectiveness and credibility of corporate responsibility (Fougère and Solitander 2009):

1. The first is a "shift away from the modernist view considering nature as a resource that can be exploited forever, which implies questioning the belief in limitless economic growth."
2. The second is that it "should be obvious that corporations cannot have it both ways, appropriating social and environmental responsibilities when it suits them and then abandoning them when there is a crisis or perceived lack of business opportunities." This means that above all corporate responsibility is a commitment and a continuous process that is not finished at any time.

What Are the Implications for the Present Study?

In the light of increasing concerns about global business ethics and corporate (environmental) responsibility, businesses need to comply with various legal obligations from different national and regional contexts and to respond to various expectations. In this respect, a parallel can be drawn between corporate responsibility (CR) and (voluntary) biodiversity offsets with regard to the expectations to contribute "to the resolution of many global and environmental ills although little is known about the social consequences and impacts of CR on society" (Halme et al. 2009). Corporate responsibility and biodiversity offsets both face expectations to help solving global environmental problems (whereas the focus of corporate responsibility is broader, biodiversity offsets specifically target to counterbalance ongoing biodiversity loss).

Similar to voluntary environmental approaches (see Sect. 3.3), corporate responsibility (and biodiversity offsets) is construed in relation to regulation and the role of government. Corporations can use activities under corporate responsibility strategically in their relations with the government, e.g., for preempting tougher government regulations or reducing the extent of monitoring. Strategic corporate responsibility also implies using it toward other stakeholders (e.g., consumers, competitors, the wider public), e.g., as part of a differentiation strategy at product, business, and corporate levels. Thus, corporate responsibility has instrumental economic value and there are potentially numerous benefits for the organizations adopting such activities, e.g., in terms of gaining competitive advantage and internal benefits. Clearly, from this a "business case" for corporate responsibility can be derived—a characteristic that has been also widely discussed to motivate the implementation of voluntary biodiversity offsets (see Sect. 4.2).

(continued)

This highlights the focus on the materiality motivation of business (as opposed to the values motivation and altruistic behavior). Consequently, business actions in terms of corporate responsibility are driven by a number of pressures or incentives arising from market, government, or civil society influence (i.e., market pressure, market failure, government pressure, government failure, and pressure from consumers and civil society). Similar pressures may also be examined for the case of biodiversity offsets.

The same is true for the interplay between contextual factors and internal business practice, which makes both corporate responsibility and biodiversity offsets increasingly complex and context dependent. Consequently, corporate responsibility is only as strong or effective as the factors shaping it. To this adds that it is not exclusively a corporate sector trend, but many constituencies such as the World Bank, OECD, EU, and UN put trust in and encourage it (again the same applies to biodiversity offsets).

In fact, the large overlap between corporate responsibility and voluntary biodiversity offsets is not surprising, because voluntary biodiversity offsets can occur as activities under the corporate responsibility framework of a business (see Sect. 6.5). However, they are not restricted to this (and there may be other cases motivated by factors other than corporate responsibility).

Thus, the discussion on corporate responsibility can provide useful hints and incentives for the analysis of voluntary biodiversity offsets, e.g., regarding the following points:

- Some initiatives are more voluntary than others due to pressures from consumers, civil society, and legal provisions.
- The recognition of "voluntary compliance" tends to blur the distinction between voluntary approaches and regulation as corporations strive to comply with norms, standards, values, and beliefs to meet stakeholder expectations.
- Not only functional and motivational, but also temporal aspects play an important role, i.e., in terms of issues maturity (cf. the four stages identified by Zadek (2004): latent stage, emerging stage, consolidating stage, and institutionalization stage). Building on this, different types of corporate responsibility or biodiversity offsets might be seen not only simultaneously, but also consecutively.
- Corporate responsibility faces similar challenges in terms of "considering nature as a resource that can be exploited forever" (Fougère and Solitander 2009), the problem of accountability (if the business itself defines its goals and responsibilities) and ensuring continuous (and stable) commitment in the long term.

(continued)

Derived potential criteria for voluntariness[10]: Motivation (3), Recompense/benefit (5), Public relations benefit (6), Influenceability (7), (Degree of) obligation (9), Community pressure/public response (10), Role of government (11), Context (14), Competitive environment (15), Social and individual commitment/societal context (16), Charitable/directed toward public good/welfare (17), Outcome/effectiveness/success (18), Flexibility (19), Lower costs/cost efficiency (20), Scope or distance (22), Ethics (25).

References

Alberini, Segerson (2002) Assessing voluntary programs to improve environmental quality. Environ Resour Econ 22:157–184

Amman H (2001) Von Freiwilligkeit sei die Rede. Ein Vorschlag zur Klärung der Begriffe

Amman (2004) Freiwilligkeit, Gemeinnützigkeit und Sozialstaat. In: Ammann (ed) Freiwilligkeit zwischen liberaler und sozialer Demokratie

Amman H (2008) Begrifflichkeiten und deren Auswirkungen auf die Forschung am Beispiel des Freiwilligen-Monitors. In: Ammann H, Hasse R, Jakobs M, Riemer-Kafka G (eds) Freiwilligkeit – Ursprünge. Erscheinungsformen, Perspektiven

Andreoni J (1988) Privately provided public goods in a large economy: the limits of altruism. J Public Econ 35:57–73

Andreoni J (1990) Impure altruism and donations to public goods: a theory of warm-glow giving. Econ J 100(401):464–477

Arora S, Cason TN (1995) An experiment in voluntary environmental regulation: participation in EPA's 33/50 program. J Environ Econ Manag 28:271–286

Arora S, Cason TN (1996) Why do firms volunteer to exceed environmental regulations? Understanding participation in EPA's 33/50 program. Land Econ 72(4):413–432

Arora S, Gangopadhyay S (1995) Toward a theoretical model of voluntary overcompliance. J Econ Behav Organ 28:289–309

Badelt C (2004) Freiwilligkeit aus Sicht der Ökonomie. In: Ammann (ed) Freiwilligkeit zwischen liberaler und sozialer Demokratie

Baker J (2001) Freedom of association and CSR, remarks of Jim Baker, ICFTU, to OECD Conference on Corporate Social Responsibility, 19 June 2001

Bakker FG, Groenewegen P, Hond F (2005) A bibliometric analysis of 30 years of research and theory on corporate social responsibility and corporate social performance. Bus Soc 44:283–317

Batson CD (2014) The altruism question. Toward a social-psychological answer. Psychology Press, New York

Benabou S (2014) Making up for lost nature?: a critical review of the international development of voluntary biodiversity offsets. Environ Soc 5:103–123

Blowfield M, Murray A (2014) Corporate responsibility. Oxford University Press, New York

Bohm P, Russell CS (1985) Comparative analysis of alternative policy instruments. In: Handbook of natural resource and energy economics, vol 1. Elsevier, pp 395–460

Boothe B (2004) Freiwilligkeit aus Sicht der Psychologie. In: Ammann H (ed) Freiwilligkeit 659 zwischen liberaler und sozialer Demokratie. Seismo, Zürich

[10]see Sect. 5.2 and refer to corresponding numbers in Table 5.1.

Bosch DJ, Cook ZL, Fuglie KO (1995) Voluntary versus mandatory agricultural policies to protect water quality: adoption of nitrogen testing in Nebraska. Rev Agric Econ 17:13–24

Brau R, Carraro C (1999) Voluntary approaches, market structure and competition. Fondazione Eni Enrico Mattei, Milan

Business for Social Responsibility (2003) Issues in corporate social responsibility. http://www.bsr.org/AdvisoryServices/Issues.cfm

Carigiet (2004) Die Stärkung des Sozialstaats durch die Zivilgesellschaft – kleine Solidaritäten als Unterstützung und nicht als Ersatz der großen Solidaritäten. In: Ammann (ed) Freiwilligkeit zwischen liberaler und sozialer Demokratie

Carraro C, Lévêque F (1999) Voluntary approaches in environmental policy. Kluwer Academic, Dordrecht

Carraro, Siniscalco (1996) Voluntary agreements in environmental policy: a theoretical appraisal. In: Xepapadeas (ed) Economic policy for the environment and natural resources. Edward Elgar, Cheltenham

Carroll AB (1983) Corporate social responsibility: will industry respond to cut-backs in social program funding? Vital Speeches Day 49:604–608

Castelló I, Lozano J (2011) Searching for new forms of legitimacy through corporate responsibility rhetoric. J Bus Ethics 100:11–29

Cranach (2008) Freiwilligkeit, Altruismus oder Egoismus? Zur Sozialpsychologie der individuellen und gesellschaftlichen Bedeutung der Freiwilligkeit. In: Ammann et al (eds) Freiwilligkeit – Ursprünge, Erscheinungsformen, Perspektiven

Dahlsrud A (2008) How corporate social responsibility is defined: an analysis of 37 definitions. Corp Soc Respon Environ Manag 15:1–13

Darbi M, Ohlenburg H, Herberg A, Wende W (2010) Impact mitigation and biodiversity offsets-compensation approaches from around the world: a study on the application of Article 14 of the CBD (Convention on Biological Diversity). BfN-Schr.-Vertrieb im Landwirtschaftsverl

Delmas MA, Terlaak AK (2001) A framework for analyzing environmental voluntary agreements. Calif Manag Rev 43(3):44–63

Dixit A, Nalebuff B (2008) Prisoners' Dilemma. The concise encyclopedia of economics. Library of economics and liberty. http://www.econlib.org/library/Enc/PrisonersDilemma.html

Doswald N, Barcellos Harris M, Jones M, Pilla E, Mulder I (2012) Biodiversity offsets: voluntary and compliance regimes. A review of existing schemes, initiatives and guidance for financial institutions. UNEP-WCMC, UNEP FI, Cambridge, Geneva

Earl G, Curtis A, Allan C (2010) Towards a duty of care for biodiversity. Environ Manag 45 (4):682–696

Egri CP, Ralston DA (2008) Corporate responsibility: a review of international management research from 1998 to 2007. J Int Manag 14:319

Ethical Performance (2003) Introduction: defining corporate social responsibility. http://www.ethicalperformance.com/bestpractice/archive/1001/introduction.html

Field A (2001) Altruistically inclined? The behavioral sciences, evolutionary theory, and the origins of reciprocity. The University of Michigan Press, Ann Arbor

Flathman RE (1992) Willful liberalism: voluntarism and individuality in political theory and practice. Cornell University Press, London

Fougère M, Solitander N (2009) Against corporate responsibility: critical reflections on context, thinking, practice, content and consequences. Corp Soc Responsib Environ Manag 16 (2009):217–227

Gordon K (2001) The OECD guidelines and other corporate responsibility instruments: a comparison

Gutmann T (2000) Freiwilligkeit als Rechtsbegriff. Beck, München

Halme M, Roome N, Dobers P (2009) Corporate responsibility: reflections on context and consequences. Scand J Manag 25(1):1–9

Harjoto MA, Jo H (2011) Corporate governance and CSR nexus. J Bus Ethics 100:45–67

Hasse R (2008) Einleitung: Freiwilligkeit in gesellschafts- und organisationswissenschaftlicher Perspektive. In: Ammann H, Hasse R, Jakobs M, Riemer-Kafka G (eds) Freiwilligkeit – Ursprünge. Erscheinungsformen, Perspektiven

Heikkurinen P (2012) Strategic corporate responsibility: a theory review and reconsideration of the conventional perspective

Hillenbrand C, Money K (2007) Corporate responsibility and corporate reputation: two separate concepts or two sides of the same coin? Corp Reput Rev 10:261–277

Howard K (2007) Voluntary biodiversity offsets: improving the environmental management toolbox. Im Internet: http://www.cortex.org/d-Cortex-%20Biodiversity%20Offsets_01Dec07.pdf

Jakobs (2008) Freiwilligkeit: Bilanz und Ausblick. In: Ammann et al (eds) Freiwilligkeit – Ursprünge. Erscheinungsformen, Perspektiven

Kahneman D, Knetsch JL (1992) Valuing public goods: the purchase of moral satisfaction. J Environ Econ Manag 22(1):57–70

Kausch (2008) Freiwilligkeit und Freiwilligenarbeit in den Kirchen. Ideeller Hat sich Anspruch und reale Bedeutung – ein Beitrag aus der Praxis kirchlicher Freiwilligenarbeit. In: Ammann et al (eds) Freiwilligkeit – Ursprünge. Erscheinungsformen, Perspektiven

Khanna (2001) Non-mandatory approaches to environmental protection. J Econ Surv 15 (3):291–324

Kilcullen M, Kooistra JO (1999) At least do no harm: sources on the changing role of business ethics and corporate social responsibility. Ref Serv Rev 27(2):158–178

Knetsch JL (1990) Environmental policy implications of disparities between willingness to pay and compensation demanded measures of values. J Environ Econ Manag 18:227–237

Kotchen (2006) Green markets and private provision of public goods. J Polit Econ 114(4):816–834

Kotchen (2007) Voluntary provision of public goods for bads: a theory of environmental offsets. Econ J R Econ Soc 119:883–899

Liebe U, Preisendörfer P, Meyerhoff J (2011) To pay or not to pay: competing theories to explain individuals' willingness to pay for public environmental goods. Environ Behav 43(1):106–130

Lyon TP, Maxwell JW (2002) "Voluntary" approaches to environmental regulation: a survey. In: Franzini M, Nicita A (eds) Economic institutions and environmental policy. Ashgate, Farnham

Maak T (2008) Undivided corporate responsibility: towards a theory of corporate integrity. J Bus Ethics 82:353–368

Maxwell, Lyon (2001) An institutional analysis of environmental voluntary agreements in the United States. In: Orts, Deketelaere (eds) Environmental contracts. Comparative approaches to regulatory innovation in the United States and Europe. Kluwer, London, pp 333–348

Maxwell JW et al (2000) Self-regulation and social welfare: the political economy of corporate environmentalism. J Law Econ 43:583

OECD (ed) (1999) Voluntary Approaches for Environmental Policy: An Assessment. Paris

OECD (ed) (2000) Foreign direct investment, development and corporate responsibility OECD, Paris

OECD (2001a) Corporate responsibility: private initiatives and public goals. OECD, Paris

OECD (2001b) OECD roundtable on global instruments for corporate responsibility, 19 June 2001, Paris

OECD (2001c) Private initiatives for corporate responsibility: an analysis. OECD working papers on International Investment, 2001/01, OECD. https://doi.org/10.1787/816826411085

Olk C (2004) Freiwilligkeit aus Sicht der Soziologie, in Ammann (ed) Freiwilligkeit zwischen liberaler und sozialer Demokratie

Orlitzky M, Siegel D, Waldman D (2011) Strategic corporate social responsibility and environmental sustainability. Bus Soc 50(I):6–27

Orsato RJ (2006) Competitive environmental strategies: when does it pay to be green? Calif Manag Rev 48(2):127–143

Paavola J, Gouldson A, Kluvánková-Oravská T (2009) Interplay of actors, scales, frameworks and regimes in the governance of biodiversity

Priller E (2008) Zivilgesellschaftliches Engagement im europäischen Vergleich. In: Ammann H, Hasse R, Jakobs M, Riemer-Kafka G (eds) Freiwilligkeit – Ursprünge. Erscheinungsformen, Perspektiven

Putnam (2000) Bowling alone. The collapse and revival of American community. Simon & Schuster, New York

Rauscher M (1997) International trade, factor movements, and the environment. Oxford University Press, Oxford

Reed D (1999) Realms of corporate responsibility: distinguishing legitimacy. Morality Ethics 21 (23–35):1999

Segerson (1999) Mandatory versus voluntary approaches to food safety. Agribusiness 15:53–70

Segerson, Li (1999) Voluntary approaches to environmental protection. Edward Elgar, Cheltenham

Segerson, Li Dawson (2001) Environmental voluntary agreements: participation and free riding. In: Orts, Deketelaere (eds) Environmental contracts. Comparative approaches to regulatory innovation in the United States and Europe. Kluwer, London, pp 369–388

Segerson, Miceli (1998) Voluntary environmental agreements: good or bad news for environmental protection? J Environ Econ Manag 36:109–130

Siegenthaler H (2008) Zur Reichweite gemeinnützigen Handelns – Von der Gruppensolidarität zum Universitalitätsprinzip. In: Ammann H, Hasse R, Jakobs M, Riemer-Kafka G (eds) Freiwilligkeit – Ursprünge. Erscheinungsformen, Perspektiven

Spurgeon J (2008) Corporate environmental strategies for economic instruments for biodiversity. Bus Law Rev 2008:221–230

Storey M, Boyd G, Dowd J (1999) Voluntary agreements with industry. In: Voluntary approaches in environmental policy. Springer, Dordrecht, pp 187–207

US Environmental Protection Agency (1992) EPA's 33/50 Program Second Progress Report: Reducing Risks through Voluntary Action

van Calster G, Deketelaere K (2001) The use of voluntary agreements in the European Community's environmental policy. In: Orts E, Deketelaere K (eds) Environmental contracts – comparative approaches to regulatory innovation in the United States and Europe. Kluwer, London

Voigt (2000) Freiwilligkeit durch Zwangsandrohung? Eine institutionenökonomische Analyse von Selbstverpflichtungserklärungen in der Umweltpolitik, in Zeitschrift für Umweltpolitik und Umweltrecht 3/2000, pp 393–407

Wu, Babcock (1999) The relative efficiency of voluntary vs mandatory environmental regulations. J Environ Econ Manag 38:158–175

Zadek S (2004) The path to corporate responsibility. Harv Bus Rev 82(12):125–132

Voluntariness of Biodiversity Offsets

4

Abstract

As voluntary biodiversity offsets are a relatively new phenomenon, most evidence and literature relate to offsets under regulatory schemes. These are rooted in numerous compensation approaches worldwide, ranging from long-standing schemes such as US Wetland Mitigation, German Impact Mitigation Regulation, and Australian offset schemes to the recent uptake of biodiversity offsets in several European countries, e.g., France, the United Kingdom, Spain, Sweden, and the Netherlands, as well as the discussion on the introduction of an offset policy at supranational (EU) level in the scope of the EU's no net loss initiative. These schemes are, on the one hand, discussed as role models and sources of information. On the other hand, they have been criticized for the inclusion of market-based mechanisms and the use of habitat banking for the implementation of offsets.

Voluntary biodiversity offsets offer a number of opportunities for developers, e.g., license to operate, reputational effects, and access to capital. However, the likely benefits of the business case can significantly vary from one case to another. Consequently, the business case is an umbrella concept that covers a variety of different types of biodiversity offsets, and how these play out hinges on context dependency and the role and motivations of stakeholders involved.

Increasingly, standards and guidance related to the choice, design, and implementation of voluntary biodiversity offsets (such as additionality, equivalence, and perpetuity) have been formulated by private and nongovernmental institutions, e.g., the Business and Biodiversity Offsets Program, the International Union for the Conservation of Nature, and the International Finance Corporation.

Keywords

Voluntary biodiversity offsets · Compensation approaches · Wetland mitigation · Impact mitigation · No net loss initiative · The business case for biodiversity offsets · Business and Biodiversity Offsets Program · Compensation agencies

M. Darbi, *Biodiversity Offsets Between Regulation and Voluntary Commitment*,
https://doi.org/10.1007/978-3-030-25594-7_4

4.1 Biodiversity Offsets and Regulation

As has been noted in Sects. 1.1 and 1.2, compensation approaches for impacts on biological diversity already existed in numerous countries worldwide before biodiversity offsets became popular (see Table 4.1, Darbi et al. 2010). Liability for

Table 4.1 Selected compensation approaches worldwide

Compensation approaches in selected countries	
Argentina	EIA, Environmental Compensation Fund
Australia	National Environmental Offset Policy based on Environment Protection and Biodiversity Conservation Act, Native Vegetation Offset Programs in Victoria, New South Wales (NSW) and Western Australia, NSW Wetland Management Policy, Biodiversity Banking and Offsets Scheme in NSW, BushTender Program and BushBroker System in Victoria
Brazil	Forest set-aside offsets, project developers' offsets and National System of Conservation Units, Environmental Compensation Fund, Proambiente Program
China	Eco-compensation (in discussion), pilot projects (road planning, land consolidation, hydropower)
Colombia	Environmental Licensing, Principles of No Net Loss and Ecological Equivalence (Manual para la asignación de compensaciones por pérdida de biodiversidad)
Egypt	EIA/Environmental and Social Impact Assessment (ESIA), sectoral guidelines for major projects
France	Biodiversity offsets regulated by 2016 Biodiversity law, Doctrine ERC and accompanying guidelines (lignes directrices), Biobanking pilots to implement the legal requirements
Germany	Impact Mitigation Regulation
India	Compensatory afforestation, Biological Diversity Rules, mitigation schemes, and wetland mitigation schemes (under development)
Madagascar	Sectoral EIA guidelines for major projects, MEC for existing facilities, biodiversity offsets for mining projects, FOREAIM Project (Bridging restoration and multi-functionality in degraded forest landscape of Eastern Africa and Indian Ocean Islands), Eco-Certification
Malaysia	Biobanking pilot
Mexico	EIA, Administration Programme of Environmental Justice
Netherlands	Compensation in road planning
South Africa	EIA, Provincial Guidelines for Biodiversity Offsets (Western Cape and Kwazulu-Natal Provinces)
South Korea	Substitute habitats (dam construction projects), eco-bridges, research/discussion about adaptation and implementation of the German impact mitigation regulation (Eingriffsregelung) (using Incheon City as a pilot project)
Spain	Avoidance of Net Loss under Law on Natural Heritage and Biodiversity, Conservation Banking under Environmental Assessment Law
Sweden	Environmental Code, Guidance on ecological compensation, legislative process under development
United States	Wetland Mitigation, Conservation Banking

Source: Modified after Darbi et al. (2010)

damages is stipulated under various sectoral laws, e.g., environment, mining, forests, waste, and water. The "Polluter Pays Principle" is widely recognized (Darbi 2010a).

One of the compensation regulations of note is the US Wetland Mitigation (1985), based on the Clean Water Act (see Sect. 4.1.1). While this compensation approach is well established in the international literature, an even older and more comprehensive compensation regulation, the German Impact Mitigation Regulation (1976), based on the Federal Nature Conservation Act (see Sect. 4.1.2), is only recently being acknowledged in the international discourse on offsets (e.g., Albrecht et al. 2014; Darbi et al. 2010; eftec, IEEP et al. 2010; Wende et al. 2018b).

In addition, Australian offset schemes have risen to prominence in recent years. At the federal level, the Environment Protection and Biodiversity Conservation Act of 1999 (the EPBC Act) sets the framework for offsets with regard to matters of national environmental significance (NES) (Commonwealth of Australia 2007). The legislative requirements to approval conditions for proposed developments/actions in general and to biodiversity offsets, in particular, are covered in Part 9 of the EPBC Act (Sections 134 and 136 in particular) (Commonwealth of Australia 2007). Moreover, many Australian states already established a more or less developed form of biodiversity offset or biodiversity banking scheme based on market statutes (eftec, IEEP et al. 2010). The two most well-known Australian offset schemes at the state level are shown in Table 4.2: BushBroker in the state of Victoria and BioBanking in the state of New South Wales.

4.1.1 US Wetland Mitigation

As the population has expanded across the United States during the past few centuries, wetlands have been drained and manipulated to accommodate human needs. This is one of the reasons why by the 1980s the wetland area in the United States was only about 53% of what it had been in the 1780s (NRC 2001). In reaction to this, in the past few decades a number of regulations were enacted to establish a regulatory framework for compensatory mitigation in order to counteract against further wetland losses as shown in Table 4.3.

The National Environmental Policy Act (NEPA) is the main environmental law in the United States. It defines the concept of environment, covers different sectors and mediums, and establishes the fundamentals of EIA and the obligation to avoid and correct environmental damages (Peters et al. 2003). After NEPA, the mitigation and compensation of wetlands according to the Clean Water Act (CWA) is the second most important approach to addressing impacts on biological diversity.

The Clean Water Act follows the goal of "no net loss" and thus, requires compensation for unavoidable impacts on wetlands (Peters et al. 2003). In principle, activities in wetlands are forbidden, if the wetland would thereby be significantly damaged or if a feasible, less environmentally harmful alternative exists. However, permissions can be granted under exceptional circumstances by the responsible authority, the US Army Corps of Engineers.

Table 4.2 The two major offsetting schemes in Australia

Bush-Broker	Initially developed in 2006 in Victoria, the scheme aims to protect and enhance native vegetation. Therefore any clearance of native vegetation is subject to permit, a process that includes finding offsets. Offsets for residual impacts are to be provided on a like-for-like-or-better basis after the mitigation hierarchy has been observed. The scheme also seeks to deliver a "net gain" in biodiversity provision. One of the principal metrics used is "Habitat Hectares" where not only the area affected by the change in condition or "quality" of the vegetation affected or preserved is measured and scored, allowing for comparison of losses and gains. Potential suppliers of offsets that meet the system's requirements post their potential offsets on the BushBroker website, which represents an offset exchange. The price is then negotiated between suppliers and developers. Developers are also able to advertise for types of credits that they are looking for. It is also interesting to note that, due to the quality of Australian biodiversity data, the credit system accounts for different types of vegetation. In effect this represents the number of markets that have been created; however, in practice only around 50–100 are actively traded.
Bio-banking	The New South Wales scheme was established in 2007, but that was several years before the first transaction was carried out. It requires any urban development to "improve or maintain" any related ecological damage, which is assessed in a "biobanking statement". Developers are obliged to meet an "improve or maintain" test. Offset providers register their credits, or biobanks, on a public registry. This "trading floor" allows buyers to find sellers. As there are currently only six biobanks it is considered too early to determine the success of the scheme. However, early assessment suggests that developers prefer to line up the credits themselves, rather than using the central registry (a similar trend was observed in the US schemes). The offset providers can charge any amount, but there are high upfront costs to participation, which may be putting participants off and limiting the number of biobanks.

Source: Extracted and modified after IEEP et al. (2012)

Wetlands under §404 are defined as "areas that are inundated or saturated by surface or ground water at a frequency and duration sufficient to support, and that under normal circumstances do support, a prevalence of vegetation typically adapted for life in saturated soil conditions. Wetlands generally include swamps, marshes, bogs, and similar areas" (U.S. EPA n.d.).

Activities regulated under §404 of the CWA include fill for development, water resource projects, infrastructure development, and mining projects. Before dredged or fill material may be discharged into waters of the United States a permit is required (U.S. EPA n.d.).

In 1990, the Department of the Army and the U.S. EPA entered into a Memorandum of Agreement (MOA). This document clarifies the "appropriate and practicable steps" required by the environmental guidelines from 1980 to offset unavoidable impacts to wetlands under the §404 regulatory program (ELI 2002).

A three-part mitigation sequence consisting of appropriate and practicable avoidance, minimization, and compensation was established to achieve the goal of no overall net loss of wetland functions and values (U.S. Corps n.d.): First, the project proponent has to avoid alteration of wetlands by using the least environmentally

Table 4.3 Steps in the history of compensatory mitigation in the United States

Year	Regulation	Content
1958	Fish and Wildlife Coordination Act	Introduction of the concept of mitigation
1969	National Environmental Policy Act (NEPA)	Establishment of the procedure for incorporating mitigation analyses into Environmental Impact Statements
1972	§404 of the Clean Water Act	Establishment of the wetlands regulatory program
1978	Council on Environmental Quality regulations for implementing NEPA	Mitigation defined as 1. Avoiding the impact 2. Minimizing the impact 3. Rectifying the impact 4. Reducing or eliminating the impact over time 5. Compensating for impacts
1990	Memorandum of Agreement between the Department of the Army and the U.S. Environmental Protection Agency (EPA)	Establishment of a three-part process for the §404 regulatory program 1. Avoid 2. Minimize 3. Mitigate
1990	Section 307 of the Water Resources Development Act	U.S. Congress instructs the U.S. Army Corps of Engineers to pursue the goal of "no overall net loss" of the nation's remaining wetlands
1995	Federal Guidance for the Establishment, Use and Operation of Mitigation Banks	Regulates off-site compensatory mitigation

Source: Information compiled from Ambrose (2000)

damaging site. This may include sites that are not owned by the proponent. Second, a plan has to be developed to minimize the adverse effects of unavoidable impacts. Finally, if after the rectification and reduction over time, impacts still remain, the proponent has to adopt appropriate compensation measures (Peters et al. 2001). This may either be done by paying monetary compensation or using the services provided by a third party, usually a private investor with a commercial interest, a so-called mitigation bank.

Concerning wetland mitigation banking, the 1990 MOA states that it "may be an acceptable form of compensatory mitigation under specific criteria designed to ensure an environmentally successful bank" (U.S. Corps and U.S. EPA 1990, cited in ELI 2002).

Generally, wetland mitigation banking refers to the principle of bundling measures that are carried out by a third party, usually a private investor with a commercial interest (Peters et al. 2003). According to Bishop (2004), wetland mitigation banking grew out of early experiments with off-site mitigation in the late 1970s and many years were spent perfecting the concept, developing guidelines and securing agreements among relevant public agencies. In the report on "Banks and Fees" of the Environmental Law Institute (2002) it is noticed that despite the established legitimacy of wetland mitigation banking through the 1990 mitigation

MOA, it was not a commonplace practice in the early 1990s because of high costs and regulatory uncertainty. However, since the middle of the 1990s mitigation banking has become a significant business in the United States (Bishop 2004).

The Issuance of the Federal Banking Guidance in 1995 brought about a proliferation of the number of banks across the United States and commercial mitigation banks became a "mainstream option" (ELI 2002).

4.1.2 German Impact Mitigation Regulation

Biodiversity offsetting in Germany is part of the mitigation hierarchy under the Impact Mitigation Regulation (IMR). This mandatory system builds on the provisions in the Federal Nature Conservation Act, which sets the overall framework that is further elaborated in the nature conservation laws of the federal states. The IMR applies to all areas, scales of impact and sectors (excluding agriculture, forestry, and fisheries). The German approach to impact mitigation provides decades of experience with the conceptual design and practical implementation of biodiversity offsets with the goal of achieving no net loss functionally and spatially (Darbi and Wende 2013; Wende et al. 2018b).

In 1976, the adoption of the Federal Nature Conservation Act, the so-called "Eingriffsregelung" (Impact Mitigation Regulation) fundamentally changed nature conservation in Germany from a static and restrictive area-related system focusing on protected areas to a more dynamic approach aimed toward the preservation of singular ecosystem functions (Wagner 2007). Since then, it has become the major landscape conservation instrument to address mitigation and compensation for impacts from developments and projects (Darbi 2010b). The overall objective is to ensure the preservation of the existing ecological situation ("no net loss") as a minimum standard (Köppel et al. 2004) by avoiding any impairment of nature and landscape in general and more specifically, of the natural assets (species and habitats, soil, water, climate, and air quality) as well as the aesthetic quality and the recreational function of the landscape, and compensating for residual unavoidable impacts (BMU 2002).

This IMR based within nature conservation legislation was later complemented by the inclusion of IMR into the Federal Building Code (Art. 1a (3) BauGB) and the Federal Spatial Planning Act (Art. 7 (2) 2 ROG). These acts respond to the specific requirements of overall spatial planning (Janssen and Albrecht 2008). They mandate the use of IMR in spatial planning in a comprehensive way, covering projects at the levels of both urban planning (e.g., development of new residential areas) and sectoral planning (e.g., construction of roads or railways).

The Impact Mitigation Regulation requires the application of a mitigation hierarchy, in order to assure the principle of full compensation. This implies following different steps for the evaluation of impacts and the elaboration of counterbalancing measures (Durner 2001), resembling a cascade, as shown in Fig. 4.1 (Wende et al. 2005).

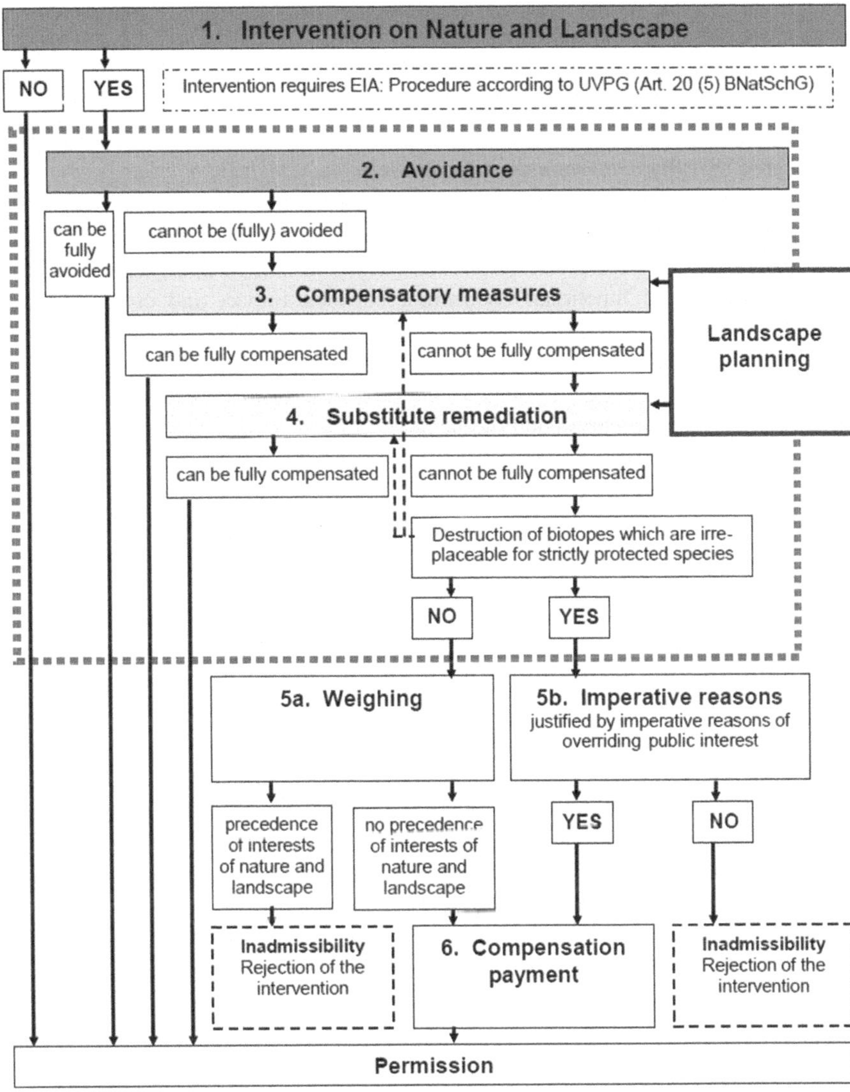

Fig. 4.1 Steps of the German Eingriffsregelung. Source: Darbi et al. (2010); modified after Köppel et al. (2004) and Bundesamt für Naturschutz (2007)

As laid down in the Federal Nature Conservation Act, the process of the IMR begins with the identification and evaluation (in terms of significance) of the impacts of a project, plan or action on nature and the landscape. Any unavoidable impairment must be compensated for through nature conservation and landscape management measures (Peters et al. 2003; Durner 2001). These include compensatory measures ("in kind" offsets, also referred to as "Ausgleich"/restoration compensation) and substitute remediation ("out of kind," also referred to as "Ersatz"/replacement compensation) and, in the case of remaining adverse impacts, a compensation payment (Peters et al. 2003).

With regard to IMR implementation, the use of pool and banking models (i.e., habitat banking) have been explored since the beginning of the 1990s (cf. Wagner 2007; Jessel et al. 2006; Spang and Reiter 2005; Böhme et al. 2005; Darbi 2010b; Wende et al. 2018b).

Amendments to the Federal Building Code in 1998 (with the aim of optimizing the enforcement and implementation of compensation measures in urban development planning) and the Federal Nature Conservation Act in 2002 (resulting in a loosened spatial and functional connection between impact and compensation) fostered the emergence of compensation pools (see BMU 2002; Bundesregierung 2009; Wende et al. 2005, 2018b; Darbi and Wende 2013). Wagner (2007) characterizes these as modern compensation instruments that build on the idea of providing compensation areas and measures.

Two types of compensation pools can be distinguished in practice:

1. "Flächenpool": Appropriate land is gathered in a pool, and is provided for implementation of compensation measures on request (pool of compensation sites).
2. "Ökokonto": Measures are provided that have already been implemented (pool of compensation measures, "Maßnahmenpool" or eco-account) (Darbi 2010b).

A statewide survey confirmed the existence of several hundreds of pools (Böhme et al. 2005).

In recent years, several of the German states have enacted ordinances in support of this practice, in addition to the framework legislation of the Federal Nature Conservation Act, thus establishing the basis for the occurrence of professional public and private providers of compensation services ("compensation agencies") (see Schöps et al. 2007; Schöps 2008; BFAD 2007).

As a result of 40 years of experience, German IMR is outstanding due to its comprehensive character and strict additionality of offsets (Herbert 2015; Albrecht et al. 2014; Darbi and Wende 2013; Darbi et al. 2010). It rests on a broad scientific base and discussion, high-quality standards and numerous detailed guides developed by the ministries of the German states and other institutions. As a result, German IMR is one of the best developed formal compensation systems worldwide and the largest one in Europe (Madsen et al. 2010).

4.1.3 Situation in the EU and No Net Loss Initiative of the European Commission

EU legislation currently creates limited demand for compensation except for damage to Natura 2000 sites. Many Member States have no additional requirements for compensation other than that required by EU legislation, so in practice, most demand for compensation arises from damage to the Natura 2000 network (IEEP et al. 2012).

The exception to this is Germany, which hosts a comprehensive impact mitigation regulation as has been presented in Sect. 4.1.2.

However, this situation is about to change. In fact, interest in offsets and experience of their application is growing in several Member States, e.g., the Netherlands, Sweden, Spain, and the United Kingdom (and thus new offset schemes and requirements are being trialed or introduced) (IEEP et al. 2012; cf. Wende et al. 2018a; Alvarez Garcia et al. 2018; Wende 2018; Van Teeffelen 2018; Baker et al. 2018; Dupont et al. 2018; Bacianskaite 2018; Boučková and Albrecht 2018). In addition, in other Member States, previously lacking enforcement of existing requirements has been strengthened, e.g., in France (Vaissière et al. 2018), which, like Germany, has a requirement to compensate for negative environmental impacts stipulated in the environmental law since 1976. However, the French compensation requirement has never been properly implemented until a set of reforms were introduced starting in 2007 as the result of a national consultative process on environmental policy (called "Grenelle de l'Environnement") (IEEP et al. 2012).

To streamline offset implementation, France has launched its own pilot habitat banking scheme and in 2008 the French ministry for the environment in cooperation with the private "CDC Biodiversité" (a specialized subsidiary of a state-owned sovereign fund) began implementing a first pilot on biodiversity offsets, by restoring natural grasslands in the Crau area in Southern France (personal communication with Delphine Morandeau 2010; Vaissière et al. 2018).

In June 2011, the French government chose to expand this pilot to several other regions and foster additional habitat banks. Furthermore, French offsetting requirements have been significantly strengthened by the issuance of a "doctrine" on how to best avoid, reduce and offset impacts on biodiversity and ecosystems as well as accompanying guidelines (lignes directrices) in 2012 and a new biodiversity law (Projet de loi pour la reconquête de la biodiversité, de la nature et des paysages) which was approved by the National Assembly in 2015. The latter includes the adoption of an important article on biodiversity offsets and introduces a no net loss/ net gain principle (Darbi 2015a).

More generally, in addition to the mentioned activities at Member State level, at the European level, the European Commission published its Biodiversity Strategy until 2020 (European Commission 2011). Target 2 of the European Biodiversity Strategy claims *to maintain and enhance ecosystems and their services* "by establishing green infrastructure and restoring at least 15% of degraded ecosystems."

Supporting actions initiated by the European Commission include the development of a strategic framework to set priorities for ecosystem restoration at subnational, national, and EU level (Action 6a), a Green Infrastructure Strategy by 2012 (action 6b), and *an initiative to ensure there is no net loss of ecosystems and their services (e.g., through compensation or offsetting schemes) by 2015 (action 7b)* (European Commission 2011).

To trial and foster the concept of no net loss, the European Commission has established a *working group on no net loss of ecosystems and their services* (European Commission 2012a, b) and contracted a report on "Policy Options for an EU No Net Loss Initiative" (Tucker et al. 2013). The report identifies more than

30 individual measures (grouped to 11 complexes) that could potentially be included in the future no loss initiative and analyses them with their expected impact compared to the business as usual scenario.

The authors concluded that the further development of an EU policy on NNL is necessary to halt biodiversity loss, and that offsetting is an important tool to achieve the NNL objective (Tucker et al. 2013).

> Therefore, the achievement of the NNL objective will require the development of a NNL policy framework that seeks to improve the implementation of existing policies *and* carefully designs and develops requisite new policy measures for offsetting (as well as other policy gaps such as relating to invasive species, soils etc.). However, policy measures will not be enough; substantial public and private support will also be needed, including commitment to awareness raising, guidance, training, capacity building and monitoring and assessment. It is therefore recommended that a comprehensive strategy and common policy framework for NNL is developed to address all stages of the mitigation hierarchy through initiatives to improve and better use existing policy instruments where feasible, complemented where necessary and appropriate by new policy measures to fill significant gaps. (Tucker et al. 2013)

As regards the implementation of offsets under an EU policy, Tucker et al. (2013) note that "there is strong evidence from practical experience of operational offset schemes reviewed in this study, that mandatory requirements for the offsetting of residual impacts would be needed to make a significant contribution to the NNL objective." They further suggest that this "could be achieved at EU level through a framework directive, or similar instrument."

However, they also note that this is "politically and technically challenging," e.g., that developers might be encouraged to forego the proper application of the mitigation hierarchy and jump straight to offsetting or that existing levels of protection under the Birds and Habitats Directive and the European Liability Directive might be weakened (Tucker et al. 2013).

Building on the findings from the policy options reports together with the work of the no net loss working group (e.g., a glossary and a scoping document) the European Commission launched a public online consultation on the planned initiative in June 2014. The consultation asked interested citizens, public authorities, businesses, and NGOs for their views on the scope and scale of the initiative, the choice of policy instruments to use and how to tackle the challenges related to offsetting, etc. (European Commission 2015a).

By closure of the consultation in October 2014, the European Commission has received more than 700 responses, mainly from individuals, NGOs, companies, and business associations (European Commission 2015a). The results were made publicly available in early 2015 and are briefly summarized below:

While more than half of respondents in principle support offsetting as part of an EU no net loss initiative, for more than two-third of this group this must be conditional on strict measures and robust safeguards being in place to prevent abuse (European Commission 2015a, b, see Fig. 4.2).

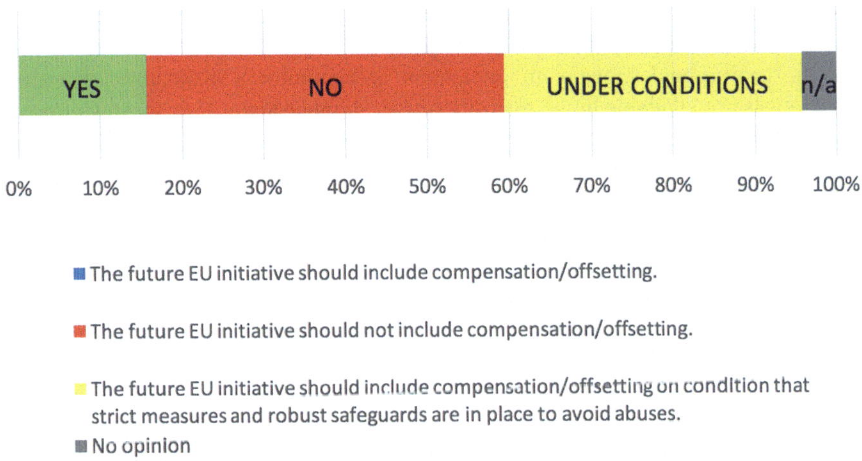

Fig. 4.2 Results of the public consultation on the planned EU no net loss initiative: Should the future EU initiative address compensation/offsetting or should this be excluded? Source: Author, building on European Commission (2015b)

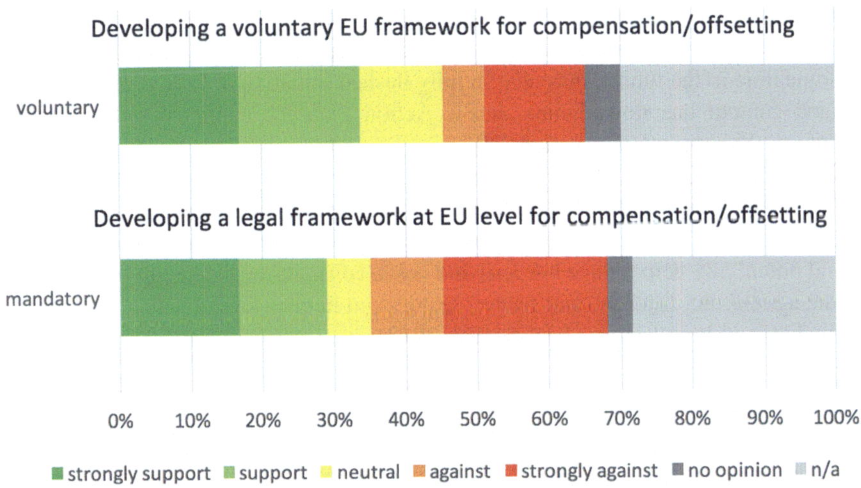

Fig. 4.3 Results of the public consultation on the planned EU no net loss initiative: voluntary vs. legal framework. Source: Author, building on European Commission (2015b)

By contrast, a (relative) majority (of 33%) has voted against the development of a legally binding compensation instrument and a legal framework for biodiversity offsets at the EU level, and rather preferred the development of a voluntary framework including technical guidance and benchmarking of good practice (European Commission 2015b, see Fig. 4.3). The reasons for this choice were based on fears that offsets were a "license to trash," that they would be abused, that the political timing was not right and that it was too complex (European Commission 2015a).

These concerns have been expressed by nature conservation NGOs in particular. In addition, businesses feared the burden of potential additional costs of offsetting and that the institutional capacity to implement and monitor a mandatory offsetting scheme was not available in the majority of EU Member States (European Commission 2015a).

The results of this consultation exemplify that no consensus exists on the appropriateness and role of biodiversity offsets in an EU context. This is particularly true with regard to the general question concerning whether or not to include offsetting in a future EU no net loss initiative (see Fig. 4.2), whereby little more than half of the respondents (52%) are in favor of inclusion (however, under strict conditions) as opposed to 43.7% who are against the inclusion of offsets as part of the initiative (European Commission 2015b).

With regard to the more specific question as to whether respondents would support the development of a legal offset framework at the EU level, this relationship is inversed, with 28.9% in support of and 33% against the introduction of a legal framework (European Commission 2015b). This shows that there is no clear tendency concerning how to proceed regarding an EU policy on offsets.

Seeing this ambiguity and the at least partial rejection of a compulsory EU-wide legal framework for biodiversity offsets, the no net loss initiative that had been initially planned for 2015 has been delayed in the course of re-examination and re-prioritization of the possible policy options and may be put on the political agenda at some time in the future. Instead of a fully fleshed initiative or even directive some related content has flowed into an EU Action Plan for nature, people, and the economy (European Commission 2017).

Nonetheless, Tucker et al. (2018) conclude that "to help advance offsetting as a key element of sustainable development, it appears necessary to create a clearer EU level no net loss policy requirement, with associated standards and guidance." This could potentially help reduce burdens and costs, consider the extension of offsetting to the agriculture, forestry, and fishery sectors and help ensure offsets are effective, equitable, and long-lasting (Tucker et al. 2018).

What Are the Implications for the Present Study?

As voluntary biodiversity offsets are a relatively new phenomenon, most evidence and literature are focused on the implementation of offsets under regulatory schemes. These are rooted in numerous compensation approaches worldwide.

The most well-known scheme, which is also widely documented in scientific and gray literature, is US Wetland Mitigation after the 1985 Clean Water Act. In addition, one of the best developed formal compensation systems worldwide, the German Impact Mitigation Regulation after the 1976 Federal Nature Conservation Act, is only recently being acknowledged in the international discourse on offsets. Both schemes follow a no net loss goal, focused on

(continued)

wetlands in the case of wetland mitigation and comprehensive (i.e., areawide) in the case of the impact mitigation regulation.

Both schemes are, on the one hand, discussed as role models and have served as a source of information and inspiration for regulatory and voluntary biodiversity offsets. On the other hand, they have been criticized for the inclusion of market-based mechanisms and the increasing use of habitat banking for the implementation of offsets (wetland mitigation banking, compensation pools, and agencies).

The same holds to be true for Australian offset schemes that have risen to prominence in recent years, which largely build on biodiversity banking (e.g., Biobanking in the state of New South Wales).

Increasingly, there has been uptake of biodiversity offsets in several European countries, e.g., France, the United Kingdom, Spain, Sweden, and the Netherlands. France has strengthened its preexisting compensation requirements by issuing a doctrine and guidelines on the implementation of the mitigation hierarchy including offsets and exploring habitat banking in a number of pilots. Similarly, the United Kingdom has explored the introduction of a regulatory offsetting scheme at national level with a number of studies and pilots.

A more recent development, that has the question of regulatory vs. voluntary biodiversity offsets at its core, is the discussion in the European Union on the introduction of an offset policy at supranational (EU) level in the scope of the EU's no net loss initiative as part of its 2020 biodiversity strategy. This has been explored by a no net loss working group, accompanied by a number of studies on behalf of the European Commission.

Finally, an online public consultation was held in 2014, which confirmed the existing controversy on the appropriateness and role of biodiversity offsets at EU level (and in general). In particular, the question as to whether a regulatory or a voluntary offset framework shall be envisaged has come to a dead end.

The ambiguity of the consultation results was also fueled by a previous biased discussion on offsets and the campaigning against biodiversity offsets in particular by several nature conservation NGOs, e.g., FERN and Friends of the Earth (cf. FERN 2014a, b; FERN and Friends of the Earth 2014; Counter balance 2014). Consequently, the no net loss initiative envisaged for 2015 has been delayed to 2016/17 in the course of re-examining and re-prioritization of the policy options to be included.

4.2 The Business Case for Biodiversity Offsets

When looking at the "business case" for biodiversity offsets, the Business and Biodiversity Offsets Program is an obvious starting point. BBOP (n.d.-b) asks:

> Why should businesses voluntarily "go the extra mile" and take on biodiversity offsets? What's in it for them?

Or, as Kerry ten Kate (Director of BBOP) puts it in a presentation she gave in 2005:

> Why should business offset the harm it causes to biodiversity?

As a conclusion from this, biodiversity offsets can be seen as the result of a cost-benefit analysis. Thus, the "business case" can be interpreted as the weighing of opportunities and risks associated with the implementation of biodiversity offsets for business, i.e., an argument or motivation why a business might engage voluntarily in offsetting their impacts.

In this line of reasoning, a study by PricewaterhouseCoopers LLP (2010) notes a gradual increase in the number of organizations and initiatives that are beginning to recognize the business benefits of biodiversity offsets, i.e., understanding potential advantages for companies, and working to integrate biodiversity considerations into business and commercial activities.

This reasoning has become mainstream in academia and business communication (Darbi 2015b). ten Kate et al. (2004) cite a number of opportunities offered by voluntary biodiversity offsets for companies and developers. Businesses can benefit from the implementation of (voluntary) compensation measures, e.g., in terms of reputation and better acceptance of their projects. Table 4.4 shows the criteria that make compensation measures from the perspective of business worthwhile. They have been named as possible motivations for businesses to implement biodiversity offsets.

However, no sufficient evidence exists that these criteria prove to be effective in practice (Darbi 2015b). Therefore, critical examination is necessary as to whether

Table 4.4 Benefits of biodiversity offsets for business

The business case for biodiversity offsets	
License to operate	Access to land and resources, speeding up approval processes and avoiding costly delays, political influence
Reputation	Good PR, improved relationship with the local population and decision makers, status of a "preferred partner"
Access to capital	Increased demands and standards of international financial institutions and donors
Efficiency	Management of risks and liability obligations
New markets	First-Mover advantage

Source: After ten Kate (2005) and Howard (2007)

these criteria provide incentives for the implementation of biodiversity offsets and whether these incentives would be positive or negative from a nature conservation point of view (e.g., license to operate might result in negative environmental outcomes).

Furthermore, the assumption that the "business case" is a simple weighing of risks and opportunities does not explain the various types of involvement of business in biodiversity offsets (Darbi 2015b).

To further explore and analyze the business case for biodiversity offsets, a number of additional questions can be raised:

- Does the "business case" imply voluntary biodiversity offsets?
- What are the motives and goals of the "business case," i.e., is it part of Public Relations, Corporate Responsibility, or Strategic Business Planning?
- What is the business in the "business case," i.e., are biodiversity offsets a business?

While only interim answers to these questions can be given at this point, some further observations will help to better understand what the "business case for biodiversity offsets" might be or what aspects it comprises.

4.2.1 Does the "Business Case" Imply Voluntary Biodiversity Offsets?

The "business case" usually refers to biodiversity offsets that are implemented without legal obligation (ten Kate et al. 2004; ten Kate 2005; Howard 2007; Darbi 2015b), i.e., voluntarily. In line with this, the business case for biodiversity offsets is used synonymously with "voluntary biodiversity offsets" in the scope of this study—despite the limitations that occur in practice.[1]

Already in 2004, Ten Kate, Bishop, and Bayon have identified "a modest, but growing number of corporate initiatives to offset unavoidable harm to biodiversity on a voluntary basis." They note that most activities are conducted on ad hoc basis and driven by the personal enthusiasm of particular leaders within companies.

This is in line with theoretical observations on corporate responsibility (as outlined in Sect. 3.6) that especially top management "tends to over-invest in CSR activities to build their own personal reputation as good global citizens (over-investment hypothesis)" and to "generate support from social and environmental activists in order to reduce the probability of CEO turnover in a future period

[1]It can be noted that while the "business case" implies voluntary biodiversity offsets, the latter do not necessarily entail business involvement in return, i.e., voluntary biodiversity offsets can be implemented, e.g., by public authorities, NGOs, and civil society groups.

Furthermore, the business case may not only be relevant to voluntary biodiversity offsets, but also to regulatory regimes, where, however, developers depending on the context conditions and likely sanctions may not see a business case for compliance with these regulatory regimes.

(strategic-choice hypothesis)" (Barnea and Rubin 2010, cited in Harjoto and Jo 2011). This example points to the importance of motivational aspects related to biodiversity offsets, i.e., biodiversity offsets may be driven by different motives and goals as is explained in the following Sect. 4.2.2.

4.2.2 What Are the Motives and Goals of the "Business Case"?

If the "business case" refers to voluntary biodiversity offsets, then the question arises why businesses implement biodiversity offsets voluntarily. As has been noted, implementing biodiversity offsets can create added value, i.e., a tangible benefit (cf. Sect. 3.2.1). The most obvious case for a business benefit arising from biodiversity is when biodiversity is the source or prerequisite for the business activities (e.g., tourism).

> The business case for biodiversity conservation is most easily made when the business in question depends directly on biodiversity to operate and survive [...] For many other businesses the case for investing in biodiversity conservation is less clear. Understanding what biodiversity means and how it affects business value is not always straightforward. (Bishop et al. 2006)

But in most cases, this dependence is not directly visible or does not necessarily require an offset from a business perspective. For example, mining builds on the extraction of natural resources, but unlike tourism or agriculture the same area can only be "used" for business purposes once.

Nevertheless, biodiversity offsets will most likely be the result of a cost-benefit analysis as has been noted above, and "make good business sense" as part of strategic business planning for a number of reasons (Darbi 2015b, see Table 4.4). Thus, the "business case" has been increasingly recognized for managing corporate impacts on the environment and led to increasingly detailed and quantified environmental goals by businesses (Rainey et al. 2015).

In the public, however, a rather negative view of biodiversity offsets (and in particular of related business activities) dominates, with businesses being accused of promoting biodiversity offsets for pure marketing and PR reasons, or even labelled "Greenwashing" by some (Darbi 2015b).

> The single greatest challenge for biodiversity offsetting comes from opposition at local level. Grassroots and NGO-led campaigns have successfully managed to communicate their negative opinions about biodiversity offsetting through on-the-ground and online campaigns, and have in the process been noticed by the mainstream media. (Ferreira 2014)

In fact, on the contrary, Ferreira (2015) finds that few companies are actually choosing offsets as a way to manage their image and to show consumers that they are environmentally responsible.

> While many companies acknowledge the potential of offsetting as a mechanism for assuring regulatory goodwill and securing social license to operate, there is very little evidence that it is being used as a mechanism for managing corporate reputations or corporate social responsibility. (Ferreira 2015)

Despite this observation, corporate (social and environmental) responsibility is possibly the strongest driver and particularly suited to examine or explain the motivation for the implementation of voluntary biodiversity offsets by business, as it combines both moral and ethical responsibilities toward the various stakeholders of a company and the added value or benefit the company may gain from doing so (Darbi 2015b, cf. Sect. 3.6).

4.2.3 What Is the Business in the "Business Case", i.e., Are Biodiversity Offsets a Business?

The business case for biodiversity offsets essentially refers to an argument or motivational state that encourages the implementation of biodiversity offsets. However, the magnitude or implications of a business case for biodiversity offsets can vary greatly, i.e., the business case may be stronger in some cases than in others.

From this, the question ultimately arises whether biodiversity offsets are more than a tool for business to address the negative externalities of their operations, but in fact open up a new business field (i.e., a market for biodiversity offsets). This question certainly touches upon one of the most debated aspects of biodiversity offsets. Opponents argue that biodiversity offsets should be targeted at achieving the best possible nature conservation outcomes and not be measured in terms of profitability.

Nevertheless, business models exist under different offsetting schemes worldwide and biodiversity offset markets have emerged to some extent, especially under Wetland Mitigation Banking in the United States (Madsen et al. 2010, 2011; Mann et al. 2014; Mann and Simons 2015; Darbi et al. 2010; Ferreira 2014, 2015; eftec, IEEP et al. 2010; ICH GHK, Bio Intelligence Service 2013, cf. Sect. 4.1).

> In the US, legislation requiring offsets for damage to wetlands and to endangered species has led to the creation of a number of new businesses. Some of these create wetland or species mitigation banks, some broker deals, some provide scientific advice and some do all of the above. The largest of these businesses have now become multi-million dollar service providers. (ten Kate et al. 2004)

While in the United States, offsets are already delivered via commercial offset providers to a large extent, this cannot be claimed for most offsetting schemes. For example, in Germany compensation pools and banks exist alongside "compensation agencies" that provide biodiversity offsets or biodiversity credits to developers (Wende et al. 2005; Darbi et al. 2010, cf. Sect. 4.1.2), but these are run by a multitude of different actors and there is only a little knowledge about the market in terms of size and homogeneity (Wende et al. 2015; Darbi 2015c).

What can be seen from this is that biodiversity offsets can provide a starting point to build a business; however, this can be risky and depends very much on the context. Worldwide, a small, but an increasing number of offset schemes exists, which may lead to the creation of a market of a biodiversity offsets services industry (cf. Madsen et al. 2010, 2011; ICH GHK, Bio Intelligence Service 2013; Darbi 2015c).

4.2.4 Summing Up: What Is the Business Case—An Approximation

Coming back to the question of what the often cited "business case for biodiversity offsets" is, it can be noted, that no "easy" answer exists—or at least none that sufficiently covers and explains what is currently labeled the "business case for biodiversity offsets" in practice.

Does the "business case" simply apply to any case where businesses voluntarily carry out biodiversity offsets that go beyond (non-)existing regulation, i.e., beyond compliance? More generally, does the business case refer to any involvement of business in designing and developing biodiversity offsets?

For some, this means that implementing biodiversity offsets (voluntarily) simply makes good business sense. For others, biodiversity offsets open up a profitable business field, i.e., habitat banks like wetland mitigation banking or compensation pools (see Sect. 4.1). Again, others might argue that the "business case" refers to all business-led initiatives that aim to implement biodiversity offsets. These, however, can be motivated by a variety of reasons, e.g., implemented in the scope of their Corporate Social Responsibility, Environmental Management and Reporting (Darbi 2015b).

Thus, in practice the answer to the posed questions will most likely vary from one case to another and "the business case" can be considered a vague concept at worst and an umbrella term for a multitude of different understandings at best. Consequently, as has been assumed in Chap. 1, voluntary biodiversity offsets and the "business case" do not represent one "case" with strictly defined conditions and particularities. Rather they cover a whole variety of different cases or types of biodiversity offsets, which are shaped by the context and conditions under which they arise. For example, offsets can be implemented on a single case basis or be based on a corporate or sectoral strategy and thus differ with regard to their scope and implications.

Consequently, the outcome of the complex and controversial task of implementing biodiversity offsets hinges on this context dependency and thus a cooperative approach, involving various stakeholders, most notably government and business and being controlled by institutions of civil society (e.g., NGOs) and the broader public (Darbi 2015b).

What Are the Implications for the Present Study?

The "business case" for biodiversity offsets and "voluntary biodiversity offsets" largely overlap and thus are used synonymously in the scope of this study. Very generally, the business case triggers offsets that are implemented without legal obligation, i.e., corporate initiatives that target the negative environmental externalities of their business operations.

However, neither the motivations for the growing number of such corporate initiatives have been explored in detail, nor does a sound definition exist that sufficiently covers and explains what is currently labelled the business case for biodiversity offsets in practice, let alone an evaluation of the outcome in terms of additional nature conservation value.

Several authors have noted that the implementation of voluntary biodiversity offsets offers a number of opportunities for companies and developers, e.g., license to operate, reputation, and access to capital.

However, only a little evidence exists that these criteria prove to be effective in practice in terms of providing pressures and incentives that drive the implementation of voluntary biodiversity offsets. For example, with regard to possible reputational benefits, Ferreira (2015) finds that there is little evidence for businesses using voluntary biodiversity offsets to improve their corporate reputation.

Moreover, in light of the controversy on biodiversity offsets in the public and the media, this seems very unlikely, in particular in a European context.

Furthermore, context is very important, and voluntary biodiversity offsets and the likely benefits of the business case can significantly vary from one case to another.

Consequently, from this arises an understanding of the business case as an umbrella concept that covers a whole variety of different cases or types of biodiversity offsets, and how these play out hinges on context dependency and the role and motivations of stakeholders involved.

4.3 Common Standards for Biodiversity Offsets

Independent of the way offsets are implemented (e.g., through regulatory, corporate, or pilot schemes), there are a number of institutions and initiatives that have discussed and formulated standards and critical points for biodiversity offsets. These include, among others:

- International Finance Corporation IFC: Performance Standards 6 on "Biodiversity Conservation and the Sustainable Management of Ecosystem Services and Living Resources" (IFC 2012).

- World Bank: World Bank Safeguard Policies, in particular, World Bank Performance Standard 6 "Biodiversity Conservation and Sustainable Management of Living Natural Resources"[2] (The World Bank 2012a, b).
- Business and Biodiversity Offsets Program (BBOP): published a whole toolkit on biodiversity offsets as is shown in Fig. 4.4, including principles, offset design handbook, implementation handbook, and cost–benefit handbook, resource papers, pilots, and case studies (BBOP 2009a, b, 2012a, b, c, 2013, cf. BBOP 2018).
- Federal Association of Compensation Agencies: (Bundesverband der Flächenagenturen in Deutschland BFAD) published quality standards for compensation service providers/habitat banks in Germany (BFAD 2007).
- International Council on Mining and Metals ICMM: published two sets of case studies on integrating mining and biodiversity conservation as well as a number of guidance documents on the application of the mitigation hierarchy and biodiversity offsets, launched commitment of its members not to mine or explore in World Heritage Sites (ICMM 2005a, b, 2006, 2010, 2016; ICMM IUCN 2012).
- International Petroleum Industry Environmental Conservation Association IPIECA: has teamed up with ICMM to launch the CSBI (see below), currently exploring the introduction of a no net loss/net gain commitment (personal communication with Jared Hardner).
- Ecosystem Marketplace: launched reports and guidance on the state of biodiversity markets, mitigation, and conservation banking in the United States (Madsen et al. 2010, 2011).
- The Equator Principles: Minimum standards, including adherence to the mitigation hierarchy (Equator Principles 2011, 2013).
- IUCN: draft biodiversity offsets policy, additional guidance policy options and technical conditions for biodiversity offsets as well as potential for the application in agriculture and forestry (IUCN 2014, 2015; Pilgrim and Ekstrom 2014; Edwards 2015; ICMM IUCN 2012).
- Cross Sector Biodiversity Initiative (CSBI): CSBI is a partnership between ICMM, IPIECA, and the Equator Principles Association, has released practical guidance on the implementation of the mitigation hierarchy for the extractive sector (The Biodiversity Consultancy 2015).
- Biodiversity for Banks Program (B4B): The Biodiversity for Banks (B4B) program, co-launched by the Equator Principles Association, World Wildlife Fund (WWF) and the Business and Biodiversity Offsets Program (BBOP), is designed to help financial institutions overcome the challenges of incorporating risks associated with biodiversity and ecosystem services, including guidance on biodiversity offsets, largely building on the work of the co-founders (Equator Principles 2011).

[2]Similar to the International Finance Corporation's Performance Standard 6.

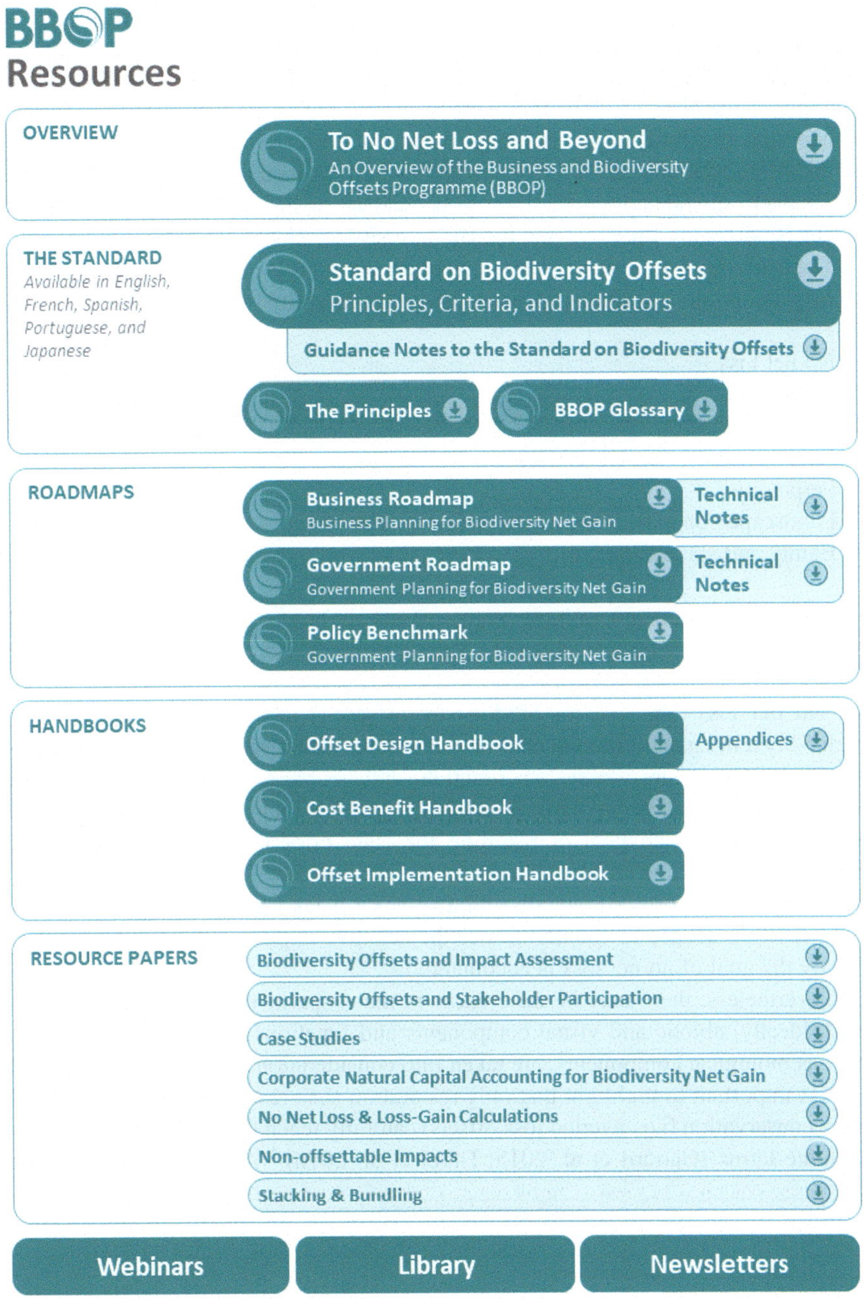

Fig. 4.4 BBOP standards on Biodiversity Offsets and associated material. Source: BBOP (2018)

As can be seen from the list above, these various standards are not independent from one another but have been mutually influencing and reinforcing. In particular, the BBOP standard has served as a source for several of the others and is referenced for example in the IUCN's draft biodiversity offsets policy (IUCN 2015), in Performance Standard 6 of the International Finance Corporation (IFC 2012) and many others. Moreover, several of the organizations have teamed up to jointly launch new initiatives to spread and transfer their own standards to new contexts, e.g., the Cross Sector Biodiversity Initiative and the Biodiversity for Banks Program.

The analysis of these standards and recommendations leads to a number of crucial aspects and points of critique associated with biodiversity offsets (cf. Fig. 4.5):

1. No net loss
2. Additionality
3. Appropriateness
4. Equivalence
5. Perpetuity
6. Landscape context
7. Equity and stakeholders

These core parameters of biodiversity offsets are briefly presented and discussed below.

(1) No net loss The goal of biodiversity offsets is to achieve no net loss and preferably a net gain in biodiversity on the ground with respect to species composition, habitat structure, ecosystem function, and people's use and cultural values associated with biodiversity (BBOP 2009a, cf. Sects. 1.1 and 1.2). In this context, no net loss refers to the goal of restoring the state prior to the impact, and thus, maintaining the same level of biodiversity, whereas net gain aims at improving biodiversity quality. While ideally, biodiversity offsets can exceed the amount of compensation needed to counterbalance the negative impacts (net gain), in practice mostly the goal of no net loss is encountered.

Nevertheless, there may be differences in the scope of the components considered. Ideally, abiotic and visual components and functions of the environment, as well as biological components, are taken into consideration.[3]

It is important to highlight that only restoration (i.e., restoration offsets) and not mere conservation (i.e., averted loss offsets) can fully reach the goal of no net loss in absolute terms (Garrard et al. 2015; Darbi et al. 2010).[4] An enhancement of the

[3]An example for a comprehensive no net loss approach is the German Impact Mitigation Regulation (cf. Sect. 4.1.2).

[4]Restoration offsets and averted loss offsets are the two most common offset strategies. While the former refers to offsets based on restoring biodiversity, the latter refers to offsets based on avoiding loss to existing biodiversity under current or expected threat of loss (Gibbons et al. 2016).

Box 2: Principles on Biodiversity Offsets supported by all the members of the BBOP Advisory Group

Biodiversity offsets are measurable conservation outcomes resulting from actions designed to compensate for significant residual adverse biodiversity impacts arising from project development* after appropriate prevention and mitigation measures have been taken. The goal of biodiversity offsets is to achieve no net loss and preferably a net gain of biodiversity on the ground with respect to species composition, habitat structure, ecosystem function and people's use and cultural values associated with biodiversity.

These principles establish a framework for designing and implementing biodiversity offsets and verifying their success. Biodiversity offsets should be designed to comply with all relevant national and international law, and planned and implemented in accordance with the Convention on Biological Diversity and its ecosystem approach, as articulated in National Biodiversity Strategies and Action Plans.

1. Adherence to the mitigation hierarchy: A biodiversity offset is a commitment to compensate for significant residual adverse impacts on biodiversity identified after appropriate avoidance, minimization and on-site rehabilitation measures have been taken according to the mitigation hierarchy.

2. Limits to what can be offset: There are situations where residual impacts cannot be fully compensated for by a biodiversity offset because of the irreplaceability or vulnerability of the biodiversity affected.

3. Landscape context: A biodiversity offset should be designed and implemented in a landscape context to achieve the expected measurable conservation outcomes taking into account available information on the full range of biological, social and cultural values of biodiversity and supporting an ecosystem approach.

4. No net loss: A biodiversity offset should be designed and implemented to achieve in situ, measurable conservation outcomes that can reasonably be expected to result in no net loss and preferably a net gain of biodiversity.

5. Additional conservation outcomes: A biodiversity offset should achieve conservation outcomes above and beyond results that would have occurred if the offset had not taken place. Offset design and implementation should avoid displacing activities harmful to biodiversity to other locations.

6. Stakeholder participation: In areas affected by the project and by the biodiversity offset, the effective participation of stakeholders should be ensured in decision-making about biodiversity offsets, including their evaluation, selection, design, and implementation and monitoring.

7. Equity: A biodiversity offset should be designed and implemented in an equitable manner, which means the sharing among stakeholders of the rights and responsibilities, risks and rewards associated with a project and offset in a fair and balanced way, respecting legal and customary arrangements. Special consideration should be given to respecting both internationally and nationally recognized rights of indigenous peoples and local communities.

8. Long-term outcomes: The design and implementation of a biodiversity offset should be based on an adaptive management approach, incorporating monitoring and evaluation, with the objective of securing outcomes that last at least as long as the project's impacts and preferably in perpetuity.

9. Transparency: The design and implementation of a biodiversity offset, and communication of its results to the public, should be undertaken in a transparent and timely manner.

10. Science and traditional knowledge: The design and implementation of a biodiversity offset should be a documented process informed by sound science, including an appropriate consideration of traditional knowledge.

* While biodiversity offsets are defined here in terms of specific development projects (such as a road or a mine), they could also be used to compensate for the broader effects of programs and plans.

Fig. 4.5 BBOP principles on biodiversity offsets. Source: BBOP (2013)

existing situation is required to effectively counterbalance the loss elsewhere occurred (BFAD 2007).

By contrast, averted loss offsets can only achieve no net loss relative to an anticipated declining baseline, also referred to as counterfactual (see Fig. 4.6), i.e., maintenance of a declining trajectory (Maron et al. 2015a). Several authors (e.g.,

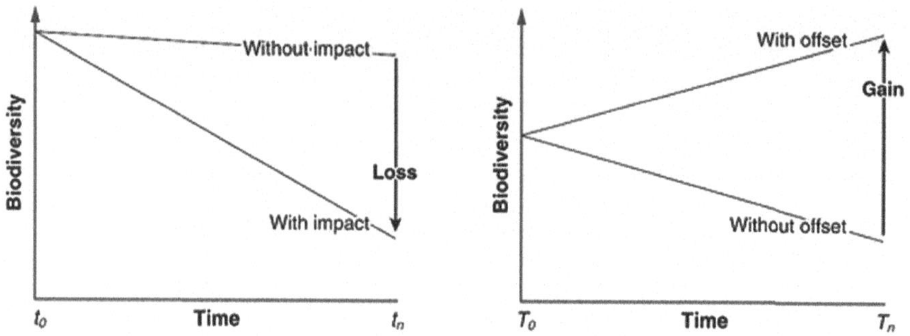

Fig. 4.6 Conceptual model for defining biodiversity loss from development and biodiversity gain from offsets relative to a declining baseline (counterfactual). Source: Gibbons et al. (2016)

Gibbons et al. 2016; Bull et al. 2013; Maron et al. 2010, 2015a) call for the use of counterfactuals in determining offsets.

> In particular, the baseline against which to measure no net loss is rarely specified. The implicit assumption is often that the biodiversity baseline is fixed at the point of the development project. However, as ecosystems are dynamic, no net loss should be defined against prevailing trends in biodiversity condition. (Bull et al. 2013)

While averted loss offsets are often less expensive and easier to obtain than restoration offsets (Maron and Gordon 2015), Gibbons et al. (2016) calculated that both restoration and averted loss offsets could theoretically lead to no net loss (cf. Maron et al. 2010). However, Poulton (2015) objects that "projecting when a site may be lost to a threat, and how much will be lost, is a more speculative exercise."

(2) Additionality Following the no net loss principle, biodiversity offsets should result in measurable compensation outcomes, beyond that which would have occurred in the absence of the offset activities (BBOP n.d.-a). Thus, biodiversity offsets are a direct result of the offset project and would not have existed otherwise (Howard 2007). This requirement of additionality implies that offsets cannot replace conservation and other obligations, for example, those of environmental authorities (Darbi et al. 2010; cf. Maron et al. 2015a, b; Maron and Gordon 2015). It must be ensured that biodiversity offsets are not mere financing tools for general nature conservation duties.

Therefore, the central question is what the (physical and financial) compensation measures are used for. Negative uses include cases where monetary compensation measures are merely used for the management and maintenance of existing protected areas, as this would not generate an additional gain to counterbalance the loss, thus undermining the no net loss principle.

> Offsets that create new protected areas, fund the management of existing protected areas and restore degraded ecosystems are already occurring in many parts of the world (Brownlie and

> Botha 2009; McKenney and Kiesecker 2010; Bos et al. 2014; Villarroya et al. 2014). Proponents of the use of offsets as a funding mechanism for achieving such protected area targets point to the slow progress of many nations toward meeting the targets (Watson et al. 2014) and the widespread inadequacy of funding for protected area management (McCarthy et al. 2012; Githiru et al. 2015; Kiesecker et al. 2015). Critics emphasize the risk of cost-shifting: using offsets to replace expenditure toward already agreed targets, thus failing the requirement that offset gains are additional, and resulting in net biodiversity loss (Pilgrim and Bennun 2014; Gordon et al. 2015). (Maron et al. 2015b)

Maron et al. (2015b) observe that biodiversity offsets are increasingly used to help meet preexisting targets in particular related to the establishment of protected areas. They conclude that this can create perverse incentives in terms of no additional conservation benefit, consequently devaluating either the offset or the preexisting commitment (Maron et al. 2015b; Maron and Gordon 2015).

(3) Appropriateness of biodiversity offsets and mitigation hierarchy (including limits to what can be offset) With regard to the appropriateness of Biodiversity Offsets, it is important to note that offsets cannot provide a justification for proceeding with projects for which the residual impacts on biodiversity are unacceptable (BBOP n.d.-a). This means that the "No Go" option has to be considered seriously and applied in cases where the destruction of unique habitats, or irreversible loss, etc. would otherwise occur (Bishop 2006). To ensure this strict adherence to the mitigation hierarchy is crucial as has been explained in Sect. 1.2 (see also Fig. 1.4).

Accordingly, adherence to the mitigation hierarchy implies that one should first seek to avoid or prevent negative impacts on the environment in general and biological diversity in particular. Second, the unavoidable impacts should be addressed through minimization and rehabilitation measures and only as a "last resort" should biodiversity offsets be established for the residual adverse impacts. This shall be done by restitution or in exceptional cases by a compensation payment (Darbi et al. 2010).

The major points of critique on biodiversity offsets focus on their appropriateness (and thus questioning the concept of offsets itself) and the lacking application of the mitigation hierarchy (cf. Sect. 1.3). One problem lies in the fact that even though the mitigation hierarchy is anchored as a theoretical principle, the practical implementation in some cases remains doubtful. When offsets replace the steps of avoidance and minimization this may lead to the approval of environmental harmful developments that would have been rejected without the offset. This is colloquially called "license to trash" (ten Kate et al. 2004, cf. Sect. 1.3).

While usually the function of biodiversity offsets as a "last resort" is formulated in many offsetting schemes, provisions for distinguishing between offsettable and not offsettable impacts are largely absent (Darbi et al. 2010). Thus, the issue of irreplaceability and the "No Go" option (non-implementation of a project due to its potential environmental impacts) are not sufficiently considered. This leads to the question for which kinds of biodiversity damage or loss an offset is admissible and effective. Building in particular on the value of biological diversity, which is difficult to determine in all its facets and the complexity of interactions, this results in a high degree of uncertainty.

With regard to criteria such as replaceability and vulnerability it is in practice hard to anticipate, e.g., the extinction of a species from implementing a specific project and to prove a causal relationship between the impact and loss of a species (and thus leading to responsibility). Therefore, biodiversity offsets require the definition of thresholds for the kind and amount of biodiversity damage that can be counterbalanced by an offset.

(4) Functional, spatial, and temporal equivalence With regard to the implementation of biodiversity offsets, a number of principles or criteria are being discussed. One of the most important issues is the functional and spatial equivalence of offsets and impacts, i.e., the question whether and how biodiversity offsets can be made comparable to the affected area and ecosystem (Bishop 2006). Regarding the functional relation, two kinds of offsets can be distinguished. Whereas "in kind" refers to measures that are equal to the lost area with respect to habitats, functions, values, or other attributes (like-for-like), "out of kind" measures (like-for-not-like) have no or only a loose functional relationship with the impacted site.

The functional relationship between impacts and offsets is closely linked to their spatial relationship. While "on site" compensation includes measures in the impact area or nearby, "off site" compensation is spatially disconnected from the impact area. Usually, "in kind" and "on site" measures are preferable to "out of kind" and "off site" measures, especially when the affected area is of high local relevance (Biodiversity Neutral Initiative 2005). However, in practice this is not always feasible or appropriate, as most projects with significant adverse impacts cause to some extent an alteration of the affected area, which makes it impossible to implement adequate functional ("in kind") compensation measures directly in the affected area (Peters et al. 2001). Furthermore, the geographical extent of what is considered to be "on site" offsetting is not properly defined and can vary largely depending on the context.

A problem with the implementation of biodiversity offsets is the time lag between the impact and the positive outcomes of the offset (e.g., newly planted trees may take decades to grow). One approach to resolve this lies in the application of an additional coefficient or ratio (more than 1:1) to calculate the amount of biodiversity that has to be gained by the offset compared to the amount of biodiversity lost by the project, taking into consideration the time lag between the impact and the maturity of appropriate offsets and the risk that these may fail (Suvantola n.d.).

A second approach regards the question of appropriate timing of offsets: It is preferable that biodiversity offsets are operational and proven prior to the occurrence of project impacts (Biodiversity Neutral Initiative 2005). This anticipatory approach may be addressed by aggregated offsets using pool and banking models. These aim at the provision or concentration of compensation areas and measures. Pools provide areas or measures, which are ready to be used. Banks are generally economic arrangements (banking, credit trading, or trust funds), which technically and financially support the implementation of compensation measures.

In some cases, the obligation to compensate for the residual adverse impacts may be linked to the possibility of a compensation payment in monetary terms instead of tangible compensation measures (Wende et al. 2005). Nevertheless, as has been noted above, general preference should be given to natural restitution.

The equivalence of impact and offset is crucial to maintain the same level of biodiversity. Therefore, critics fear that an adequate quantitative and qualitative relation (functional and spatial relation, compensation ratio) can hardly be achieved because of difficulties in evaluating and balancing impact and offset (Christensen 2007). The reasons are various: First, the value of biological diversity cannot be measured objectively as different people assign different values to it (Rajvanshi 2006). Second, even the best techniques will never be able to predict and cover all impacts that may arise in the long term, especially with respect to indirect and cumulative effects. This becomes even more obvious when biodiversity data is incomplete and less detailed (BBOP 2009a). Last, the evaluation and balancing process consumes work, time, and costs. As a result of these facts, it will always be a compromise between the aim for an exact and detailed evaluation on the one hand and the need for pragmatism on the other hand.

Therefore, the necessary evaluation and balancing methods are diverse and built on different procedures and criteria (Biodiversity Neutral Initiative 2005). For Germany alone more than 40 evaluation approaches can be distinguished (Bruns 2007; Darbi and Tausch 2010) and BBOP assumes more than hundred methods worldwide (BBOP 2009a). This existing multitude of evaluation and balancing methods is critical as it impedes to achieve comparability and transparency, which are extremely important (Darbi and Tausch 2010). Furthermore, these methods face a number of deficiencies, for example:

- Inappropriate use of simplified area-based balancing approaches (value points)
- Negligence of components of the natural balance, in particular the abiotic components soil, water, climate, air, and the landscape scenery.
- Lacking cross-sectional evaluation and consideration of the natural balance and landscape as a complex.
- No consideration of cumulative effects.
- Lacking distinction between object-level and value-level (Spang and Reiter 2005).

A lack of equivalence does not only concern the ecological situation but may also have socioeconomic effects, e.g., a development may cause disadvantages to the local population (e.g., restriction of recreational use of an area) which cannot be offset in the area, thus resulting in social inequality (Darbi et al. 2010; ten Kate et al. 2004).

(5) Perpetuity Temporal aspects represent another important group of criteria, most notably the duration of impacts. There are impacts that last only a very short time and others that may endure many years or even centuries. Alongside these temporary impacts, other impacts may be permanent. In this respect the question of reversibility is important and the feasibility and costs of remediation.

Furthermore, timing must be considered. Some effects may appear immediately, while others may only become visible after many years. The latter are difficult to address by appropriate measures as noted before. Furthermore, in the long term biodiversity offsets may fail when they are not or not properly implemented or do not fulfill all functions. Even though biodiversity offsets should last as long as the impairments persist, i.e., usually in perpetuity, in practice different time frames are applied (e.g., 20, 25, 30 years in German Impact Mitigation) which is of course seen critically. Thus, even more emphasis must be placed on the importance of monitoring, follow up and long-term environmental management plans in tackling the lack of implementation of offsets and in ensuring their long-term effectiveness.

As a general principle, offsets must be designed for sustainability, aiming at long-term success (Escorcio Bezerra 2007; Bishop 2006; BBOP 2007). Therefore, to assure effectiveness, offsets should be subject to monitoring and follow up (Peters et al. 2001).

(6) Landscape Context An important issue to consider is the scale of the offset, i.e., that "small sites can require a disproportionate effort of management in order to maintain their ecological interest and mitigation costs for separate developments can also be higher where administration and management are replicated both spatially and over time in separate commissions for the same target habitat or species" (Latimer and Hill 2008). Therefore, the potentials of aggregated offsets and banking have to be highlighted. Thus, conservation pools and mitigation banks can be effectively used to implement biodiversity offsets with a minimum of land consumption, bureaucracy, and costs for the acquisition of land and the maintenance of measures and a maximum of nature conservation (Darbi and Voss 2014; Darbi et al. 2014).

From an ecological point of view, landscape and habitat pattern (e.g., core sites, green corridors, and ecological networks) play an important role in ensuring efficient and adequate compensation measures (Darbi et al. 2010; Latimer and Hill 2008). Through bundling and matching of measures and their concerted management habitat banks/compensation pools can contribute to the creation of protected areas networks with sound natural dynamics and larger and functionally more effective habitats. This is at the same time more cost-efficient due to reduced administrative effort.

Thus, the ecosystem approach "taking into account available information on the full range of biological, social and cultural values of biodiversity" (BBOP n.d-a) is being widely supported, i.e., biodiversity offsets can be integrated into the overall planning concept at landscape level (Landscape Level Planning), e.g., building on the provisions of landscape plans and other planning instruments (BBOP n.d-a, 2013). Thereby, offsets (and possibly habitat banks/compensation pools) can take a development function and be used even to enhance poor landscapes.

(7) Stakeholders and Equity A biodiversity offset should be designed and implemented in an equitable manner, which means the sharing among stakeholders of the rights and responsibilities, risks, and rewards associated with a project and offset in a fair and balanced way, respecting legal and customary arrangements (BBOP 2010). Special consideration should be given to respecting both internationally and nationally recognized rights of indigenous peoples and local communities.

In areas affected by the project and by the biodiversity offset, the effective participation of stakeholders should be ensured in decision-making about biodiversity offsets, including their evaluation, selection, design, implementation, and monitoring. The design and implementation of a biodiversity offset, and communication of its results to the public, should be undertaken in a transparent and timely manner (BBOP 2010).

What Are the Implications for the Present Study?

Increasingly, standards and guidance on biodiversity offsets are being formulated by a number of institutions and initiatives mostly from the private and nongovernmental sector, e.g., the Business and Biodiversity Offsets Program (BBOP), the International Union for the Conservation of Nature, the International Council on Mining and Metals (ICMM), and the International Finance Corporation.

This existing multitude of standards, however, appears to be largely coherent. In fact, there are strong links between the various standards, which are the result of the cooperation of a number of key actors (e.g., the Cross Sector Biodiversity Initiative is a partnership between ICMM, IPIECA, and the Equator Principles Association). Thus, the standards are mutually influencing and reinforcing, with major emphasis on BBOP, whose standard and principles were not only the first for a global context (i.e., beyond recommendations and guidelines at the national level), they are also now the common point of reference in practice and in the literature and most widely accepted.

The standards and principles formulate crucial aspects related to the choice, design, and implementation of biodiversity offsets, such as additionality, equivalence, and perpetuity. These have been increasingly uptaken and discussed in research.

Nevertheless, a number of aspects remain debated and no consensus could be reached, e.g., regarding the question of appropriate loss-gain calculation methods to determine equivalence or the relation of biodiversity offsets and other nature conservation obligations, particularly protected areas (Are they mutually reinforcing or are perverse incentives in terms of cost shifting and reduced overall nature conservation outcome most likely to occur?).

References

Albrecht J, Schumacher J, Wende W (2014) The German impact-mitigation regulation—a model for the EU's no-net-loss strategy and biodiversity offsets? Environ Policy Law 44(3):317–325

Alvarez Garcia D, Darbi M, Wende W (2018) Spain. In: Wende W, Tucker GM, Quétier F, Rayment M, Darbi M (eds) Biodiversity offsets: European perspectives on no net loss of biodiversity and ecosystem services. Springer, Cham

Ambrose RF (2000) Wetland mitigation in the United States: Assessing the success of mitigation policies. Wetlands (Australia) 19:1–27. Available at: http://www.mitigationactionplan.gov/ambrose_wetlandmitigationinus.pdf

Bacianskaite V (2018) Lithuania. In: Wende W, Tucker GM, Quétier F, Rayment M, Darbi M (eds) Biodiversity offsets: European perspectives on no net loss of biodiversity and ecosystem services. Springer, Cham

Baker J, Papadopoulou L, Sheate W (2018) United Kingdom. In: Wende W, Tucker GM, Quétier F, Rayment M, Darbi M (eds) Biodiversity offsets: European perspectives on no net loss of biodiversity and ecosystem services. Springer, Cham

BBOP (2007) Biodiversity offsets: a tool for conservation and development. An open discussion. Available at: http://www.forest-trends.org/biodiversityoffsetprogram/Presentations/BBOP5/June%2018/BBOP5_K_ten_Kate_summing_up%20_BBOP_LN.pdf

BBOP (2009a) Biodiversity offset cost-benefit handbook. BBOP, Washington, DC. Available at: www.forest-trends.org/biodiversityoffsetprogram/guidelines/cbh.pdf

BBOP (2009b) Biodiversity offset implementation handbook. BBOP, Washington, DC. Available at: www.forest-trends.org/biodiversityoffsetprogram/guidelines/oih.pdf

BBOP (2010) Principles on biodiversity offsets supported by the BBOP Advisory Committee. Available at: http://bbop.forest-trends.org/guidelines/principles.pdf

BBOP (2012a) Standard on biodiversity offsets. BBOP, Washington, DC. Available at: http://bbop.forest-trends.org/guidelines/Standard.pdf

BBOP (2012b) Guidance notes to the standard on biodiversity offsets. BBOP, Washington, DC. Available at: http://bbop.forest-trends.org/guidelines/Standard_Guidance_Notes.pdf

BBOP (2012c) Glossary. BBOP, Washington, DC. 2nd updated edition. Available at: http://bbop.forest-trends.org/guidelines/Updated_Glossary

BBOP (2013) To no net loss and beyond: an overview of the Business and Biodiversity Offsets Programme (BBOP). Washington, DC. Available at: http://www.forest-trends.org/documents/index.php?pubID=3319

BBOP (2018) Working for biodiversity net gain: an overview of the Business and Biodiversity Offsets Programme (BBOP) 2004–2018. Forest Trends, Washington, DC. Available at: https://www.forest-trends.org/bbop_pubs/overview2018

BBOP (n.d.-a) Biodiversity offsets and the Business and Biodiversity Offsets Programme (BBOP). A draft consultation paper for discussion and comment. Available at: http://www.forest-trends.org/biodiversityoffsetprogram/documents/UNEP/BBOP-UNEP-CBD-COP-9-Inf-29.pdf

BBOP (n.d.-b) Biodiversity offsets: the business case. Available at: http://bbop.forest-trends.org/pages/business_case

BFAD (2007) (Bundesverband der Flächenagenturen in Deutschland e. V.): Qualitätsstandards für Flächenpools. Available at: http://www.verband-flaechenagenturen.de/Downloads/BFAD_Qualitaetsstandards_Flaechenpools.pdf

Biodiversity Neutral Initiative (2005) Environmental offset policies, principles, and methods: a review of selected legislative frameworks. Available at: http://www.forest-trends.org/biodiversityoffsetprogram/library/new/Environmental%20Offset%20Legislative%20Frameworks%20DRAFT%20March%2 030.pdf

Bishop J (2004) Producing and trading habitat or land development as a source of funding for biodiversity conservation: a review of mitigation and conservation banking in the USA and its implications for global biodiversity conservation (Draft). Available at: http://www.eftec.o.uk/

UKNEE/Seminars/2004_documents/Seminar1_2004_Producing_and_Trading_Habitat_paper.pdf

Bishop J (2006) Biodiversity offsets: introduction and context. The role of biodiversity offsets in conservation. An open roundtable discussion. CBD COP8, Curitiba, 21 Mar 2006. Available at: http://www.forest-trends.org/biodiversityoffsetprogram/documents/LN2/LNppt6.ppt

Bishop J, Kapila S, Hicks F, Mitchell P (2006) Building biodiversity business: report of a scoping study. Available at: http://www.globalnature.org/bausteine.net/f/6644/Building_Biodiversity_Business_-_Draft_Report_%28Oct_2006%29.pdf?fd=2

BMU (eds) (2002) Translation federal nature conservation act of 25 March 2002. Available at: http://www.bmu.de/files/pdfs/allgemein/application/pdf/bundnatschugesetz_neu060204.pdf. Accessed 23 Mar 2009

Böhme C, Bruns E, Bunzel A, Herberg A, Köppel J (2005) Flächen- und Maßnahmenpools in Deutschland, Bundesamt für Naturschutz (ed) Schriftenreihe "Naturschutz und Biologische Vielfalt" Heft 6, Bonn – Bad Godesberg

Boučková A, Albrecht J (2018) The Czech Republic. In: Wende W, Tucker GM, Quétier F, Rayment M, Darbi M (eds) Biodiversity offsets: European perspectives on no net loss of biodiversity and ecosystem services. Springer, Cham

Bruns E (2007) Bewertungs- und Bilanzierungsmethoden in der Eingriffsregelung. Analyse und Systematisierung von Vorgehensweisen des Bundes und der Länder. Dissertation. Available at: http://opus.kobv.de/tuberlin/volltexte/2007/1565

Bull JW, Suttle KB, Gordon A, Singh NJ, Milner-Gulland EJ (2013) Biodiversity offsets in theory and practice. Oryx 47(03):369–380

Bundesamt für Naturschutz (Hrsg.) (2007) Bestimmung des Verhältnisses von Eingriffsregelung, FFH-VP, UVP und SUP im Vorhabensbereich. Endbericht des gleichnamigen Forschungs- und Entwicklungsvorhabens

Bundesregierung (2009) Gesetz zur Neuregelung des Rechts des Naturschutzes und der Landschaftspflege, Entwurf v. 11.03.2009

Christensen M (2007) Biodiversity offsets—an overview of selected recent developments: New Zealand—where to from here? Anderson Lloyd Lawyers, Apr 2007

Commonwealth of Australia (2007) Use of environmental offsets under the environment protection and biodiversity conservation act 1999—discussion paper. Australian Government Department of the Environment and Water Resources, Canberra

Counter Balance (2014) Biodiversity offsetting, making dreams come true. Available at: http://www.counter-balance.org/biodiversity-offsetting-making-dreams-come-true/

Darbi M (2010a) Biodiversity offsets—a tool for environmental management and biodiversity conservation. In: Sanchez Bengoa D, Powell D (eds) TOP biodiversity 2010. Conference Proceedings. Intercollege-Larnaca, Larnaca, S. 289–301

Darbi M (2010b) Compensation agencies in Germany—a tool to facilitate development and to safeguard environmental protection and restoration. In: Davy B et al (eds) Program and abstracts of the fourth international academic conference on planning, law, and property rights. Technische Universität, Dortmund, S. 86–87

Darbi M (2015a) Biodiversity law (including biodiversity offsets) under development in France. Biodiversity Offsets Blog. Available at: http://www.biodiversityoffsets.net/biodiversity-law-under-development-in-france/

Darbi M (2015b) What is the business case for biodiversity offsets? SQ-raising the Sustainability Quotient (Bombay Chamber of Commerce and Industry) 4 (1 Jan–March 2015):3–5

Darbi M (2015c) Biodiversity are not a business—why I don't think we'll have a commercial platform for biodiversity offsets somewhere in Europe, soon. Biodiversity Offsets Blog. Available at: http://www.biodiversityoffsets.net/biodiversity-offsets-are-not-a-business-why-i-dont-think-well-have-a-commercial-platform-for-biodiversity-offsets-somewhere-in-europe-soon/

Darbi M, Tausch C (2010) Loss-gain-calculations in German Impact Mitigation. Available at: http://www.forest-trends.org/documents/index.php?pubID=2404

Darbi M, Voss J (2014) Key lessons from the 'Burgberg Zschaitz' habitat bank in Saxony. Using government metrics and ensuring permanent protection for biodiversity offsets under the German impact mitigation regulation. Presentation. "To no net loss of biodiversity and beyond".

Zoological Society of London, Regent's Park, 3–4 Jun 2014 Available at: http://www.forest-trends.org/documents/files/doc_4594.pdf
Darbi M, Wende W (2013) Biodiversity offsets and German impact mitigation regulation. In: Japan-German bio web city/region symposium: new trend of landscape design 2—enhancing ecological services and biodiversity in urban, suburban and rural areas. Nagoya, S. 5–17
Darbi M, Wende W, Voss J (2014) How is a Habitat Bank established and maintained under the German Impact Mitigation Regulation? The example of the Burgberg Zschaitz Habitat Bank in Saxony. Webinar in the scope of the first global conference on approaches to avoid, minimise, restore, and offset biodiversity loss: "TO NO NET LOSS OF BIODIVERSITY AND BEYOND", 3–4 June, 2014, Zoological Society of London, Regent's Park. Available at: http://www.biodiversityoffsets.net/wp-content/uploads/2014/08/2014_Darbi_Webinar-Habitat-Bank_NNL-Conference.pptx
Darbi M, Ohlenburg H, Herberg A, Wende W (2010) Impact mitigation and biodiversity offsets-compensation approaches from around the world: a study on the application of Article 14 of the CBD (Convention on Biological Diversity). BfN-Schr.-Vertrieb im Landwirtschaftsverl
Dupont V, Schoukens H, Born C-H (2018) Belgium. In: Wende W, Tucker GM, Quétier F, Rayment M, Darbi M (eds) Biodiversity offsets: European perspectives on no net loss of biodiversity and ecosystem services. Springer, Cham
Durner W (2001) Kompensation für Eingriffe in Natur und Landschaft nach deutschem und europäischem Recht. Natur und Recht 23 (11/2001):601–610
Edwards S (2015) A balancing act. IUCN. Available at: http://www.iucn.org/?18817
eftec, IEEP et al (2010) The use of market-based instruments for biodiversity protection—the case of habitat banking. Technical Report. For European Commission DG Environment
ELI (Environmental Law Institute) (2002) Banks and fees. The status of off-site wetland mitigation in the United States
Equator Principles (2011) Biodiversity for banks program (B4B). Available at: http://www.equator-principles.com/index.php/b4b. Accessed 24 Feb 2016
Equator Principles (2013) The equator principles. June 2013. A financial industry benchmark for determining, assessing and managing environmental and social risk in projects. Available at: http://www.equator-principles.com/resources/equator_principles_III.pdf
Escorcio Bezerra LG (2007) Biodiversity offsets in national (Brazil) and regional (EU) mandatory arrangements: towards an international regime? Available at: http://www.forest-trends.org/biodiversityoffsetprogram/library/new/Dissertation%20Biodiversity%20Offsets%20LGB%20IUCN%20BBOP.doc
European Commission (2011) The EU biodiversity strategy to 2020, COM(2011) 244 endg
European Commission (2012a) Draft mandate for the EU working group on no net loss of ecosystems and their services, unpublished draft
European Commission (2012b) Scoping document for the first meeting of the working group on no net loss of ecosystems and their services, unpublished draft
European Commission (2015a) No net loss of biodiversity public consultation. Summary overview of the responses. Available at: http://ec.europa.eu/environment/nature/biodiversity/nnl/pdf/report.pdf
European Commission (2015b) Public consultation on the future EU initiative on no net loss of biodiversity and ecosystem services. Statistics. Available at: http://ec.europa.eu/environment/nature/biodiversity/nnl/pdf/NoNetLoss%20statistics%20analysis.pdf
European Commission (2017) Communication from the Commission to the European Parliament, the Council, the European Economic and Social Committee and the Committee of the Regions. An Action Plan for nature, people and the economy. COM(2017) 198 final. Available at: https://ec.europa.eu/environment/nature/legislation/fitness_check/action_plan/communication_en.pdf
FERN (2014a) Briefing note 2: what is biodiversity offsetting and why is it problematic? Available at: http://www.fern.org/sites/fern.org/files/Biodiversity2_EN.pdf. Accessed 24 Feb 2016
FERN (2014b) Briefing note 3: biodiversity offsetting in practice. Available at: http://www.fern.org/sites/fern.org/files/Biodiversity3_EN.pdf. Accessed 24 Feb 2016

FERN and Friends of the Earth (2014) Case studies of biodiversity offsetting: voices from the ground. Available at: http://www.fern.org/sites/fern.org/files/Offset%20stories%20-%20Final.pdf

Ferreira C (2014) Biodiversity offset markets: current challenges and prospective development: PhD thesis—executive summary. Coventry University

Ferreira C (2015) Opinion biodiversity offsets as corporate responsibility: opportunity or paradox? Available at: http://www.ecosystemmarketplace.com/pages/dynamic/article.page.php?page_id=10666&section=news_articles&eod=1

Garrard G et al (2015) Offset policies don't work. So maybe we should be weighing up the alternatives. Decision Point Online 91/August 2015

Gibbons P, Evans MC, Maron M, Gordon A, Le Roux D, von Hase A et al (2016) A loss-gain calculator for biodiversity offsets and the circumstances in which no net loss is feasible. Conserv Lett 9(4):252–259

Harjoto MA, Jo H (2011) Corporate Governance and CSR Nexus. J Bus Ethics 100:45–67

Herbert, M (2015) German impact mitigation regulation—national, European, international. A Comparison from the Viewpoint of the Federal Agency for Nature Conservation. UVP-Report 29(3):149–151. Eingriffsregelung; Umweltverträglichkeitsprüfung; Strategische Umweltprüfung; Biologische Vielfalt

Howard K (2007) Voluntary biodiversity offsets: improving the environmental management toolbox. Im Internet: http://www.cortex.org/d-Cortex-%20Biodiversity%20Offsets_01Dec07.pdf

ICH GHK, Bio Intelligence Service (2013) Exploring potential demand for and supply of habitat banking in the EU and appropriate design elements for a habitat banking scheme

ICMM (2005a) Biodiversity offsets: a proposition paper from The International Council on Mining and Metals. ICMM, London

ICMM (2005b) Biodiversity offsets—a briefing paper for the mining industry. ICMM, London

ICMM (2006) Good practice guidance for mining and biodiversity. ICMM, London

ICMM (2010) Mining and biodiversity. A collection of case studies—2010 edition. ICMM, London

ICMM (2016) International council on mining and metals. Homepage. Available at: https://www.icmm.com/

ICMM IUCN (2012) Independent report on biodiversity offsets. Prepared by The Biodiversity Consultancy. Available at: www.icmm.com/biodiversity-offsets

IEEP et al (2012) Policy options for an EU no net loss initiative. A tender for the European Commission. Unpublished

IFC (2012) Performance standard 6. Biodiversity conservation and sustainable management of living natural resource. Available at: http://www.ifc.org/wps/wcm/connect/bff0a28049a790d6b835faa8c6a8312a/PS6_English_2012.pdf?MOD=AJPERES. Accessed 24 Feb 2016

IUCN (2014) Biodiversity offsets technical study paper. IUCN, Gland, 65pp

IUCN (2015) Biodiversity offsets policy: draft for public consultation. Available at: http://cmsdata.iucn.org/downloads/iucn_biodiversity_offsets_draft_policy___for_consultation.pdf

Janssen G, Albrecht J (2008) Umweltschutz im Planungsrecht - Die Verankerung des Klimaschutzes und des Schutzes der biologischen Vielfalt im raumbezogenen Planungsrecht. Dessau-Roßlau : Umweltbundesamt, 2008, S.146 (Texte / UBA; 10/08)

Jessel B, Schöps A, Gall B, Szaramowicz M (2006) Flächenpools in der Eingriffsregelung und regionale Landschaftsentwicklung am Beispiel der Mittleren Havel. Abschlussbericht zum E+E-Vorhaben „Entwicklung und modellhafte Umsetzung einer regionalen Konzeption zur Bewältigung von Eingriffsfolgen am Beispiel der Kulturlandschaft Mittlere Havel. – Naturschutz und Biologische Vielfalt 33

Köppel J, Peters W, Wende W (2004) Eingriffsregelung, Umweltverträglichkeitsprüfung, FFH-Verträglichkeitsprüfung (No. UTB 2512)

Latimer W, Hill D (2008) Mitigation banking: securing no net loss to biodiversity? Available at: http://www.forest-trends.org/biodiversityoffsetprogram/library/new/Latimer_mitigation_banking.doc

Madsen B, Moore Brands K, Carroll N (2010) State of biodiversity markets: offset and compensation programs worldwide
Madsen B, Carroll N, Kandy D, Bennett G (2011) Update: state of biodiversity markets. Forest Trends, Washington, DC
Mann C, Simons A (2015) Local emergence and international developments of conservation trading systems: Innovation dynamics and related problems. Environ Conserv 42(4):00–00
Mann C, Voß J-P, Simons A, Amelung N, Runge T (2014) Challenging futures of biodiversity offsets and banking. Critical issues for robust forms of biodiversity conservation. A report based on an interactive, anticipatory assessment of the dynamics of governance instruments, 19 Apr 2013. Technische Universität (TU) Berlin, Berlin
Maron M, Gordon A (2015) Could well-designed offset policies actually increase loss? Decision Point Online 91/August 2015
Maron M, Dunn PK, McAlpine CA, Apan A (2010) Can offsets really compensate for habitat removal? The case of the endangered red-tailed black-cockatoo. J Appl Ecol 47(2):348–355
Maron M, Bull JW, Evans MC, Gordon A (2015a) Locking in loss: baselines of decline in Australian biodiversity offset policies. Biol Conserv 192:504–512
Maron M, Gordon A, Mackey BG, Possingham HP, Watson JE (2015b) Interactions between biodiversity offsets and protected area commitments: avoiding perverse outcomes. Conserv Lett 9:384–389
NRC (2001) Compensating for wetland losses under the clean water act. National Academy Press, Washington, DC. Available at: http://www.nap.edu/nap-cgi/report.cgi?record_id=10134&type=pdfxsum
Peters W, Siewert W, Szaramowicz M (2001) Eingriffsfolgenbewältigung im internationalen Vergleich. Kurzfassung des Endberichts für das F+E Vorhaben. Bundesamt für Naturschutz
Peters W, Siewert W, Szaramowicz M (2003) Folgenbewältigung von Ein-griffen im internationalen Vergleich. Endbericht zum F+E Vorhaben. BfN-Skripten 82:220p
Pilgrim JD, Ekstrom JMM (2014) Technical conditions for positive outcomes from biodiversity offsets. An input paper for the IUCN Technical Study Group on Biodiversity Offsets. IUCN, Gland, 46 pp
Poulton D (2015) Key issues in biodiversity offset law and policy: a comparison of six jurisdictions. Report for Ontario Nature
PricewaterhouseCoopers LLP (2010) Biodiversity offsets and the mitigation hierarchy: a review of current application in the banking sector. Study completed on behalf of the Business and Biodiversity Offsets Programme and the UNEP Finance Initiative. Available at: http://www.unepfi.org/fileadmin/documents/biodiversity_offsets.pdf
Rainey HJ et al (2015) A review of corporate goals of no net loss and net positive impact on biodiversity. Oryx 49:232–238
Rajvanshi A (2006) Reconciling conservation and development. The concept of biodiversity offsets. Wildlife institute of India, Dehradun, Presentation IAIA '06, Pre-meeting training course, Stavanger
Schöps A (2008) Erfahrungen und Konzepte des BFAD und seiner Mitglieder, presentation at the conference "Kompensation durch Kooperation". Frankenhausen 2008-09-24
Schöps A, Szaramowicz M, Busch D, Geßner J (2007) Flächenpools und Flächenagenturen: Handbuch für die Praxis, Bundesamt für Naturschutz (ed) Naturschutz und Biologische Vielfalt Heft 55, Bonn – Bad Godesberg
Spang W, Reiter S (2005) Ökokonten und Kompensationsflächenpools in der Bauleitplanung und der Fachplanung. Anforderungen. Erfahrungen. Handlungsempfehlungen. Beiträge zur Umweltgestaltung A 160, Erich Schmidt Verlag, Berlin
Suvantola L (n.d.) Environmental offset arrangements in biodiversity conservation. Available at: http://www.neln.life.ku.dk/NELN-events/Past-events/~/media/migration%20folder/upload/neln/docs/pdf/suvantola%20environmental%20offsets.pdf.ashx
ten Kate K (2005) Biodiversity offsets: good for business and biodiversity? Available at: https://www.cbd.int/financial/offsets/g-offsetsbusiness.pdf

ten Kate K, Bishop J, Bayon R (2004) Biodiversity offsets: views, experience, and the business case. IUCN, Insight Investment, Gland, London

The Biodiversity Consultancy (2015) A cross-sector guide for implementing the mitigation hierarchy. Available at: http://www.icmm.com/document/9460

The World Bank (2012a) Safeguard policies. Available at: http://go.worldbank.org/WTA1ODE7T0

The World Bank (2012b) Performance standard 6. Biodiversity conservation and sustainable management of living natural resources. Available at: http://siteresources.worldbank.org/INTSAFEPOL/Resources/PerformanceStandard6.pdf

Tucker G, Allen B, Conway M, Dickie I, Hart K, Rayment M, Schulp C, van Teeffelen A (2013) Policy options for an EU no net loss initiative. Report to the European Commission. Institute for European Environmental Policy, London

Tucker G, Darbi M, Wende W, Quétier F, Rayment M (2018) Conclusions: lessons from biodiversity offsetting experiences in Europe. In: Wende W, Tucker GM, Quétier F, Rayment M, Darbi M (eds) Biodiversity offsets: European perspectives on no net loss of biodiversity and ecosystem services. Springer, Cham

U.S. Army Corps of Engineers (U.S. Corps) (ed) (n.d.) Final environmental assessment, finding of no significant impact, and regulatory analysis for the compensatory mitigation regulation. Available at: http://www.epa.gov/owow/wetlands/pdf/MitRule_Regulatory_Analysis.pdf

U.S. EPA (n.d.) U.S. Environmental Protection Agency (ed) The wetland fact sheet series—wetland regulatory authority. Available at: http://www.epa.gov/owow/wetlands/pdf/reg_authority_pr.pdf

Vaissière A-C, Quétier F, Bas A, Calvet C, Gaucherand S, Hay J, Jacob C, Kermagoret C, Levrel H, Malapert A, Pioch S, Scemama P (2018) France. In: Wende W, Tucker GM, Quétier F, Rayment M, Darbi M (eds) Biodiversity offsets: European perspectives on no net loss of biodiversity and ecosystem services. Springer, Cham

Van Teeffelen A (2018) The Netherlands. In: Wende W, Tucker GM, Quétier F, Rayment M, Darbi M (eds) Biodiversity offsets: European perspectives on no net loss of biodiversity and ecosystem services. Springer, Cham

Wagner S (2007) Ökokonten und Flächenpools. Die rechtlichen Grundlagen, Möglichkeiten und Grenzen der Flächen- und Maßnahmenbevorratung als Ausgleichsmethoden im Rahmen der Eingriffsregelung im Städtebaurecht, Schriften zum Umweltrecht Vol. no. 153, Duncker & Humblot Berlin

Wende W (2018) Other EU member states. In: Wende W, Tucker GM, Quétier F, Rayment M, Darbi M (eds) Biodiversity offsets: European perspectives on no net loss of biodiversity and ecosystem services. Springer, Cham

Wende W, Herberg A, Herzberg A (2005) Mitigation banking and compensation pools: improving the effectiveness of impact mitigation regulation in project planning procedures. Journal for Impact Assessment and Project Appraisal, Guildford, Surrey, 23 (2005) 2, pp 101–111

Wende W, Darbi M, Stein C (2015) Annex 4 evidence of the costs of offsetting in Germany. Case study provided by Prof. Wolfgang Wende, Marianne Darbi and Christian Stein (Technische Universität Dresden). In: The Institute for European Environmental Policy (IEEP), Biotope, eftec and ICF (eds) Supporting the elaboration of the impact assessment for a future EU initiative on no net loss of biodiversity and ecosystem services. 2nd Draft Interim/Task 2 Report. Unpublished draft

Wende W, Tucker GM, Quétier F, Rayment M, Darbi M (eds) (2018a) Biodiversity offsets: European perspectives on no net loss of biodiversity and ecosystem services. Springer, Cham

Wende W, Albrecht J, Darbi M, Herbert M, May A, Schumacher J, Szaramowicz M (2018b) Germany. In: Wende W, Tucker GM, Quétier F, Rayment M, Darbi M (eds) Biodiversity offsets: European perspectives on no net loss of biodiversity and ecosystem services. Springer, Cham

5 Development of a Typology of Biodiversity Offsets

Abstract

A typology of biodiversity offsets was developed using four consecutive steps: (1) The starting point was an impressionistic classification of types ("intuitive types") that was refined in several steps throughout the research process and over time. (2) Building on the theoretical analysis of the concept of voluntariness, attributes, or criteria were extracted from the literature that provided the theoretical frame of reference to analyze the degree or intensity of voluntariness of the types presented in this work. These criteria were grouped and rearranged into different categories (with distinctive functions), notably threshold criteria, descriptive criteria, performance criteria, and quality criteria. Thus, through a process of elimination two core criteria were identified: influence and initiators, and motivation and ultimate goal. (3) These core criteria, and the respective value categories assigned to them, were used to construct the underlying attribute space, i.e., a 3 × 7 matrix. (4) Finally, the impressionistic types that have been developed over time (building on the author's own observation and experience) and the constructed attribute space (building on criteria derived from the literature) were integrated to obtain a more refined typology of seven types.

Furthermore, a screening of biodiversity offset cases was carried out to empirically underpin these types. This generated an exploratory list of 90 biodiversity offset cases worldwide. A number of these were rejected, e.g., due to poor information quality or project failure/abandonment. The remaining 72 cases were analyzed and assigned to the seven types.

Keywords

Impressionistic classification of types · Impressionistic types · Attribute space · Mandatory biodiversity offsets · Voluntary biodiversity offsets · Biodiversity offsets examples

M. Darbi, *Biodiversity Offsets Between Regulation and Voluntary Commitment*,
https://doi.org/10.1007/978-3-030-25594-7_5

As outlined in Sect. 2.2, the development of a typology of biodiversity offsets in the scope of this study consists of four consecutive steps:

1. Development of an impressionistic classification of types.
2. Derivation (and reduction) of relevant attributes/criteria for voluntariness from the theory.
3. Substruction of the underlying attribute space and combinations of attributes.
4. Transformation (rectification) of the impressionistic types and analysis of meaningful correlations.

The application of each of these steps to the given research subject is described in the following subchapters.

5.1 Development of an Impressionistic Classification of Types

Following Kluge (1999) and Lazarsfeld (1937) the development of a typology of biodiversity offsets started with an impressionistic classification of types ("intuitive types") that has been refined in several steps throughout the research process and over time as is documented in the following subchapters.

5.1.1 Step 1: Dichotomy Between Mandatory and Voluntary Biodiversity Offsets[1]

A general distinction can be made between two kinds of offsets: on the one hand, mandatory biodiversity offsets are required by legal and planning provisions and on the other hand, voluntary biodiversity offsets can either exceed these requirements or exist completely independent from them (Darbi 2010a, b). This distinction between the two kinds of biodiversity offsets is underlined by BBOP as follows (BBOP Secretariat 2010):

(a) Voluntary biodiversity offsets, which a developer undertakes in circumstances where there is no legal requirement to do so, because it perceives a business advantage (such as license to operate, reputational benefits, competitive advantage, market share, etc.).
(b) Regulatory/mandatory biodiversity offsets, which are required by law.

Figure 5.1 illustrates in which cases voluntary biodiversity offsets may emerge. They are especially important where the legal basis and general guidance are lacking. However, voluntary biodiversity offsets have also been observed as

[1]cf. Darbi (2010b).

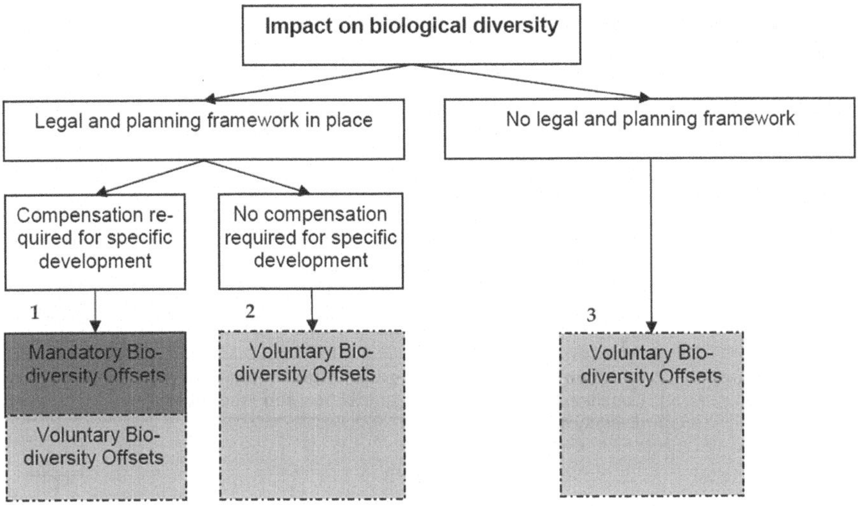

Fig. 5.1 Mandatory biodiversity offsets and voluntary biodiversity offsets. Source: Darbi (2010a, b)

complements to mandatory approaches in some cases. This can be interpreted as the first indicator of the need for a more diversified typology of biodiversity offsets.

5.1.2 Step 2: Typology with Four Biodiversity Offset Types (Building on Pressures and Incentives)[2]

As has been noted, the distinction into two categories is rather simplified and does not represent actual practice. Instead, there is a whole range of biodiversity offsets between two poles: a high degree of regulation on the one side and a high degree of voluntary commitment on the other. In this continuum, evidence from practical experience can be clustered to identify a (non-exhaustive) number of specific cases. Therefore, in order to understand why biodiversity offsets are implemented a number of questions need to be asked. Figure 5.2 suggests a cascade of four questions leading to a set of four biodiversity offset categories (cases 1 and 2 driven by pressures, case 3 driven by incentives, and case 4 without influence):

- *Case 1 Regulatory biodiversity offsets*: The question whether a legal requirement to implement offsets exists, is alone not sufficient. In fact, mandatory or regulatory biodiversity offsets only exist if there is law enforcement.
- *Case 2 Conditional biodiversity offsets*: Besides the legal obligation, there may be other requirements that, while not being mandatory, make biodiversity offsets a

[2]cf. Darbi (2014).

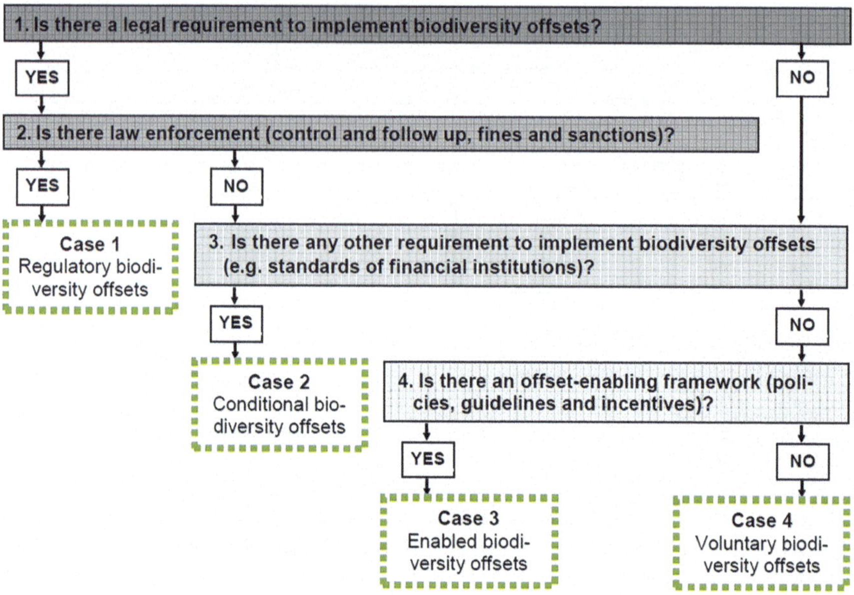

Fig. 5.2 Cascade of biodiversity offsets (4 types). Source: Author

condition when the project proponent wants to obtain for example financial support (e.g., Performance Standards 6 of the International Finance Corporation, IFC). While not obliged to do so, the project proponent will usually comply with these standards and implement biodiversity offsets (which from a business point of view will usually be preferable to losing the financial support).

- *Case 3 Enabled biodiversity offsets*: Even if there is no kind of legal or other requirement presents there may still be an "offset-friendly" environment. An offset-enabling framework may include policies, guidelines, and forms of incentives from public authorities and civil society.
- *Case 4 Voluntary biodiversity offsets*: In contrast to regulatory biodiversity offsets pure voluntary biodiversity offsets emerge and are qualified as such when neither any forms of requirements nor incentives exist.

5.1.3 Step 3: Updated Typology with Six Biodiversity Offset Types (as of August 2014)[3]

Already in 2004, ten Kate et al. described a "conservation case", a "regulatory case," and a "business case" for biodiversity offsets. Bishop et al. (2006) distinguish a

[3]cf. Darbi (2015).

"business case for biodiversity", a "conservation case for biodiversity business," and a "development case for biodiversity business."

Starting from this, the "business case for biodiversity offsets" is explored in a new attempt to refine the interim typology. The business case is usually established as the opposite of "mandatory" biodiversity offsets that are required by some kind of regulation (Howard 2007; Ten Kate 2005; Escorcio Bezerra 2007; Darbi 2010a; Albrecht et al. 2014). In this sense, it is used synonymously to what is commonly referred to as voluntary biodiversity offsets. Nonetheless, "business case" can be considered a more appropriate term than "voluntary biodiversity offsets" to describe the whole spectrum of biodiversity offset cases that lies beyond legal compliance[4] and which is implemented by businesses due to a variety of different motivations and drivers. Building on these motivations or drivers, the typology is further refined. As suggested in Fig. 5.3, a cascade of six consecutive questions can be used, leading to a set of six biodiversity offset cases or categories.

These different cases can vary significantly regarding their design and implementation. Despite this, the large majority (cases 2–6) would be covered under the narrow definition of the "business case" as has been outlined above (core business case for biodiversity offsets). What can be concluded from this is that it makes more sense to distinguish different cases within the "business case" than to distinguish the "business case" against regulatory offsetting.

5.1.4 Results of the Impressionistic Classification

Finally, as a result of the impressionistic classification, six types of biodiversity offsets have been extracted from the literature with the accompanying empirical observations:

1. *Regulatory biodiversity offsets*: Required by law and enforced
2. *Conditional biodiversity offsets*: Required by financial institutions (World Bank, International Finance Corporation. . .)
3. *Enabled biodiversity offsets*: Fostered by governments (and NGOs) through pilot schemes, guidance, etc.
4. *Sectoral biodiversity offsets*: Take part in a voluntary self-commitment of a sector (e.g., mining)
5. *Corporate biodiversity offsets*: Resulting from a voluntary self-commitment of a corporation
6. *Local biodiversity offsets*: Single offsets, that are most likely developed at local level in a consensual process

[4]It should be noted, however, that neither do regulatory biodiversity offsets exclude business (in fact all offsets target harmful activities by business, industry, and public developers), nor is the "business case" exempt from all kind of government involvement (see Fig. 5.3 with regard to the core business case and extended business case for biodiversity offsets).

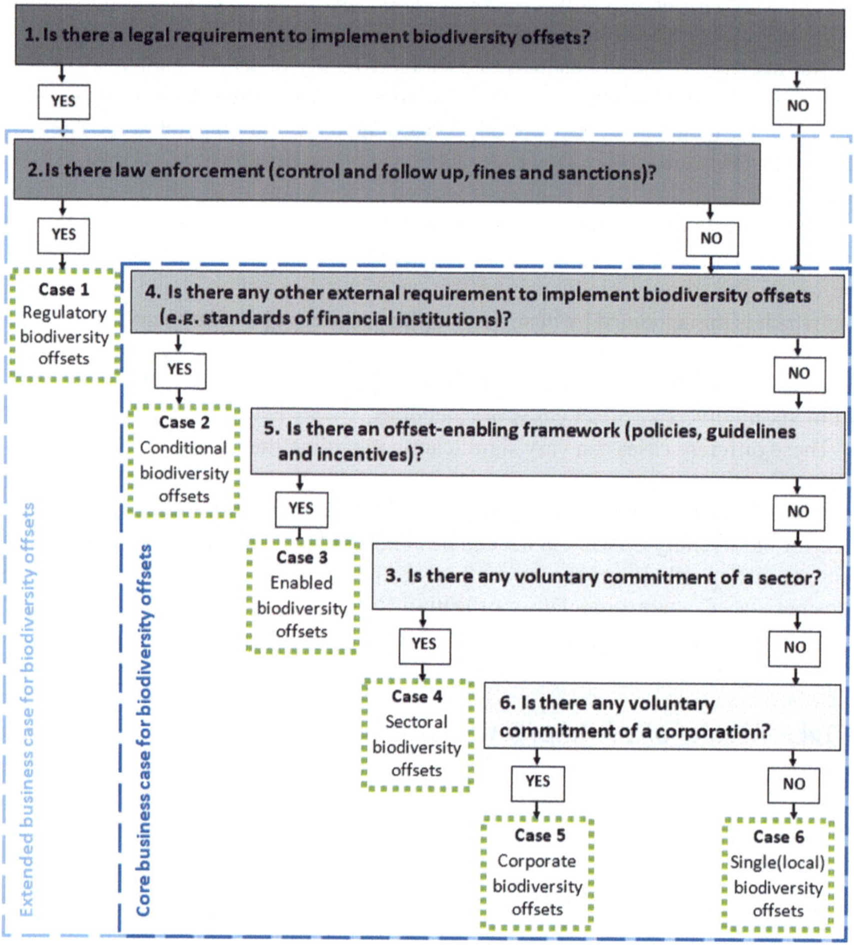

Fig. 5.3 Cascade of biodiversity offsets (6 types). Source: Author

5.2 Derivation of Relevant Attributes/Criteria for Voluntariness from the Theory

Building on the theoretical analysis of the concept of voluntariness (see Chap. 3), attributes or criteria can be extracted from the literature that will provide the theoretical frame of reference to analyze the degree of voluntariness of the types or cases presented in this work. This shall then be used to judge whether or not an action is voluntary, or more precisely by which degree or intensity of voluntariness it is characterized.

5.2.1 Degree or Intensity of Voluntariness

Voluntariness and coercion have not only categorizing functions, but can also differ in degree or intensity (i.e., quantity). However, quantifying this intensity is a difficult task and remains mostly speculative. As a result, different approaches are applied to quantify factors that diminish the voluntariness of an action. Simple approaches focus either on the forced action or on the likely consequences, e.g., by measuring the intensity of coercion with the negative use function (i.e., how much an action is undesired). Three main problems are associated with determining the degree of voluntariness or coercion. First, the relevance of quantifying the degree of coercion is very much context dependent, i.e., it differs depending on the specific case or for normative reasons. Second, it can only be evaluated normatively (what is rational or reasonable?). Third, while voluntariness or coercion may be gradually determined, the (legal) consequences of an action are most likely binary decisions (yes/no). This means that in legal terms for specific actions and situations thresholds have to be defined to judge whether an action is voluntary or compliance (Gutmann 2000).

What can be seen from this is that the degree or intensity of voluntariness is a superordinate criterion that itself includes many aspects. It is, therefore, necessary to specify a variety of sub-criteria that are operationalizable. A large number of potentially suitable criteria have been extracted from the literature. How significant is each of these criteria? Which criteria are crucial and which are expendable?

Table 5.1 lists 25 criteria with the respective source and a short judgment as to the suitability of the criterion in the scope of this work and the envisaged action.

As can be seen from the table the theoretically grounded criteria can be grouped and rearranged to a set of criteria that fall into different categories (with distinctive functions). These are explained by category in the following subchapters.

5.2.2 Threshold Criteria

Threshold criteria are defined as those that clearly distinguish voluntary from involuntary action. These binary criteria (Yes/No) do not allow for qualifying characterization, but need to be fulfilled for an action to be considered voluntary.

Intentionality

Intentionality is a major driver of voluntary action, i.e., complex intentions are the prerequisite for communicative action under vis compulsiva (Gutmann 2000). By contrast, overcompliance can be the result from using uniform abatement technologies in environmental regulation, hence being incidental and not intentional (Arora and Cason 1995). Thus, in an overcompliance case offsets are a by-product of raw practices. If intentionality is defined as a prerequisite for an action to be considered voluntary, then these cannot be considered voluntary offsets.

However, firms may also make an effort to change products, processes, and services, thus intentionally overcomplying (Arora and Cason 1995).

Table 5.1 Criteria for voluntariness

No.	Criterion	Source (not exclusive)	Comment/action
1.	Intentionality	Arora and Cason (1995)	Use as threshold criterion
2.	Additionality	Nadai (2004)	Use as threshold criterion
3.	Motivation	Siegenthaler (2008) and Cranach (2008)	Group together: motivation and ultimate goal, Use as core criterion
4.	Ultimate goal	Batson (2014)	
5.	Recompense/ benefit	Ammann (2004) and Boothe (2004)	Group together: Recompense or benefit, Use as secondary criterion to motivation and ultimate goal
6.	Public relations benefits	Segerson (1999) and Caswell and Henson (1997)	
7.	Influenceability	Priller (2008)	Group together: Influence and initiators, use as core criterion
8.	Driven by actors and actions of civil society	Olk (2004)	
9.	(Degree of) obligation	Olk (2004) and Ammann (2004)	
10.	Community pressure/public response	Alberini and Segerson (2002) and Segerson (1999)	
11.	Role of government		
12.	Choice, eligibility, or alternatives	Gutmann (2000) and Benn (1982)	Group together: choice or eligibility, Use as core criterion
13.	Causality	Gutmann (2000)	
14.	Context	Priller (2008)	Group together: Context, use as descriptive criterion
15.	Competitive environment	Segerson (1999)	
16.	Social and individual commitment, societal context	Draser (2013) and Paavola et al. (2009)	
17.	Charitable, directed towards public good/ welfare	Siegenthaler (2008), Priller (2008), and Ammann (2004)	Group together: Outcome and charitableness Use as quality criterion
18.	Outcome/ effectiveness / success	Altorfer (2004), Boothe (2004), Alberini and Segerson (2002), and Segerson (1999)	
19.	Flexibility	Alberini and Segerson (2002) and Segerson and Li Dawson (2001)	Use as performance criterion
20.	Lower costs/cost-efficiency	Alberini and Segerson (2002) and Segerson and Li Dawson (2001)	Use as performance criterion
21.	Know-how/ professionalism/ competence	Ammann (2004), Badelt (2004), and Boothe (2004)	Reject (minor relevance)

(continued)

Table 5.1 (continued)

No.	Criterion	Source (not exclusive)	Comment/action
22.	Scope or distance	Siegenthaler (2008)	Reject (minor relevance)
23.	Free choice/one's own initiative	Ammann (2004)	Reject (minor relevance)
24.	Taking over responsibility	Boothe (2004)	Reject (minor relevance)
25.	Ethics	Boothe (2004)	Reject (minor relevance)

Source: Author

As has been noted, cases are grouped according to the intentionality criterion into two categories:

1. Yes
2. No

Additionality

As Nadai (2004) notes, for the individual, voluntary action is additional, i.e., not meant to make a living from it. This may also be generalized and transferred to developers, business entities, or any legal person.

In this regard, it is important to note that voluntariness needs to be put in relation to the acting person, i.e., the same action may be considered mandatory if undertaken by one person and voluntary if undertaken by another. This is particularly true for environmental compensation and biodiversity offsets. Not infrequently is the person responsible for an impact different from the one implementing an offset. For example, a habitat banking provider will usually establish an offset in the course of his/her regular business activities to earn profit from it (and thus not voluntarily), while a developer may choose to voluntarily implement an offset (which he is not otherwise obliged to do).

Thus, the additionality of an offset (in terms of "not for profit") can only be judged in relation to the impact and the person responsible to offset this impact, i.e., whether there is a legal or any other requirement or incentive for the developer to implement an offset. The way this offset is then delivered, i.e., whether the responsible person chooses to contract a third party for this purpose or buys credits from a habitat bank, is yet another issue tied to the implementation of offsets.

As has been noted, cases are grouped according to the additionality criterion into two categories:

1. Yes
2. No

5.2.3 Core Criteria

Those criteria that greatly influence the voluntariness of an offset are marked as core criteria. They are prominently represented in the literature and also comprise the underlying fundamentals of the impressionistic classification of types of biodiversity offsets.

Motivation and Ultimate Goal

When Cranach (2008) asks why humans act voluntarily and how free voluntary action is, this leads to one of the core criteria of voluntariness: the motivation behind a voluntary action.

Taking the intentionality of an action as a starting point, any action is driven by a specific motivation and directed toward a certain goal. While the motivation can (and in reality mostly will) be a combination of different motivations, there is usually one ultimate goal toward which the voluntary action is oriented. This can be either selfless (altruistic) or selfish (for one's own benefit or to prevent damage to the own person).

However, an action can be considered voluntary even if it is not purely altruistic and accompanied by selfish motivations (Siegenthaler 2008). Voluntary actions can be both intrinsically and extrinsically motivated (Cranach 2008). Regarding extrinsic motivations, rewards and punishments can be distinguished. As Fig. 5.4 illustrates, motivation and ultimate goal are closely related, i.e., a reward usually provides an extrinsic motivation to pursue one's own benefit, whereas a punishment or sanction provides an extrinsic motivation to prevent damage to one's own person. An intrinsic motivation usually targets a selfless goal (however, it may often be accompanied by weaker forms of selfishness, e.g., striving for self-fulfillment and personal satisfaction).

Consequently, for the case of biodiversity offsets three different motivations are distinguished:

1. The pressure to comply with certain requirements to prevent damage (extrinsically motivated).
2. The incentive to gain an additional benefit from it (extrinsically motivated).
3. The altruism motivation based on perceived responsibility (intrinsically motivated, in practice at least partly also extrinsically motivated in terms of gaining acceptance).

Sub-criterion: Recompense or Benefit

The potential benefits of an offset are crucial—not only in cases where they are the ultimate goal and reason why the developer implements an offset. Despite the fact that voluntariness is unpaid, i.e., not rewarded (Ammann 2004; Boothe 2004), businesses may gain (additional) benefits from implementing biodiversity offsets voluntarily. Above all, this relates to the public relations benefits.

As Segerson (1999) notes, the benefit that a firm has from applying a voluntary approach "will reflect not only net revenue from the sale of its product, including

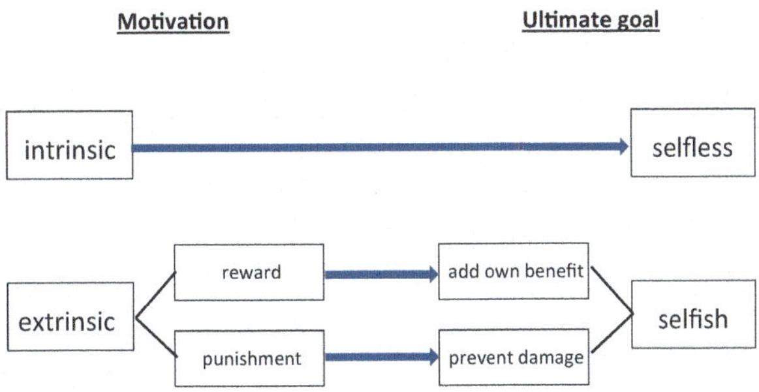

Fig. 5.4 Relation of motivation and ultimate goal. Source: Author

safety or increased shelf-life, but also any public relations benefit (e.g., increased current or future demand because of increased "good will" toward the firm)." Similarly, Caswell and Henson (1997) argue "that the loss of reputation and market sales are likely to be of more importance to firms than the direct damage costs imposed through liability". In addition, a firm is not likely to receive the same public relations benefits from meeting a mandatory standard than from a voluntary action, i.e., it would be rational for a firm to implement voluntary measures if a threat of a certain regulation is credibly imposed (Caswell and Henson 1997).

Influence and Initiators

As Priller (2008) and Olk (2004) note, voluntary commitment is influenceable and customizable by actors and actions of civil society. It is always business or a single firm that decides to comply with a certain rule or not, i.e., to implement biodiversity offsets. However, the originator of this rule can be very different—either intrinsic (the single firm/project) or from various extrinsic sources. In general terms, government influence and legal pressures on the one hand and community pressure and related public response on the other hand are the two most commonly encountered extrinsic influences.

Sub-criterion: Government Influence and Legal Pressures

Theoretically, voluntariness is not binding (Olk 2004). Nevertheless, the degree of obligation can vary, shaped by the legal boundaries and requirements, through governmental regulation and enforcement (Ammann 2004; Olk 2004).

The role of government and the related legal framework are important aspects for the emergence and implementation of biodiversity offsets. As has been shown in Sect. 5.1.1 (Step 1: Dichotomy between mandatory and voluntary biodiversity offsets) this has led to a categorization of two classes—mandatory and voluntary biodiversity offsets (the former being driven by government and legal pressures). Historically, many offset schemes are tied to a high degree of government involvement and regulation (e.g., US Wetland Mitigation and German Impact Mitigation Regulation, see Sect. 4.1).

This is further emphasized by observations of offset development in recent years, e.g., in France, the United Kingdom, and Spain. Driven by the French government, existing but unenforced legal provisions have been further specified and accompanied by new legislation. Similarly, in Spain amendments to the law aim for the introduction of offsetting and habitat banking. In the United Kingdom, the government has proclaimed the aim to put in place an offsetting system and commissioned a number of pilots.

While this underlines the vital role of government for offsets, this role is not limited to that of regulator, but also a facilitator in an increasingly intertwined globalized network of different actors.

Sub-criterion: Community Pressure/Public Response

Seeing the increasing diversity and drivers of actors involved in offsetting, pressure to implement offsets is no longer restricted to government but can originate from different sources—hence making the distinction between mandatory offsets required by governmental regulation and voluntary offsets (all cases exempt from governmental regulation) not adjusted to the diversity of practice.

In particular community pressure and public response (i.e., as part of public relations of a business) increasingly play a role. In the absence of regulatory programs and enforcement, communities may exert significant formal and informal pressures on firms to pollute less (Alberini and Segerson 2002). If a firm is considered liable for a certain contamination, the severity of the (anticipated) public response (e.g., reduced future purchases or consumer boycotts) can influence a firm's behavior (i.e., enhance voluntary commitment) (Segerson 1999).

For the specific case of biodiversity offsets, the following sources of influence have been identified:

1. Government
2. Financial institutions
3. Business sector
4. Single business/corporation
5. Local communities/stakeholders
6. NGOs/global organizations (BBOP, IUCN, BirdLife)
7. Without extrinsic driver (altruism)

Choice or Eligibility

Concepts of choice/eligibility, wishes, goals, preferences, or interest influence the voluntariness of an action (Gutmann 2000). Choice is a criterion for voluntariness that builds on the concept of freedom of action. Benn and Weinstein (1971) describe the latter as a state of nonrestricted options.

Thus, only those drivers or influences on the actions of an individual are limiting his/her freedom that reduces his/her possible choices. Unfree in the sense of involuntary means a restriction of options. The limitation or removal of an option for action is, independently from the intentions of the acting individual, seen as limiting the freedom or voluntariness of the individual's actions (even if this option was not his preference). Consequently, possible courses of action are independent from the action preferences of a person (Gutmann 2000).

In the case of vis compulsiva the consequences of the different options are changed, so that they are no longer practically eligible (Benn 1982). The concept of freedom of action implies that voluntariness is based on the choice or eligibility rather than the possibility of options (Benn 1982). According to Benn, an option is not eligible if for a rational person it would be inappropriate to choose (Gutmann 2000).

The choice of options or eligibility of possible alternatives corresponds very much to the degree of voluntariness; however, it faces the same problem. From the criteria identified it has possibly the highest normativity, which makes any generalization highly speculative. For this reason, choice and eligibility are not used to build the attribute space for the typology, but are documented and discussed separately.

In the course of this study, the following categories have been (normatively) set regarding the question whether there is an influence/change to the eligibility of options for a rational person:

1. None (willful voluntary action)
2. Marginal (coercive offer)
3. Significant (vis compulsiva)
4. Strong (vis absoluta)

5.2.4 Descriptive Criteria

Descriptive criteria are meant to explain or describe the setting and circumstances or conditions under which offsets are implemented, i.e., the case-specific context.

Context

Not one criterion, but a whole complex shapes the situation in a specific case. It is, therefore, important to analyze the context of biodiversity offsets to identify the strongest driver(s). This criterion cannot be captured using some keywords, but only by means of a detailed description.

Considering the context needs to emphasize questions of scale both with regard to geographical and political location (regional, national, international). As Priller (2008) notes, voluntary commitment is mostly shaped at national level, but one can also differentiate at the regional or sometimes transnational level.

Furthermore, how voluntariness is societally embedded explains existing national differences to a large degree (Priller 2008). This also means that different societal spheres, e.g., regulation, culture, and the market environment, need to be considered.

> Although these initiatives are voluntary and private, they are influenced in various ways by the broader environment—cultural, social, legal, economic and political—from which they emerge. (Kolk 2000)

Voluntariness is a social, cultural, and economic phenomenon (Priller 2008). These two attributes (scale and spheres) are also interrelated: the single business is influenced by national and international regulation and business codes as well as the overarching "moral environment" of the society within which it operates (ING Group and Corporate Responsibility, Pieter Kroon 2001).

Sub-criterion: Domestic Legislation in the Home Country

As has been noted (cf. Sub-criterion: Government influence and legal pressures) the role of government and the related legal framework are important aspects for the emergence and implementation of biodiversity offsets.

Generally, a public policy backed by national (and or regional) legislation forms an important part of the institutional framework influencing voluntary initiatives of business.

This is underpinned by practical evidence from a survey of company environmental programmes in North America, Europe, and Asia, which lists domestic legislation in the home country and legal sanctions as the most important influences on these initiatives (Kolk 2000).

Sub-criterion: Competitive Environment

As mandatory controls or mechanisms are likely to be imposed on an industry as a whole, firms that fail to implement adequate environmental improvements can impose negative externalities on other firms. Thus, the interaction between firms[5] and the expected behavior of other firms in the industry influences the strategies and voluntary actions of a firm (Segerson 1999). Competition and market dynamics can influence or incentivize offsets (e.g., within a sector or within a region).

Sub-criterion: Cultural and Social Context: Social and Individual Commitment

As Paavola et al. (2009) note "human behavior is governed by a much wider range of institutions than those that are embodied in or are enabled by the state." This refers in particular to social expectations and cultural norms as major drivers not only for the individual but also for business behavior. They exert a binding character and apply even without formal pressure (Draser 2013).

5.2.5 Performance Criteria

Performance criteria are those that do not refer to the drivers or to the outcome of offsets, but the way they are delivered, in terms of increased flexibility and cost efficiency.

---

[5]In game theory this could be modelled as a Nash equilibrium, i.e., "the optimal outcome of a game is one where no player has an incentive to deviate from his or her chosen strategy after considering an opponent's choice. Overall, an individual can receive no incremental benefit from changing actions, assuming other players remain constant in their strategies" (source: http://www.investopedia.com/terms/n/nash-equilibrium.asp).

Flexibility and Cost Efficiency

One of the core advantages of using voluntary approaches as an alternative to regulation is that they provide polluters with greater flexibility in meeting set environmental quality goals, thus opting for the option with the least costs, i.e., increasing financial or cost efficiency (Alberini and Segerson 2002; Segerson and Li Dawson 2001). Another form of cost efficiency could also be to search for the best outcome within a fixed budget.

The flexibility and cost efficiency criterion are asked as a simple relevance question in the scope of this study (does the type or case of biodiversity offsets allow for greater flexibility and thus increase efficiency?):

1. Yes
2. No

5.2.6 Quality Criteria

Quality criteria serve the purpose of evaluating the quality of an offset, in terms of the outcome of a voluntary commitment and the associated voluntary action. This is based on the premise that a voluntary action has to make an impact in order to be relevant. As the focus of this work is intentionally not on evaluation, these criteria (however, crucial) will be considered to be of minor relevance for this study.

Outcome and Charitableness

It is an inherent characteristic of voluntary commitment that an action should serve the public by creating or contributing to a public good (Siegenthaler 2008; Priller 2008), i.e., voluntary commitment requires that the individual provides goods that are important to the general public (workforce, money, goods/infrastructure, prestige) (Ammann 2004).

The provision of goods or services can be judged by its effectiveness. In fact, any voluntary action is meaningless and irrelevant if it is not effective, i.e., it should lead to a perceived real outcome (that is, the result of this action). Segerson (1999) assumes that "the outcome of the voluntary measures (if adopted) is the same as the outcome of the mandatory standards that might be imposed", i.e., the target level of performance shall be equal to the target level set by the government. However, this might be different in negotiated voluntary agreements (Alberini and Segerson 2002; Segerson 1999).

This indicates that the success of a voluntary approach has to be measured in terms of outcome compared to a baseline, i.e., what would the level of environmental quality be in the absence of the voluntary approach and what alternative (policy) might have been used and with what outcome (Altorfer 2004; Boothe 2004). This is referred to as the counterfactual. The counterfactual can be either static or dynamic. If the counterfactual is not explicitly mentioned, it is usually implicitly assumed to be

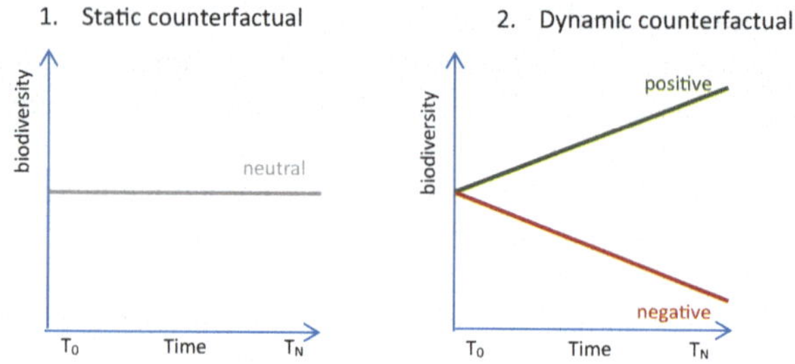

Fig. 5.5 Static and dynamic counterfactuals relative to voluntary environmental approaches. Source: Author

static. However, more precisely, three general trends can be distinguished for the counterfactual as shown in Fig. 5.5:

1. Static (neutral)
2. Dynamic:
   (a) Biodiversity increase (positive)
   (b) Biodiversity decline (negative)

Biodiversity offsets, on the one hand, should be directed toward public welfare (i.e., result in a measurable effect). On the other hand, due to the fact that offsets are assigned to a prior impact, the concept of biodiversity offsets usually can ensure no net loss at best and thus not be welfare enhancing. The latter may only be applicable if a "net positive impact" is to be achieved.

The outcome and charitableness of biodiversity offsets can be split into four categories:

1. Net Gain: Positive balance, i.e., gain > loss
2. No Net Loss: Neutral balance, i.e., gain = loss
3. Reduced loss: Minimum balance, i.e., 0 < gain < loss
4. (Complete) loss: Not effective, i.e., gain = 0

As has been said, theoretically, only no. 1 would be considered welfare enhancing. In practice, however, even no. 2 and 3 could add to the public good if the counterfactual was a complete loss. Therefore, it is necessary to judge the outcome and charitableness in relation to the counterfactual as has been defined above.

5.2.7 Rejected Criteria

A number of criteria that have been extracted from the literature on voluntariness have been rejected in the scope of this study because they are of no or minor relevance for biodiversity offsets under the given research premises. These include:

- *Free choice*: Ammann (2004) defines voluntariness as the contribution of an individual to the public welfare that the individual has chosen freely.

- *Scope or distance of voluntary action*: Voluntary action can exist at different levels (Siegenthaler 2008), building on close social bonds (i.e., caring for family and friends) or a more generalized system of (global) responsibility (e.g., fighting climate change).
- *Know-how/professionalism*: In order to be effective, voluntary action requires appropriate know-how (Ammann 2004; Badelt 2004; Boothe 2004).
- *Taking over Responsibility*: Voluntariness means to take over a certain responsibility (Boothe 2004).
- *Ethics*: Voluntariness is not per se ethically good (Boothe 2004).

5.2.8 Measurement and Value Categories of the Criteria

Table 5.2 summarizes the different types of criteria and the respective value categories for each criterion.

Table 5.2 Grouping and measurement of criteria for voluntariness

Type of criteria	Criteria	Value categories
Threshold criteria	Intentionality	1. Yes 2. No
	Additionality	1. Yes 2. No
Core criteria	*Motivation and ultimate goal*	1. Pressure/compliance 2. Incentive/cost benefit 3. Altruism or responsibility (gain acceptance)
	• Recompense or benefit	1. Yes 2. No
	Influence and initiators	1. Government 2. Financial institute(s) 3. Sector 4. Corporation 5. Local community 6. NGO(s) 7. None
	Choice or eligibility	1. No influence (willful voluntary action) 2. Marginal influence (coercive offer) 3. Significant influence (vis compulsiva) 4. Strong influence (vis absoluta)

(continued)

Table 5.2 (continued)

Type of criteria	Criteria	Value categories
Descriptive criteria	Context	
	• Domestic legislation in the home country	No value categories, descriptive criteria
	• Competitive environment	
	• Cultural and social context	
Performance criteria	Increased flexibility	1. Yes 2. No
	Cost efficiency	1. Yes 2. No
Quality criteria	Outcome and charitableness	
	• Counterfactual	1. Static (neutral) 2. Dynamic (a) Positive (biodiversity increase) (b) Negative (biodiversity decline)
	• Outcome and charitableness relative to the counterfactual	1. Positive balance (net gain) 2. Neutral balance (no net loss) 3. Minimum balance (reduced loss) 4. Not effective (loss)

Source: Author

5.3 Substruction of the Underlying Attribute Space and Combinations of Attributes

The central question for the creation of the attribute space for a typology of biodiversity offsets is: *Which of the criteria are suitable for constructing the underlying attribute space from?* As has been shown (see Section 5.2 Derivation of relevant attributes/criteria for voluntariness from the theory) the core criteria can be derived through a process of elimination in different steps:

1. *Rejected criteria*: A number of criteria that have been extracted from the literature on voluntariness are of no or minor relevance for biodiversity offsets under the given research premises and thus are rejected.
2. *Threshold criteria*: Intentionality and additionality are defined as threshold criteria which need to be fulfilled for an action to be considered voluntary, but do not allow for qualifying characterization.
3. *Performance criteria*: Flexibility and cost efficiency are another set of subordinate criteria. They refer to the way offsets are delivered, i.e., performance related.
4. *Quality criteria: Outcome and charitableness of* serving the purpose of evaluating the quality of an offset. However, as the focus of this work is intentionally not on evaluation, quality criteria are not used to build the attribute space of the typology upon.

Table 5.3 Substruction of the underlying attribute space of the impressionistic typology

influence and initiators \ motivation and ultimate goal	Pressure (compliance)	Incentive (cost-benefit)	Altruism / responsibility
Government	Case 1 (regulatory)	Case 3 (enabled)	?
Financial institute	Case 2 (conditional)		?
Sector	Case 4 (sectoral)	?	?
Corporation	Case 5 (corporate)	?	?
Local community	Case 6 (local)		?
NGOs	?	Case 3 (enabled)	?
Without major influence	?	?	?

Source: Author

5. *Remaining core criteria:* After this process five criteria remain. Context is a descriptive criterion, which cannot be qualified by different values and hence not suitable as categorizing criterion. Recompense or benefit is a secondary criterion, which is closely tied to motivation and ultimate goal. Finally, choice and eligibility as described in the scope of this study is a highly normative concept. As this would prevent any transferability or generalization of the typology, it is also excluded from the attribute space.

A combination of two suitable criteria remains:

1. Influence and initiators
2. Motivation and ultimate goal

In a first step, these two criteria and the respective value categories that have been assigned to them (see Sect. 5.2) are used to build a 3 × 7 matrix, leading to 21 fields as shown in Table 5.3.

In a second step, the built impressionistic types (see Sect. 5.1) are sorted into the matrix. One type or case can occupy one or several fields. The result is a filled-in matrix. However, several fields (highlighted in gray) remain empty. This makes verification and possibly rectification of the impressionistic types necessary in order to align them with the attribute space (see Sect. 5.4).

5.4 Transformation (Rectification) of the Impressionistic Types and Analysis of Meaningful Correlations

The integration of the impressionistic types that have been developed over time (building on the author's own observation and experience) and the constructed attribute space (building on criteria derived from the literature) shows overlaps (thus mutually confirming, i.e., strong parts), but also unveils empty fields of the attribute space (open questions, i.e., weak parts). These empty fields need to be analyzed in light of the following questions:

- Did the impressionistic typology miss any types? If yes, which?
- Can there be blank fields? If yes, why?
- Are there any cases that could theoretically exist, but are practically improbable?
- Are the borders between the cases adequate or is there a need to refine the scope of the single cases (i.e., more extended or narrow)?

Table 5.4 illustrates the transformation of the impressionistic types and the underlying attribute space.

The head of the matrix lists the formulated three value categories for the motivation of a (more or less) voluntary biodiversity offset. As in practice, an action may often be driven by a multitude of motivations, here the dominating motivation is considered (which may be accompanied by others). What can be noted is that among the three value categories "altruism and responsibility" is different from the other two as in its pure form it can only exist without major influence of any form. This means that it is theoretically logical to exclude fields no. 3, 6, 9, 12, 15, and 18 as empty spaces in the attribute space. In practice, however, a perceived cost–benefit motivation induced by some form of incentive may be combined with the sincere conviction to act responsibly (see fields 2 and 3, 14 and 15, 17 and 18).

Similarly, *fields no. 19 and 20 can be excluded* because without major influence there can neither be a pressure nor an incentive to implement an offset.

Field no. 1 is characterized by a governmental (or legal) pressure to comply with an offsetting requirement. It is thus clearly determined by *case 1: regulatory biodiversity offsets*.

Field no. 2 relates to a case where government or public bodies facilitate or provide incentives to implement offsets, i.e., an offset enabling framework as described in *case 3: enabled biodiversity offsets*. As has been noted this may be equally driven by or combined with an altruistic motivation, thus being extended to *field no. 3*. Similar to government, NGOs (e.g., IUCN and others) may provide an offset enabling framework that extends over *fields no. 17 and 18*. This leads to a subdivision of *case 3a: enabled biodiversity offsets—by government* and *case 3b: enabled biodiversity offsets—new global norms*.

Table 5.4 Transformation of the impressionistic types and underlying attribute space

influence and initiators \ motivation and ultimate goal	Pressure (compliance)	Incentive (cost-benefit)	*Altruism / responsibility*
Government	Case 1 (regulatory)	Case 3a (enabled: government)	
Financial institute	Case 2 (conditional)		
Sector	Case 4 (sectoral)		
Corporation	Case 5 (corporate)		
Local community	Case 6a (local: reputational risk)	Case 6b (local: license to operate)	
NGOs		Case 3b (enabled: new global norms)	
Without major influence			*Case 7 (pure altruism)*

Source: Author

Financial institutes (such as the World Bank) provide at the same time pressure to comply with their requirements (especially Performance standard 6 on biodiversity offsets) and an incentive in terms of providing financing for development projects that are in compliance with these. *Case 2: conditional biodiversity offsets* thus corresponds to *fields no. 4 and 5* of the matrix.

A sector may also make a voluntary commitment to implement biodiversity offsets (e.g., via its representative organization as in the case of the International Council on Mining and Metals) thus exerting pressure to comply on its members, which may also see a reputational benefit in participating. *Case 4: sectoral biodiversity offsets* thus corresponds to *fields no. 7 and 8.*

A corporation may exert pressure on its single operations worldwide by formulating a biodiversity strategy, environmental management requirements or no net loss/net gain strategies (e.g., Rio Tinto's biodiversity strategy). This is described by *case 5: corporate biodiversity offsets* (covering field *no. 10*). Due to its direct impact, a corporation usually does not provide incentives for offsetting, thus leaving *field no. 11 blank.*

Cases 2, 4, and 5 have in common that they are driven by the pressure to comply with a certain requirements, thus being incompatible with an altruistic motivation and consequently *fields no. 6, 9, and 12 remain empty.*

At the local level, offsets may be required in response to threatened reputational damage by the local stakeholders (often supported or mobilized by NGOs). This is described as *case 6a: local biodiversity offsets—reputational risk* and refers *to fields no. 13 and 16*. Local biodiversity offsets may also be used as a tool to gain acceptance with local stakeholders, e.g., to speed up lengthy participation and permitting procedures. *This case 6b: local biodiversity offsets—license to operate* refers to *fields no. 14 and 15* (as it may be combined with an altruistic motivation).

As a result of the substruction process the typology can be updated to contain the following seven types (including four subtypes):

1. *Regulatory biodiversity offsets*: required by law and enforced
2. *Conditional biodiversity offsets*: required by financial institutions (WB, IFC, . . .)
3. *Enabled biodiversity offsets*: fostered by governments (*case 3a: enabled biodiversity offsets—by government*) and NGOs (*case 3b: enabled biodiversity offsets—new global norms*) through pilot schemes, guidance, etc.
4. *Sectoral biodiversity offsets*: take part in a voluntary self-commitment of a sector (e.g., mining)
5. *Corporate biodiversity offsets*: resulting from a voluntary self-commitment of a corporation
6. *Local biodiversity offsets*: single offsets, that are most likely developed at local level in a consensual process, including *case 6a: local biodiversity offsets—reputational risk* and *case 6b: local biodiversity offsets—license to operate*
7. *Altruistic biodiversity offsets*: truly voluntary offsets that are driven by the altruistic motivation to make a positive impact

These types are described in more detail and illustrated by examples in Chap. 6.

Voluntariness of biodiversity offsets is a gradual continuum. The graduation from case 1 to case 7 does not necessarily imply an increase in voluntariness (i.e., from less voluntary to more voluntary), but holds account of the different degrees of institutionalization (from high to low). Consequently, the cascade as presented in Fig. 5.3 could be presented in a different order: For example, the pressure that a corporation exerts on the local or project level (case 5) may in some instances more directly and more strictly affect the specific case than the regulative pressure in case 1.

5.5 Screening of Biodiversity Offset Cases Worldwide and Grouping of Cases According to the Built Types

The screening of offset cases was at the core of the netnographic research approach and evolved over the years. It thus builds on a puzzle of participant observational networking measures in biodiversity offset experts circles (i.e., via the Business and Biodiversity Offsets Programme, the Biodiversity Offsets Blog, LinkedIn, scientific conferences and review of scientific publications, news articles, policy and gray literature, and personal communication). This exploratory list of worldwide biodiversity offset cases (which by no means intends to be exhaustive, nearly exhaustive, or representative) was then checked for adequate and simple accessible (online) information quality (using mainly the Internet search engines Google and Google scholar). Additional information that was retrieved included:

- Scientific articles and books
- Government regulations and guidelines
- Reports and case studies by international organizations and NGOs (BBOP, ICMM, TNC)
- Websites of operations that implement offsets
- Reports and strategies related to offsets by corporations or at project level
- Presentations at conferences and workshops

In the scope of the screening 90 possible biodiversity offset cases were gathered to underpin the typology as shown in Appendix 1, Chap. 10. A number of these have been rejected due to poor information quality or other reasons (e.g., project failure/abandonment). The remaining 72 cases are displayed on a world map in as shown in Fig. 5.6.

The table in Appendix 1, Chap. 10 gives a brief overview for each of the encountered cases on the following aspects:

1. No.: The cases are consecutively numbered. When referred to in the text these numbers are given for easier and explicit identification.
2. Name/brief description (and impact type): This includes the name of the project or operation, usually the operator or consortium and the type of the impact (e.g., open pit gold mine).
3. Country/location: The country and (if applicable) region or province is given. This is of particular importance to appropriately judge the context (e.g., in terms of existing legislation).
4. Influence and initiators: The two core criteria are displayed with their respective values to tick and possibly some short explanation (e.g., government influence, US Wetland Mitigation following Clean Water Act). One or several boxes can be ticked building on what can be interpreted from the available sources (these may, however, be incomplete or give a biased view, e.g., shaped by the developer).
5. Motivation and ultimate goal: see above.
6. Type: Based on items 4 and 5, this is then translated into the different types. This is a normative judgment that builds on the information available. One case can be characteristic for one type or represent a combination of two or more types. Types in bold type represent primary (major) attribution to a type, non-bold secondary attribution.
7. Sources and notes: The sources that have been identified in a quick scan under the given research premises (Internet-based survey) are documented. In addition, comments on particularities of the specific case may be given.
8. Information quality: The screening serves as a basis for the identification of illustrative case studies for the types. This identification is based on the representativeness of a case for a type (see item 6) and the availability and quality of

information. Four categories of information quality are built—very good (++) represents the ideal category (from which case studies ideally should be chosen), fair (+) refers to an acceptable information quality (case studies can also be chosen from this category), sparse (–) means that very basic information or just a few sources are available (cases form this category do not qualify as a case study, but may underpin the overall description of the types) and poor (– –) refers to cases that have been mentioned, but no additional or more precise information could be identified (their function is simply to show the variety of cases).

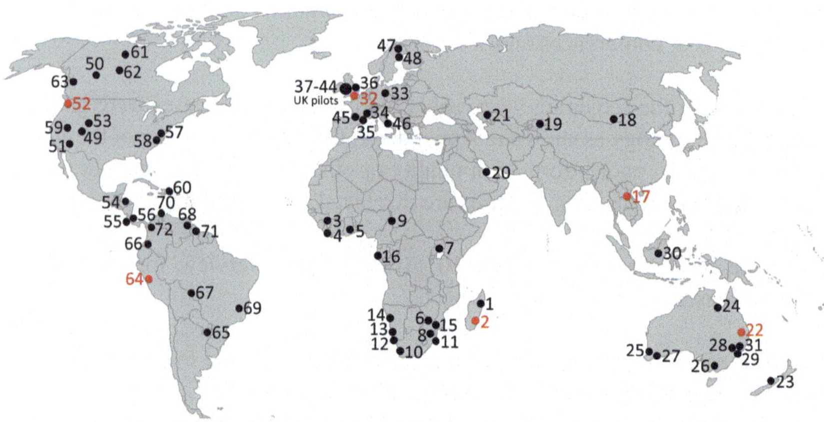

Africa
1. Ambatovy BBOP Pilot
2. QIT Mining Madagascar QMM
3. Simandou
4. Piloting aggregated offsets and mining in Liberia
5. BBOP Pilot Akyem Gold Mining Project
6. Potgietersrust platinums limited (PPRust) BBOP Pilot
7. Bujagali Energy Limited
8. Anglo Coal Isibonelo colliery
9. Petroleum Development and Pipeline Project
10. Mount Royal Golf Estate
11. Pulp United Pulp Mill
12. De Beers Marine Kleinzee and Alexander Bay Sea Areas
13. Namdeb Diamond Corporation
14. Rossing Uranium Mine
15. Palabora
16. Shell Foundation offset on Gamba Complex

Asia
17. Nam Theun 2 Hydropower
18. Oyu Tolgoi LLC
19. Kumtor Gold Mine
20. PEARL GTL QATAR BBOP Pilot
21. Petroleum and Natural Gas Development in the Ustyurt

Australia
22. Koala Offset in South East Queensland
23. Strongman Coal Mine BBOP Pilot
24. Mc Arthur River Mine Open Cut Project
25. Newmont Boddington mine
26. Basslink Undersea Power Cable
27. Raventhorpe Nickel Project
28. Karuah bypass offset
29. Cumberland Plain Woodland of Western Sydney
30. IndometCoal Project
31. Warkworth Coal Mine

Europe
32. Thameslink Railway offsets
33. Burgberg Zschaitz habitat bank
34. CNM high speed rail bypass
35. Operation Cossure
36. Bardon Hill Quarry
37. South Devon UK pilot
38. North Devon UK pilot
39. Essex UK pilot
40. Doncaster UK pilot
41. Development Project in Dudley
42. Project Pinewood
43. Nottinghamshire UK pilot
44. Warwickshire UK pilot
45. A65 Langon-Pau motorway
46. Apennine Wind Farms
47. Openpit Mining Mertainen
48. Ume Delta River Railway

North & Central America
49. Kennecott Utah Copper Mine
50. Suncor Biodiversity Offset
51. Resolution Copper Mining
52. Bainbridge Island BBOP pilot
53. Jonah Natural Gas Field
54. Via San Lorenzo – Olanchito
55. Hidroelectrica Reventazon
56. Establishment of the Braulio Carrillo National Park for the construction of a highway
57. Tygart Valley River Stream and Wetland Bank
58. South Fork Hoppers Creek Mitigation
59. Newmont Long Canyon Project
60. Pueblo Viejo gold mine expansion
61. NGTL pipeline proposals in Horne River area
62. Total Joslyn oil-sands mine
63. Enbridge Northern Gate-way pipelines

South America
64. Antamina Mine
65. Hidroelectrica Yacyreta
66. Pasto Mocoa road
67. Gasoducto Gasbol
68. Brisas Gold & Copper Project
69. AngloGold Ashanti Brasil Mineracao
70. Cerrejón mine
71. Amaila Fall Hydro Project
72. Gramalote Mining Project

Fig. 5.6 Screening of biodiversity offset cases worldwide (nonrepresentative and not exclusive) (Case studies highlighted in red). Source: Author

References

Alberini A, Segerson K (2002) Assessing voluntary programs to improve environmental quality. Environ Resour Econ 22(1):157–184

Albrecht J, Schumacher J, Wende W (2014) The German impact-mitigation regulation—a model for the EU's no-net-loss strategy and biodiversity offsets? Environ Policy Law 44(3):317–325

Altorfer H (2004) Zusammenfassende Thesen. In: Ammann H (ed) Freiwilligkeit zwischen liberaler und sozialer Demokratie. Seismo, Zürich

Ammann H (2004) Freiwilligkeit, Gemeinnützigkeit und Sozialstaat. In: Ammann H (ed) Freiwilligkeit zwischen liberaler und sozialer Demokratie. Seismo, Zürich

Arora S, Cason TN (1995) An experiment in voluntary environmental regulation: participation in EPA's 33/50 program. J Environ Econ Manag 28:271–286

Badelt C (2004) Freiwilligkeit aus Sicht der Ökonomie. In: Ammann H (ed) Freiwilligkeit zwischen liberaler und sozialer Demokratie. Seismo, Zürich

Batson CD (2014) The altruism question. Toward a social-psychological answer. Psychology Press, New York

BBOP Secretariat (2010) Biodiversity offsets: a tool for CBD parties to consider and a briefing on the Business and Biodiversity Offsets Programme (BBOP). Draft, unpublished

Benn SI (1982) Community as a social ideal, vol 1. Edward Arnold, London, p 1

Benn SI, Weinstein WL (1971) Being free to act, and being a free man. Mind 80(318):194–211

Bishop J, Kapila S, Hicks F, Mitchell P (2006) Building biodiversity business: report of a scoping study. Available at: http://www.globalnature.org/bausteine.net/f/6644/Building_Biodiversity_Business_-_Draft_Report_%28Oct_2006%29.pdf?fd=2

Boothe B (2004) Freiwilligkeit aus Sicht der Psychologie. In: Ammann H (ed) Freiwilligkeit zwischen liberaler und sozialer Demokratie. Seismo, Zürich

Caswell JA, Henson SJ (1997) Interaction of private and public food quality control systems in global markets. In: Proceedings of the European Association of agricultural economists conference on globalization of the food industry: policy implications, pp 217–36

Cranach (2008) Freiwilligkeit, Altruismus oder Egoismus? Zur Sozialpsychologie der individuellen und gesellschaftlichen Bedeutung der Freiwilligkeit. In: Ammann H et al (eds) Freiwilligkeit—Ursprünge, Erscheinungsformen, Perspektiven. Seismo, Zürich

Darbi M (2010a) Biodiversity offsets—a tool for environmental management and biodiversity conservation. In: Sanchez Bengoa D, Powell D (eds) TOP biodiversity 2010. Intercollege-Larnaca, Cyprus. Conference proceedings. Intercollege-Larnaca, Larnaca, S. 289–301

Darbi M (2010b) Voluntary Biodiversity Offsets: Freiwillige Kompensationsmaßnahmen für Eingriffe in die Biodiversität – eine neue Perspektive für den Erhalt und die Wiederherstellung der biologischen Vielfalt weltweit und in Deutschland? In: Feit U, Korn H (Bearb) Treffpunkt Biologische Vielfalt IX – Interdisziplinärer Forschungsaustausch im Rahmen des Übereinkommens über die biologische Vielfalt. BfN, Bonn, S. 199–204

Darbi M (2014) Early draft on a typology of biodiversity offsets. Biodiversity offsets blog. Available at: http://www.biodiversityoffsets.net/typology-of-biodiversity-offsets/

Darbi M (2015) What is the business case for biodiversity offsets? In: "SQ-raising the sustainability quotient" (Bombay chamber of commerce and industry), IV(1): 3–5

Draser (2013) Freiwillig nur unter Zwang. Nachhaltig verpflichtet – aber wodurch. factor y. Magazin für nachhaltiges Wirtschaften

Escorcio Bezerra LG (2007) Biodiversity offsets in national (Brazil) and regional (EU) mandatory arrangements: towards an international regime? Available at: http://www.forest-trends.org/biodiversityoffsetprogram/library/new/Dissertation%20Biodiversity%20Offsets%20LGB%20IUCN%20BBOP.doc

Gutmann T (2000) Freiwilligkeit als Rechtsbegriff. Beck, München

Howard K (2007) Voluntary biodiversity offsets: improving the environmental management toolbox. Available at: http://www.cortex.org/d-Cortex-%20Biodiversity%20Offsets_01Dec07.pdf

ING Group and Corporate Responsibility, Pieter Kroon (2001) Guidelines for multinational enterprises. Roundtable on Global Instruments for Corporate Responsibility, Paris, 19 Jun 2001. Available at: http://www.oecd.org/investment/mne/1898195.ppt

Kluge S (1999) Empirisch begründete Typenbildung. Zur Konstruktion von Typen und Typologien in der qualitativen Sozialforschung. Leske + Budrich Verlag, Opladen

Kolk A (2000) Economics of environmental management. Financial Times-Prentice Hall, Harlow

Lazarsfeld PF (1937) Some remarks on the typological procedures in social research. Am Political Sci Rev 99(3):435–452

Nadai E (2004) Der Lohn der Tugend. Zur Rationalisierung von Freiwilligenarbeit. In: Ammann H (ed) Freiwilligkeit zwischen liberaler und sozialer Demokratie. Seismo, Zürich

Olk C (2004) Freiwilligkeit aus Sicht der Soziologie. In: Ammann H (ed) Freiwilligkeit zwischen liberaler und sozialer Demokratie. Seismo, Zürich

Paavola J, Gouldson A, Kluvánková-Oravská T (2009) Interplay of actors, scales, frameworks and regimes in the governance of biodiversity. Environ Policy Gov 19(3):148–158

Priller E (2008) Zivilgesellschaftliches Engagement im europäischen Vergleich. In: Ammann H, Hasse R, Jakobs M, Riemer-Kafka G (eds) Freiwilligkeit – Ursprünge, Erscheinungsformen, Perspektiven. Seismo, Zürich

Segerson K (1999) Mandatory versus voluntary approaches to food safety. Agribusiness 15 (1):53–70

Segerson K, Li Dawson N (2001) Environmental voluntary agreements: participation and free riding. In: Orts EW, Deketelaere K (eds) Environmental contracts. Comparative approaches to regulatory innovation in the United States and Europe. Kluwer Law International, Boston, pp 369–388

Siegenthaler H (2008) Zur Reichweite gemeinnützigen Handelns – Von der Gruppensolidarität zum Universitalitätsprinzip. In: Ammann H, Hasse R, Jakobs M, Riemer-Kafka G (eds) Freiwilligkeit – Ursprünge, Erscheinungsformen, Perspektiven. Seismo, Zürich

ten Kate K (2005) Biodiversity offsets: good for business and biodiversity? Available at: https://www.cbd.int/financial/offsets/g-offsetsbusiness.pdf

6 Illustrative Case Studies for Different Types of Biodiversity Offsets

Abstract

A typology of biodiversity offsets has been developed. Seven types (including four subtypes) were identified: (1) Regulatory biodiversity offsets: required by law and enforced; (2) Conditional biodiversity offsets: required by financial institutions; (3) Enabled biodiversity offsets: fostered by governments (case 3a: enabled biodiversity offsets—by government) and NGOs (case 3b: enabled biodiversity offsets—new global norms) through pilot schemes, guidance, etc.; (4) Sectoral biodiversity offsets: taking part in a voluntary self-commitment of a sector (e.g., mining); (5) Corporate biodiversity offsets: resulting from a voluntary self-commitment of a corporation; (6) Local biodiversity offsets: single offsets that are most likely developed at local level in a consensual process, including case 6a: local biodiversity offsets—reputational risk and case 6b: local biodiversity offsets—license to operate; and (7) Altruistic biodiversity offsets: truly voluntary offsets that are driven by the altruistic motivation to make a positive impact.

These types are described and analyzed both in general and using practical examples in the form of illustrative case studies for each type. The presentation of these case studies is structured and characterized by criteria for voluntariness that have been identified from the literature: threshold criteria, descriptive criteria, core criteria, performance criteria, and quality criteria.

Keywords

Biodiversity offsets typology · Biodiversity offsets types · Biodiversity offsets case studies · Regulatory biodiversity offsets · Conditional biodiversity offsets · Enabled biodiversity offsets · Sectoral biodiversity offsets · Corporate biodiversity offsets · Local biodiversity offsets · Altruistic biodiversity offsets

M. Darbi, *Biodiversity Offsets Between Regulation and Voluntary Commitment*,
https://doi.org/10.1007/978-3-030-25594-7_6

6.1 Type 1: Regulatory Biodiversity Offsets

6.1.1 General Description

Regulatory biodiversity offsets are based on requirements set by governmental bodies at different levels (e.g., national, state, or regional level) to compensate for impacts on biodiversity and the natural environment arising from development activities. The more mature approaches include Wetland Mitigation after the US Clean Water Act, offsetting under the Australian Environmental Protection and Biodiversity Conservation Act, and the German Impact Mitigation Regulation. The latter two are also examples of a varied offsetting practice according to the different provisions of federal states (under an overarching national framework). Other long-standing offset regimes include offsets after the Brazilian SNUC (National System of Conservation Units) Act and the Forest Code, Canadian fish habitat compensation, impact mitigation regulation under the jurisdiction of the Austrian federal states and at European supra-national level appropriate assessment under the EU birds and habitats directive.

These regulatory offset regimes are of very different natures, ranging from restricted approaches focusing on selected protected or threatened species or ecosystems, e.g., wetlands under US Wetland Mitigation, to an ecosystem approach under German Impact Mitigation Regulation. Furthermore, they can differ with regard to scale and mechanisms to deliver offsets, e.g., through the use of habitat banking.

Regulatory biodiversity offsets have in common that they are (usually legally) bound to the approval of an envisaged development activity and enforced by governmental bodies. This actual enforcement is crucial and builds on significant administrative capacity. In Germany, for instance, more than a thousand full-time employees in the lower nature conservation authorities are concerned with the approval of offsets under the Impact Mitigation Regulation (Wende et al. 2015). This is in contrast to the case of France where offsetting provisions have been legally fixed for decades, but were hardly enforced. Only recently, the government has pushed toward the implementation of offsetting through guidance and pilot projects. This highlights that apart from enforcement capacity, implementation guidance is a crucial part of regulatory biodiversity offsets.

Regulatory biodiversity offsets appear to be a growing case with increasing regulatory or government interest in offsetting in recent years. New compensation schemes have arisen in particular in Latin American countries, e.g., Colombia and Peru (Lopez Arbeláez and Quintero Sagre 2015). Spain has introduced offsetting and habitat banking as part of the new Law 2172013 about environmental assessment (Mercados de Medio Ambiente 2013, 2014). Furthermore, compensation requirements by Swedish municipalities and cities are increasing (Enetjärn et al. 2015; Naturvårdsverket 2015), and British Columbia (Canada) has set up an Environmental Mitigation Policy (British Columbia Ministry of Environment 2014).

This trend is also backed by many experts, who now see a strong regulatory system as the (only) way forward (e.g., Albrecht et al. 2014; Herbert 2015; Darbi et al. 2016; eftec, IEEP et al. 2010; IEEP et al. 2012). This refers particularly to the different levels of control and the possibility of sanctions. Possibly the most prominent example is the discussion about the introduction of a mandatory scheme (i.e., through an EU directive or regulation) in the scope of 2015 no net loss initiative (Working group on no net loss of ecosystems and their services 2013). As of November 2015, this now seems very unlikely due to huge opposition expressed during the public consultation to the No Net Loss Initiative (Darbi et al. 2016). This example shows how difficult it is to set up a regulatory offsetting system. However, once in place and properly implemented it is a powerful tool to safeguard the integrity of the natural environment.

6.1.2 Case Study for Regulatory Biodiversity Offsets: Koala Offset in South East Queensland[1] (Tables 6.1, 6.2, 6.3, 6.4, 6.5, and 6.6)

Table 6.1 General description for the case study for regulatory biodiversity offsets: Koala Offset in South East Queensland

General description	
Location	South East Queensland, Australia
Development activity	Residential subdivision
Environmental baseline and impact	Clearing of vegetation, loss of koala habitat
Impact size	67.14 ha
Offset size	161.11 ha
Loss-gain calculation	The area of the offset required was calculated using the EPBC Offset Assessment Guide and the "How to use the Offset Assessment Guidelines" which gives guidance to the input figures for the spreadsheet.
Goal of the offset and surrogate	The koala is the single species targeted for the offset—i.e., the offset is to provide increased habitat and forage trees for the koala population. The management regime proposed for the offset area is to enhance the level of protection afforded to existing koala habitat through the exclusion of land management practices that are incompatible to achieving a net gain in koala habitat quality.
Type and scope of offset measures	The measures aim to improve the condition of the *Eucalyptus* open forest. This includes, in particular, the exclusion of forestry operations or any clearing of native vegetation, a weed and fire management plan as well as reduction of grazing and awareness raising through signage.
Involved parties	1. Australian Government (Department of Environment) 2. The developer 3. A private landholder 4. (a consultancy)
Timeframe	Started: 2014, planning: 07-11/2014, implementation: 01/2015, completed: 2025 (expected)

Source: Author

[1]Information based on Key and Darbi (2015) and personal communication with Alan Key. For more information see: http://www.environment.gov.au/epbc/publications/epbc-act-environmental-offsets-policy

Table 6.2 Threshold criteria for the case study for regulatory biodiversity offsets: Koala Offset in South East Queensland

Threshold criteria			
Intentionality	Yes	No	Comment
	X		–
Additionality	Yes	No	Comment
	X		–

Source: Author

Table 6.3 Descriptive criteria (context) for the Case study for regulatory biodiversity offsets: Koala Offset in South East Queensland

Descriptive criteria: context	
Domestic legislation in the home country	Australia has advanced offsetting legislation, both at the federal and at the state level. At the national level, the Environmental Protection and Biodiversity Conservation Act (EPBC Act, 1999) requires offsetting for impacts on "Threatened Ecological Communities" and "Threatened Species". In addition, all federal states and territories have specific provisions that go beyond these requirements. In this case, the offset was required due to a loss of koala habitat, which is listed as an endangered species of national significance under the EPBC Act. Thus, approval was bound to the securing of an Offset Area Management Plan (OAMP) prior to commencement of on-the-ground work. The developer is required by the Environmental Authority to implement the management plan. The conditions of the development, agreed to between the regulator (DoE) and the developer, require that the agreed Offset Management Plan be implemented. This is the legal agreement between the Australian Government and the developer for what needs to occur for the development to proceed. The landholder is bound by a legal contract with the developer with regard to the monetary payment amount and schedule, for implementing the agreed management actions within the Offset Area Management Plan. This Offset Area Management Plan is bound legally to the property and is therefore binding on current and future owners.
Competitive environment	–
Cultural and social context	–

Source: Author

Table 6.4 Core criteria for the Case study for regulatory biodiversity offsets: Koala Offset in South East Queensland

Core criteria			
motivation and ultimate goal / influence and initiators	**Pressure (compliance)**	**Incentive (cost-benefit)**	***Altruism / responsibility***
Government	**Case 1 (regulatory)**	Case 3a (enabled: government)	
Financial institute	Case 2 (conditional)		
Sector	Case 4 (sectoral)		
Corporation	Case 5 (corporate)		
Local community	Case 6a (local: reputational risk)	Case 6b (local: license to operate)	
NGOs		Case 3b (enabled: new global norms)	
Without major influence			*Case 7 (altruism)*
Comments	The offset is a legal arrangement between three parties: the Australian Government (major driver) represented by the Department of Environment (DoE), the developer and a private landholder on whose property the offset is secured. A consultancy supports and carries out the planning of the offset, negotiates with the regulators and between the parties with regard to management actions, price and contractual arrangements. The offset is also secured on the property by a legally binding instrument administrated by the Queensland (State) Government by the Land Act 1994 and the Land Titles Act 1994 in compliance with the management plan enforceable under the Vegetation Management Act 1999. The motivation of the developer is to comply with the legal requirements to obtain approval for the envisaged development by the environmental authority. Motivations for landholders vary. In this case, the landholder recognized the property needed rehabilitation due to the clearing of timber and overgrazing over an extended period resulting in a degraded ecosystem, weed invasion and habitat degradation. The use of the property for offsets has enabled the landholder to start improving the condition and rehabilitating the ecosystem and habitat with the contract between the landholder and the developer providing the money necessary for this. Compliance is undertaken each year by an independent third party to verify that the offset management plan is being undertaken/adhered to and that the offset is progressing or improving as measured against ecological data collected at the start of the offset. These annual reports are sent to the regulators, developer and the landholder.		

(continued)

Table 6.4 (continued)

	No influence (willful voluntary action)	Marginal influence (coercive offer)	Significant influence (vis compulsiva)	Strong influence (vis absoluta)	Comment
Choice or eligibility				X	The offset was clearly mandated by law (EPBC Act) and enforced by the Australian and Queensland Governments (approval bound to implementation of the OAMP).
Recompense or benefit	Yes	No	Comment		
		X	-		

Source: Author

Table 6.5 Performance criteria for the Case study for regulatory biodiversity offsets: Koala Offset in South East Queensland

Performance criteria			
Increased flexibility and cost efficiency	Yes	No	Comment
	X		With a thorough process, understanding of the technical, legal, and financial processes and variables and the determination from the start of the process as to the outcomes required by each of the parties involved, offsets can be negotiated and secured within a reasonable period of time (under 6 months) and be legally and financially responsible to the parties involved.

Source: Author

Table 6.6 Quality criteria (outcome and charitableness) for the case study for regulatory biodiversity offsets: Koala Offset in South East Queensland

Quality criteria: outcome and charitableness					
Counterfactual	Static (neutral)	Dynamic		Comment	
		Positive	Negative		
	x				
Outcome and charitableness relative to the counterfactual	Positive balance (net gain)	Neutral balance (no net loss	Minimum balance (reduced loss)	Not effective (loss)	Comment
	X				Achieving a net gain in koala habitat quality

Source: Author

6.2 Type 2: Conditional Biodiversity Offsets

6.2.1 General Description

Large development projects are highly dependent on lending financial institutions and the conditions that these provide or require. In the past decade, lenders have increasingly adopted social and environmental standards. Public and private financial institutions such as the World Bank, the International Finance Corporation (IFC), the European Bank for Reconstruction and Development, and the European Investment Bank (EIB) are making biodiversity offsetting part of their standards and practices (see Table 6.7).

The most important standards are the Equator Principles (Equator Principles 2011, 2013), the World Bank Safeguard Policies (The World Bank 2012) and International Finance Corporation Performance Standards (IFC 2012).

The Equator Principles (EPs) is a global intersectoral risk management framework, which has been adopted by more than 80 financial institutions, covering over 70% of international Project Finance debt in emerging markets (Equator Principles 2013). It is used for determining, assessing, and managing environmental and social risks of development projects (with capital costs exceeding ten million US Dollars) and ensure minimum standards are met. This includes adherence to the mitigation hierarchy:

> [N]egative impacts on project-affected ecosystems, communities, and the climate should be avoided where possible. If these impacts are unavoidable they should be minimised, mitigated, and/or offset. (Equator Principles 2013)

The International Finance Corporation exists since 1956 as the private sector arm of the World Bank. Following its (environmental and social) Safeguard Policies (1998), IFC committed to Performance Standards in 2006, which have been updated in 2012 (IFC 2012, 2015). IFC's Environmental and Social Performance Standards define IFC clients' responsibilities for managing their environmental and social risks throughout the life of investment by IFC. The eight performance standards, accompanied by Guidance and Interpretation Notes, form the operational basis for the Equator Principles.

Since the 2012 revision of the IFC Performance Standards biodiversity offsets are for the first time explicitly mentioned as part of the mitigation hierarchy in PS 6 on "Biodiversity Conservation and Sustainable Management of Living Natural Resources" (IFC 2012, 2015). Core provisions are set out in paragraphs 10 and 19 (see Fig. 6.1).

IFC PS 6 distinguishes and sets clear provisions for impacts on modified habitats, natural habitats, critical habitats, and protected areas (in order of increasing gravity).

Table 6.7 Examples of offset requirements of some development banks

Institution	Offset requirements
International Finance Corporation (IFC)	The revised Performance Standard 6 (PS6) on Biodiversity Conservation and Sustainable Management of Living Natural Resources states that clients "should seek to avoid impacts on biodiversity and ecosystem services. When avoidance of impacts is not possible, measures to minimize impacts and restore biodiversity and ecosystem services should be implemented." In addition, it states that: "For the protection and conservation of biodiversity, the mitigation hierarchy includes biodiversity offsets, which may be considered only after appropriate avoidance, minimization, and restoration measures have been applied. A biodiversity offset should be designed and implemented to achieve measurable conservation outcomes that can reasonably be expected to result in no net loss and preferably a net gain of biodiversity; however, a net gain is required in critical habitats."
Asian Development Bank (ADB)	The ADB 2009 Safeguard Policy Document prioritizes avoidance, minimization, and mitigation, but acknowledges the use of offsets as a last resort (ADB 2009, paragraph 24).
European Bank for Reconstruction and Development (EBRD)	The EBRD 2008 environmental and social policy explicitly recognizes the mitigation hierarchy, with one of the stated objectives of Performance Requirement 6 on Biodiversity Conservation and Sustainable Management of Living Natural Resources being "to avoid, minimize and mitigate impacts on biodiversity and offset significant residual impacts, where appropriate, with the aim of achieving no net loss or a net gain of biodiversity."
European Investment Bank (EIB)	EIB follows the principles of the EU Nature Conservation Policy, including the Birds and Habitats Directives, and Natura 2000. They also support the use of the mitigation hierarchy. They state, "promoter may propose biodiversity offsets, where appropriate" (EIB 2009, paragraph 70).
Inter-American Development Bank (IDB)	The IDB Environmental and Safeguards Compliance Policy encourages the use of the mitigation hierarchy, including offsets (IDB 2006, section B).
Chinese Development Bank (CDB)	The CDB has not yet committed to the mitigation hierarchy but does state that "banks shall effectively identify, assess, monitor, control, and mitigate environmental and social risks."
Equator Principles Financial Institutions15	Equator Principle 3 states that: "For projects located in non-OECD countries, and those located in OECD countries not designated as High-Income, as defined by the World Bank Development Indicators Database, the Assessment will refer to the then applicable IFC Performance Standards [i.e., PS6] and the then applicable Industry Specific EHS Guidelines."

Source: ICMM IUCN (2012)

IFC Performance Standard 6: Biodiversity Conservation and Sustainable Management of Living Natural Resources

10. For the protection and conservation of biodiversity, the mitigation hierarchy includes **biodiversity offsets**, which may be considered only after appropriate avoidance, minimization, and restoration measures have been applied. A biodiversity offset should be designed and implemented to achieve **measurable conservation outcomes** that can reasonably be expected to result in no net loss and preferably a net gain of biodiversity; however, a net gain is required in critical habitats. The design of a biodiversity offset must adhere to the **"like-for-like or better" principle** and must be carried out in alignment with best available information and current practices. When a client is considering the development of an offset as part of the mitigation strategy, external experts with knowledge in offset design and implementation must be involved.

> ***Biodiversity offsets*** *are measurable conservation outcomes resulting from actions designed to compensate for significant residual adverse biodiversity impacts arising from project development and persisting after appropriate avoidance, minimization and restoration measures have been taken.*
>
> ***Measurable conservation outcomes*** *for biodiversity must be demonstrated in situ (on-the-ground) and on an appropriate geographic scale (e.g., local, landscape-level, national, regional).*
>
> *The* ***principle of "like-for-like or better"*** *indicates that biodiversity offsets must be designed to conserve the same biodiversity values that are being impacted by the project (an "in-kind" offset). In certain situations, however, areas of biodiversity to be impacted by the project may be neither a national nor a local priority, and there may be other areas of biodiversity with like values that are a higher priority for conservation and sustainable use and under imminent threat or need of protection or effective management. In these situations, it may be appropriate to consider an "out-of-kind" offset that involves "trading up" (i.e., where the offset targets biodiversity of higher priority than that affected by the project) that will, for critical habitats, meet the requirements of paragraph 17 of this Performance Standard.*

19. In instances where biodiversity offsets are proposed as part of the mitigation strategy, the client must demonstrate through an assessment that the project's significant residual impacts on biodiversity will be adequately mitigated [...].

Fig. 6.1 Extract from the IFC Performance Standard 6, specifying the application of biodiversity offsets Source: IFC (2012)

In areas of a **modified habitat** that include significant biodiversity value, impacts should be minimized and appropriate mitigation measures be implemented (paragraphs 11 and 12).

In areas of **natural habitat**, mitigation measures are targeted to achieve **no net loss** of biodiversity, including a number of appropriate actions as specified in paragraph 15.

In case of likely impacts on **critical habitat**, a number of preconditions or requirements need to be fulfilled before a development project can be pursued. In this case, mitigation measures are targeted to achieve **net gain** of biodiversity values (paragraphs 16–18).

If a development project likely affects a **legally protected area or an internationally recognized area**, in addition to the aforementioned provisions, a number of additional requirements need to be met as outlined in paragraph 20.

In October 2015, the Banking Commission of the United Nations Environment Programme Finance Initiative (UNEP FI), which includes over 100 banks from across the globe, has issued a "Positive Impact Manifesto," calling for a new financial paradigm:

> Positive Impact Finance is that which verifiably produces a positive impact on the economy, society or the environment once any potential negative impacts have been duly identified and mitigated. (UNEP Finance Initiative 2016)

The Manifesto includes a Roadmap to Positive Impact Finance by 2017, aiming particularly to develop a set of Positive Impact Principles.

This recent development is in line with the aforementioned initiatives and standards and underlines the dominating role of lenders as drivers for offset requirements in large development projects worldwide that hinge on financial loans.

6.2.2 Case Study for Conditional Biodiversity Offsets: Nam Theun 2 Hydropower Project in Laos[2] (Tables 6.8, 6.9, 6.10, 6.11, 6.12, and 6.13)

[2]Information for the case study was derived from the following sources: BBOP (2009b), Serusclat (2004), European Investment Bank (2015), Kormos and Kormos (2011), The World Bank (2010), IAG et al (2005).

Table 6.8 General description for the case study for conditional biodiversity offsets: Nam Theun 2 Hydropower Project in Laos

General description	
Location	The project is located in the mountainous center of Laos, on the Nakai Plateau through which the Nam Theun River flows.
Development activity	Hydropower project in the scope of a 3000-MW electricity export program between Lao PDR and Thailand. Building of a 48-m high and 325-m long dam and power station, creating a 45,000-ha water reservoir.
Environmental baseline and impact	The Nakai plateau is part of the Northern Annamite Rain Forest Ecoregion in southern Laos. It harbors a high biodiversity, including 38 mammal, 17 bird, and 10 reptile species, listed on the IUCN Red List of Threatened Animals (e.g., tigers, Asian elephants, and clouded leopards). However, large parts of the area were in a degraded condition: poor quality agricultural land and degraded forest, with some fertile zones, cultivated by the local population; furthermore some permanent and some seasonal wetlands and areas of untouched rainforest. The project dams the Nam Theun river, resulting in three major impact areas: 1. Nakai Plateau: Inundation of mixed vegetation, including substantial areas of seminatural forest habitat, with loss of animal habitats and biodiversity, in particular affecting the habitats of elephants on and around the plateau. 2. Downstream of the power station and Xe Bang Fai river: Riverbank erosion and changes in water quality (biomass, solar radiation in shallow water, mixing of the water column, aquatic weeds). 3. Downstream of Nakai dam: Disturbance of the natural riverbed and cutting off most of the flow of the Nam Theun river will impact aquatic plants, fish and fisheries. Summarizing, the Nam Theun 2 project has considerable and to some extent cumulative effects on the physical, biological, and human conditions. The direct impacts include land clearance for construction works, reservoir inundation and general degradation, and ecosystem disturbance. Further indirect impacts will arise from the migration of workers to the area and infrastructure development facilitating access to the area.
Impact size	Given numbers vary from 45,000 ha (inundated area) to 98,020 ha directly and 32,568 ha indirectly impacted habitats.
Offset size	393,618 ha (Nakai Nam Theun National Biodiversity Conservation Area and two associated corridors).
Goal of the offset	Mitigation of environmental impacts whenever technically feasible, or offset (goal: protection of the watershed), contribute to the conservation of the largest protected area in Laos and one of the remaining regions of outstanding biodiversity in Southeast Asia, the Nakai Nam Theun National Biodiversity Conservation Area (NBCA) and the goal of achieving world heritage status for the NBCA.
Loss–gain calculation	A Natural Habitats Accounting approach was applied to determine and asses the significance of project impacts on biodiversity and to evaluate the compensatory conservation area in relation to these impacts.

(continued)

Table 6.8 (continued)

General description	
	This approach quantified the hectares, quality and ecosystem function of each habitat type directly impacted by the project, and evaluated the significance of habitat in terms of either its national or subnational conservation value. The area of each habitat type that would be converted, degraded, or lost as a result of the development was then compared with the area of each habitat type in the existing Nakai Nam Theun Biodiversity Conservation Area (NBCA), to determine the extent to which habitats "lost" would be represented within the NBCA.
Type and scope of offset measures	Mitigation measures include sound catchment management, riverbank protection as well as operational restrictions during flood events and a regulating dam and pond downstream of the power station. Furthermore, to improve water quality a number of measures are applied: removal of biomass and drawing riparian releases from the warmer more oxygenated upper layers of the water column and aerating it through a cone valve, a weir and further aerating devices. The loss of the inundated area is offset by the protection of an ecologically similar area, the Nakai Nam Theun National Biodiversity Conservation Area (NBCA)—a globally ranked protected area (which is more extensive, more diverse, and more pristine → trade up). The NBCA is the largest single protected area in Laos, located in the Annamite Mountains, a center of high biodiversity and species endemism.
Involved parties	4. Government of Laos, Mekong River Commission Secretariat 5. Developer "Nam Theun 2 Power Company" (NTPC): A joint venture of Electricité de France, the Electricity Generating Public Co Ltd of Thailand, Government of Laos, and the Italian-Thai Development Public Co Ltd. 6. Lenders: World Bank, European Investment Bank, Asian Development Bank, Nordic Investment Bank, bilateral agencies and syndicates of international, and Thai banks 7. International Advisory Group to the World Bank 8. Consultants
Timeframe	Environmental and social studies carried out since 1991, consultation with affected population since 1996, construction began in 2005, operation since 2010. The management of the offset will be funded over the 30 years concession period (with one million dollars annually).

Source: Author

Table 6.9 Threshold criteria for the case study for conditional biodiversity offsets: Nam Theun 2 Hydropower Project in Laos

Threshold criteria			
Intentionality	Yes	No	Comment
	X		–
Additionality	Yes	No	Comment
			Additionality can be questioned, as the developer ensures protection of the area over the 30-year concession period because it covers around 95% of the catchment for the Nam Theun 2 Hydropower Project.

Source: Author

Table 6.10 Descriptive criteria (context) for the case study for conditional biodiversity offsets: Nam Theun 2 Hydropower Project in Laos

Descriptive criteria: context	
Domestic legislation in the home country	No legal offsetting provisions are in place. Nevertheless, an Environmental Impact Assessment was carried out in accordance with the Laos Environment Protection law. The project responded to an observed (or expected?) lack of capacity and/or resources for the effective long-term management of protected areas.
Competitive environment	The project makes a significant contribution to the development of Laos: it provides power to the Lao grid and is part of a 3,000-MW electricity export program between Laos and Thailand (with assured revenues through two 25-year fixed-price power purchase agreements. With an estimated 1.9 billion US dollars it will be one of the most significant sources of foreign income.
Cultural and social context	Laos is one of Asia's poorest and least developed countries, which has limited strategic options to develop revenues for poverty reduction. It consists of diverse indigenous groups, e.g., the highly vulnerable Vietic-speaking people. These people rely on the use of natural resources, e.g., traditional forest product-gathering on the plateau. They cannot be expected to give up these traditional practices unless alternatives are available and working. Thus, illegal hunting, wildlife trade, and unsustainable use of non-timber forest products, etc. threaten the natural environment. Furthermore, an urgent need for environmental capacity building in environmental agencies can be observed. Based on estimates, around 75,000 people are affected by the project. On this background, the developer has accepted that Nam Theun 2 is a multipurpose project that had to react to emerging requirements and changing priorities, in particular, the greater emphasis on poverty alleviation and outbalancing conservation and development to ensure livelihood conditions.

Source: Author

Table 6.11 Core criteria for the case study for conditional biodiversity offsets: Nam Theun 2 Hydropower Project in Laos

Core criteria			
motivation and ultimate goal / influence and initiators	**Pressure (compliance)**	**Incentive (cost-benefit)**	***Altruism / responsibility***
Government	Case 1 (regulatory)	Case 3a (enabled: government)	
Financial institute	**Case 2 (conditional)**		
Sector	Case 4 (sectoral)		
Corporation	Case 5 (corporate)		
Local community	Case 6a (local: reputational risk)	Case 6b (local: license to operate)	
NGOs		Case 3b (enabled: new global norms)	
Without major influence			*Case 7 (altruism)*
Comments	The 1.45 billion dollar project (funded mainly by World Bank and Asian Development Bank) was promoted by the World Bank as a model of social and environmental food practice. The lending organisations conducted a due diligence process, requiring extensive and stringent project and structural conditionality to secure the implementation of important mitigation and compensation measures. The biodiversity offset (i.e. the conservation of the NBCA) to protect the watershed above the reservoir was an important condition for project approval. The project conforms broadly with the requirement of the World Bank's safeguard policies (in particular OP 4.04 on natural habitats) and in some aspects exceeds them. The project supports the European Investment Bank's environmental policy in terms of tackling climate change and promoting the sustainable use of renewable natural resources.		

	No influence (willful voluntary action)	Marginal influence (coercive offer)	Significant influence (vis compulsiva)	Strong influence (vis absoluta)	Comment
Choice or eligibility			X		NTPC is contractually responsible for implementing and covering all costs of mitigation and compensation measures during construction and operation.

	Yes	No	Comment
Recompense or benefit	x		Conserving the watershed also helps to prolong the life of the dam's operation and to sustain water quality.

Source: Author

Table 6.12 Performance criteria for the Case study for conditional biodiversity offsets: Nam Theun 2 Hydropower Project in Laos

Performance criteria			
Increased flexibility and cost efficiency	Yes	No	Comment
			n/a

Source: Author

Table 6.13 Quality criteria (Outcome and charitableness) for the case study for conditional biodiversity offsets: Nam Theun 2 Hydropower Project in Laos

Quality criteria: Outcome and charitableness					
Counterfactual	Static (neutral)	Dynamic		Comment	
		Positive	Negative		
			X	The impact–offset analysis was based on an analysis of the expected degradation and conversion expected in the NBCA, which is threatened by regional demand for wildlife products and timber resources (leading to poaching and illegal logging). A newly established agency, the Watershed Management, and Protection authority is responsible for reducing these threats.	
Outcome and charitableness relative to the counterfactual	Positive balance (Net Gain)	Neutral balance (No Net Loss	Minimum balance (reduced loss)	Not effective (loss)	Comment
	X				The lenders expect the project to generate a net environmental and social benefit both at the local and global levels: the loss of the inundated area is offset by the protection of an ecologically similar area, which is more extensive, more biologically diverse and more pristine. The project strengthens the use of renewable energy and the reduction of greenhouse gas emissions.

Source: Author

6.3 Type 3: Enabled Biodiversity Offsets

6.3.1 General Description

As has been noted, biodiversity offsets are often required by some form of compliance regime (either imposed by government, donors, or corporate commitments), however, they can also be triggered by incentives or enabling schemes, e.g., through technical guidance, capacity building, and implementation support. Two tendencies can be noted in terms of the originator of this support:

1. Governments (case 3a: enabled biodiversity offsets—by government).
2. Nature conservation organizations, NGOs, and other civil society groups (case 3b: enabled biodiversity offsets—new global norms).

Governments may not only enforce offsets, but they can also provide an offset enabling framework. The case of the United Kingdom is currently the probably most well-known example of a government enabled or promoted offset scheme, however, with the ultimate intention of introducing a regulatory offset scheme. In fact, this offset category is usually of temporary nature as a precursor to developing a government framework and binding legislation (pilot phase).

In response to the UK government's commitments at the COP10 to the Convention on Biological Diversity in Nagoya and the EU biodiversity commitments, the England Biodiversity Strategy was published in 2011 (Defra 2011). This involved as one of four policy actions the establishment of a new voluntary approach to biodiversity offsetting outside of protected areas.

Consequently, an exploration phase was underway in six county pilots over the period 2012–2014 (see Appendix 1), accompanied by detailed guidance provided by the government (for different stakeholders, including guiding principles and a technical paper on the metrics). The Environment Bank, a private company that acts as an independent broker, facilitates the implementation of several offsets. In autumn 2013, the Department for Environment, Food and Rural Affairs (Defra) launched a public consultation on its "Offsetting" Green Paper, which sets out proposals for biodiversity offsetting and how it might be introduced in England (Collingwood Environmental Planning Limited and Tucker et al. 2013; Defra 2013).

French biodiversity policy has been improved and strengthened since 1976, when the "Law on Nature Protection" provided the first mention of the mitigation hierarchy and stipulated that all development projects required an EIA (Darbi 2015a). In particular, in recent years the French Ministry for the Environment has multiplied its efforts to improve impact mitigation and foster biodiversity offsets, namely by establishing a doctrine on the mitigation hierarchy as well as guidelines for the implementation (Ministère de l'Écologie, du Développement durable, des Transports et du Logement 2012; Commissariat Général Au Développement Durable. Direction de l'eau et de la Biodiversité 2013; Darbi 2014). Furthermore, local initiatives and projects exploring options for habitat banking are being fostered by the government.

Similar to the situation in France, the Spanish government has aimed to halt biodiversity loss from projects and works that impact the environment and consequently improve impact mitigation. One of the measures of the "Strategic Plan for Natural Heritage and Biodiversity" is to improve environmental compensation for impacts on protected habitats or species. This involved exploration of habitat banking that has been included in the recent amendment of the law on environmental impact assessment (Mercados de Medio Ambiente 2013, 2014).

In contrast to the situation in the United Kingdom, which so far has remained voluntary, government action enabling offsetting in France and Spain has been translated to legal action (thus being transformed to type 1: regulatory biodiversity offsets, Darbi 2015a and Mercados de Medio Ambiente 2013, 2014).

At the transnational level, the attempts of the European Commission to explore offsetting as part of its planned No Net Loss Initiative can be interpreted as offset enabling. For this purpose, the EU commissioned a study on policy options to reach the NNL goal and set up a working group (Tucker et al. 2013; Darbi et al. 2016). Finally, the initially favored option of (mandatory) offsetting was abandoned after public opposition and the largely negative feedback received in the course of the related public consultation. The expected result is an offset enabling scheme that includes capacity building and exchange of knowledge and best practice on offsets among the EU member states (Darbi et al. 2016).

New global standards for offsets (case 3b) are increasingly driving many offset projects. Doswald et al. (2012) claim that offsets are usually carried out by companies with the support of conservation organizations, consulting firms or through partnerships with various organizations. The Business and Biodiversity Offsets Programme (BBOP) is the most extensive and most prominent example in this field, but many others now exist, e.g., The Nature Conservancy's "Development by Design" program or the partnerships of BirdLife International, Fauna and Flora International, and Conservational International with business (ICMM 2010; BBOP 2009a; BBOP n.d.).

More recently, the International Union for Conservation of Nature (IUCN) adopted a resolution at the World Conservation Congress 2012 to develop an IUCN general policy on biodiversity offsets (Darbi 2015b). To this end, a working group was established and a technical study paper was prepared in the first phase of the process (Pilgrim and Ekstrom 2014; IUCN 2014). In 2015, IUCN has launched its Draft Biodiversity Offsets Policy for public consultation (Darbi 2015c). This draft was presented and reviewed during the World Conservation Congress in autumn 2016 in Hawaii. As a result, IUCN Members have agreed on a policy on biodiversity offsets, emphasizing that priority must be given to avoid biodiversity loss, and leaving offsets must a measure of last resort.

BBOP has discussed and developed standards for offsets as well as extensive implementation guidance in a collaborative process (BBOP 2010). BBOP has also fostered a small number of pilot projects to explore biodiversity offsets in different settings and to implement the BBOP standards and methodology (BBOP 2016). Currently, it actively maintains two pilots: the Ambatovy Project, a nickel mine in Madagascar and the Strongman coal mine in New Zealand, operated by Solid Energy (von Hase 2014; von Hase et al. 2014).

Furthermore, it has initiated a "Community of Practice" to share practical experiences, skills, and lessons learned among a broad network of professionals. It hosts conferences, workshops, discussion groups, and webinars, in addition to planning and running training courses for capacity building of companies, banks, governments, representatives from civil society, and their respective consultants (BBOP 2016). One example for this is the training program "Biodiversity for Banks" (B4B) which was launched in 2011 together with the Equator Principles Association and the World Wildlife Fund to help banks incorporate the value of nature into their lending decisions (Equator Principles 2011).

This highlights the overarching role of BBOP not only as a promoter of standards and best practices for biodiversity offsets, but also as an influencer of financial institutions, governments, companies, and NGOs, thus acting as a catalyzer for different offset schemes (cf. type 1 regulatory biodiversity offsets, type 2 conditional biodiversity offsets and type 5 corporate biodiversity offsets). As can be seen from the screening of biodiversity offset cases (see Appendix 1) many projects exist where BBOP—even if not the driving force—has provided additional motivation and guidance (either directly or indirectly) to implement offsets.

6.3.2 Case Study for Enabled Biodiversity Offsets: Blakely Harbour Residential and Public Road Development on Bainbridge Island in the United States (BBOP and City of Bainbridge Biodiversity Offset Pilot)[3] (Tables 6.14, 6.15, 6.16, 6.17, 6.18, and 6.19)

Table 6.14 General description for the case study for enabled biodiversity offsets: Blakely Harbour residential and public road development on Bainbridge Island in the United States (BBOP and City of Bainbridge biodiversity offset pilot)

General description	
Location	Southeastern shoreline of Blakely Harbour on Bainbridge Island (in the central Puget Sound Basin), Washington State, United States
Development activity	12 unit single-family residential development and public infrastructure project (relocation of a shoreline road)
Environmental baseline and impact	The island supports 73% forest cover and many areas of high biodiversity, including terrestrial and intertidal habitats. The upland forests surrounding Blakely Harbour provide one of the largest remaining tracts of forest canopy in the area. They support habitat and water quality functions and are highly valued for their aesthetic value and contribution to the area's sense of place.

(continued)

[3]Information based on Burgin (2008), City of Bainbridge Island (2009), ten Kate (2007), Bainbridge Island Land Trust (2015), BBOP (n.d.) and OECD (2013).

Table 6.14 (continued)

General description	
	The coastal ecosystem has been very severely impacted by past developments. Rapid growth and residential development have led to reduced forest cover, altering intertidal habitats, fragmenting wildlife corridors, and allowing invasive species to compromise native vegetation. The current pilot projects will involve approximately 12 ha of direct impact on second growth forest, forested wetland and highly disturbed urban meadow. The **public infrastructure project** involves moving a road inland, away from the shoreline. This would impact marine riparian area and a small (0.4 ha) forested wetland (the latter subject to regulatory wetland mitigation). The **residential development** is located on a hillside along the shoreline. The mixed deciduous and coniferous forest has a medium to high sensitivity. Of the 16 ha, a total of 3.24 ha are cleared for 12 home sites, an access road and stormwater facilities. The impacts of these developments include the clearing of vegetation (forest), additional impervious surfaces, and noise and water quality impacts.
Impact size	12 ha
Offset size	12,4 ha + linear infrastructures (see Type and scope of measures below)
Goal of the offset	The overall goal is to relocate the shoreline road and to restore significant amounts of shoreline habitat, particularly to create shallow nearshore refuges and foraging and migration habitats for salmonids, while reducing the risk of erosion/failure of the road. Salmon recovery and restoration of Puget Sound habitat is a top priority in the region and in Blakely Harbour in particular.
Loss–gain calculation	Quantification of loss and gain uses an accounting approach, which builds on a combination of techniques. This includes the Washington State Wetland Functional Analysis and the Australian Habitat Hectares method as adapted by BBOP. According to this, with the use of benchmark sites a multi-attribute scoring index was created against which losses and gains are measured. This was modified to incorporate data collected from research on the ecology of old-growth forests in the region, including Douglas Fir and Mixed-Conifer Forests in the Pacific Northwest and California.
Type and scope of offset measures	Priority offset areas lie in the intertidal and coastal habitats as well as in the degraded areas of the forest. The offset for the residential project focuses on restoring and protecting forest habitat and the offset for the road relocation involves restoration on the site of the removed road as well as restoration of intertidal habit.

(continued)

Table 6.14 (continued)

General description	
	Offset measures for **shoreline restoration** include: • Decommission 700 m of road and utilities with a new road. • Location beyond the risk of erosion/failure and habitat restoration area. • Removal of 300 m of bulkhead and of intertidal debris. • Restoration of areas of upper intertidal sediment suitable for forage fish spawning. • Restoration of 700 m of riparian forest. • Place suitable pieces of large woody debris to provide habitat complexity • Interpretative signage (raising public awareness and engaging community volunteers). Offset measures for **upland forest restoration** include: • Removal of invasive species. • Replanting of appropriate native conifers. • Preservation in perpetuity of potentially developable property (12.4 ha), either through conservation easements or transfer of development rights.
Involved parties	1. City of Bainbridge Island Planning and Engineering Departments 2. Upper Blakely LLC (the developer for residential development, a family partnership with a long history in the community) 3. Private landowner (on whose property the shoreline road is located) 4. Bainbridge Island Forestry Commission 5. Bainbridge Island Land Trust (responsible for long term conservation success) 6. Washington SeaGrant 7. Washington State Department of Ecology 8. US Army Corps of Engineers
Timeframe	Start of project 2005, development permit 2006, clearing started in 2007, implementation of the proposed offsets for the residential project occurred simultaneously with site development, planting of native conifer species since late 2008.

Source: Author

Table 6.15 Threshold criteria for the case study for enabled biodiversity offsets: Blakely Harbour residential and public road development on Bainbridge Island in the United States (BBOP and City of Bainbridge biodiversity offset pilot)

Threshold criteria			
Intentionality	Yes	No	Comment
	X		–
Additionality	Yes	No	Comment
	X		–

Source: Author

Table 6.16 Descriptive criteria (context) for the case study for enabled biodiversity offsets: Blakely Harbour residential and public road development on Bainbridge Island in the United States (BBOP and City of Bainbridge biodiversity offset pilot)

Descriptive criteria: Context	
Domestic legislation in the home country	Offsets specifically addressing biodiversity loss are not commonly required by local, state or federal regulations in the United States. Regulations at all levels of government require that development projects complete environmental assessments, detailing predicted impacts to habitat and addressing the mitigation hierarchy. Offsets are required for impacts to wetlands and the habitat of State or Federally listed species; however, offsets are rarely required for noncritical habitats and do not always consider indirect impacts to biodiversity. Biodiversity offsets present an opportunity to achieve greater and more specific conservation of habitats such as mature forests or intertidal habitat that are not otherwise protected by existing law. Therefore, in order to encourage development projects to incorporate biodiversity offsets for other habitat types (e.g., native forests, shoreline habitat) in their design process, the City of Bainbridge Island is exploring a variety of policy incentives: • Expedited permit processing timelines • Density bonuses • Flexible infrastructure and zoning requirements The Blakely Harbour Pilot Project offset design is influenced by federal, state, and local regulations and by local regulations protecting critical areas: • NEPA: Requires projects with federal funding to process an EIA and an ALTERNATIVES analysis. • Federal Clean Water Act: Requires permits and mitigation for impacts to waters of the United States, i.e., wetlands and coastal waters. • SEPA: State law requiring all projects to consider environmental impacts and propose mitigation. • State Shoreline Management Act. Regulates impacts within shoreline and riparian zones. • Local Shorelines and Critical Areas Regulations: Requires mitigation for impacts to critical areas such as wetlands. Restricts development on or near steep slopes and shorelines.
Competitive environment	There is an increasing demand for housing development on Bainbridge Island (in particular through the expansion of Seattle).
Cultural and social context	Bainbridge Island is an urban island community west of the Seattle metropolitan area. Due to this situation, the island needs to accommodate a growing population and development, in particular housing.

Source: Author

Table 6.17 Core criteria for the case study for enabled biodiversity offsets: Blakely Harbour residential and public road development on Bainbridge Island in the United States (BBOP and City of Bainbridge biodiversity offset pilot)

Core criteria			
motivation and ultimate goal / **influence and initiators**	**Pressure (compliance)**	**Incentive (cost-benefit)**	***Altruism / responsibility***
Government	Case 1 (regulatory)	**Case 3a (enabled: government)**	
Financial institute	Case 2 (conditional)		
Sector	Case 4 (sectoral)		
Corporation	Case 5 (corporate)		
Local community	Case 6a (local: reputational risk)	Case 6b (local: license to operate)	
NGOs		**Case 3b (enabled: new global norms)**	
Without major influence			*Case 7 (altruism)*
comment	The pilot programme is a cooperation between a private landowner and the City of Bainbridge, supported by the Business and Biodiversity Offsets Programme as part of their offset portfolio. The City of Bainbridge Island has fostered policy development and the establishment of pilot projects to test and demonstrate biodiversity offset design for terrestrial and intertidal offsets in the context of US environmental policy. Two pilots include the Blakely Harbour Project and a small residential development. The Blakely Harbour site includes 700 metres of shoreline road placed on bulkheads. Due to extensive recurrent repair works, the City of Bainbridge Island and the adjacent private landowners are both interested in relocating the road and restoring shoreline habitat. The landowner is in the process of developing a portion of the upland forest on the pilot site. Offsets are proposed to compensate for impacts resulting from clearing and construction for the residential development. The landowner will donate right of way on this property so that the City can relocate an adjacent public shoreline road away from the shoreline. For the landowner, collaborating with the City on the biodiversity offset would facilitate smooth progress with the project development and help secure the long term quality of the local environment (to the benefit and enjoyment of those who will live in the property development). The developer (a local family partnership) has a long-term interest in preserving the natural environment in and around their property for themselves and future generations. For the City, the offset would result in more varied and higher priority conservation than the developer would otherwise undertake. Restoration of the intertidal zone is of high priority throughout Puget Sound and this project is an opportunity to demonstrate offsets in the context of near-shore restoration. This is in accord with the biodiversity preservation policies that the City of Bainbridge Island developed in 2007, including: • Creating a programme with effective mechanisms to offset impacts on biodiversity • Considering establishing a habitat bank • Providing incentives for developers to offset unavoidable impacts to biodiversity in high-quality natural habitats		

(continued)

Table 6.17 (continued)

Choice or eligibility	No influence (willful voluntary action)	Marginal influence (coercive offer)	Significant influence (vis compulsiva)	Strong influence (vis absoluta)	Comment
		X			
Recompense or benefit	Yes	No	Comment		
	X		The private landowners on whose property the shoreline road is located are asking no incentives for conserving a large portion of their land, but would welcome a part of the shoreline being preserved as a public park.		

Source: Author

Table 6.18 Performance criteria for the case study for enabled biodiversity offsets: Blakely Harbour residential and public road development on Bainbridge Island in the United States (BBOP and City of Bainbridge biodiversity offset pilot)

Performance criteria			
Increased flexibility and cost efficiency	Yes	No	Comment
			–

Source: Author

Table 6.19 Quality criteria (Outcome and charitableness) for the case study for enabled biodiversity offsets: Blakely Harbour residential and public road development on Bainbridge Island in the United States (BBOP and City of Bainbridge biodiversity offset pilot)

Quality criteria: Outcome and charitableness					
Counterfactual	Static (neutral)	Dynamic		Comment	
		Positive	Negative		
	X			n/a	
Outcome and charitableness relative to the counterfactual	Positive balance (Net Gain)	Neutral balance (No Net Loss	Minimum balance (reduced loss)	Not effective (loss)	Comment
		X			No net loss of biodiversity is addressed in the offset options. This shall be reached by combining the restoration of forest habitat on the project site with additional offsets of shoreline habitat.

Source: Author

6.4 Type 4: Sectoral Biodiversity Offsets

6.4.1 General Description

Biodiversity offsets are not only globally diverse, but also implemented in a variety of different sectors. As can be seen from the screening of biodiversity offset cases (see Appendix 1), examples exist for the following sectors:

- Mining
- Oil and gas
- Renewable energy: Hydropower and wind power
- Road and railway infrastructure
- Real estate and commercial development
- Agriculture and forestry
- Tourism
- Water supply

Although the screening of biodiversity offset cases is clearly not representative in quantitative terms, it can be observed that these cases are not evenly spread across sectors. Although several examples are observed in the infrastructure, energy, and housing sectors, the extractive industries (mining, oil and gas) are fairly dominant globally. This is substantially due to their great scale, the magnitude of associated impacts, and the corresponding public visibility or controversy. Evidence on no net loss-type commitments as shown in Fig. 6.2 supports these findings.

> For the extractive industries sector, demonstration of good biodiversity performance and risk management is important in gaining access to land and resources, obtaining a "licence to operate", maintaining regulatory goodwill and demonstrating responsible company performance to investors and other stakeholders. (F&C Asset Management 2004, cited in ICMM IUCN 2012).

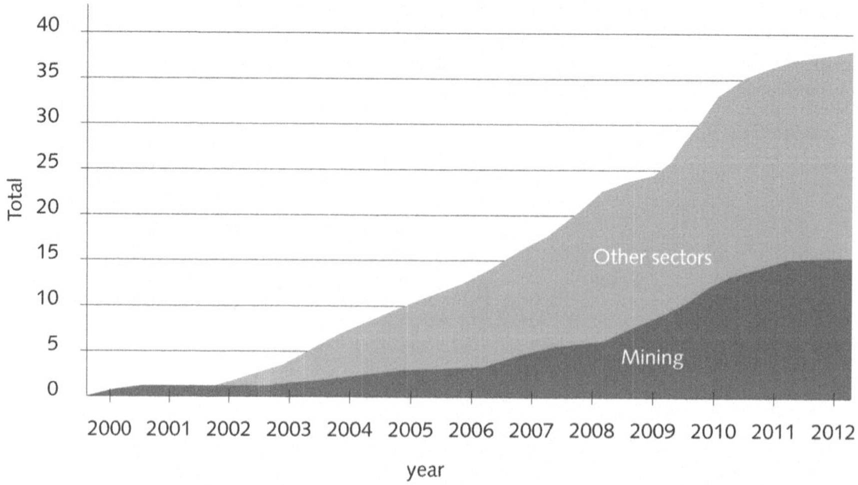

Fig. 6.2 The rise in No Net Loss-type commitments in the private sector 2000–2012. Source: ICMM IUCN (2012)

Thus, it is certainly no coincidence that this sector is particularly active in the field of biodiversity offsets, most notably through the International Council on Mining and Metals (ICMM), the International Petroleum Industry Environmental Conservation Association (IPIECA), the Cross-Sector Biodiversity Initiative (CSBI) and smaller associations such as Canada's Oil Sands Innovation Alliance (COSIA).

Driven by higher expectations from stakeholders and investors with regard to responsible mining practice, the mining industry has acknowledged the challenge of addressing its social, economic, and environmental footprint and started a dialogue with environmental and social NGOs in the mid-1990s (ICMM 2016). As a result, the International Council on Mining and Metals was formed in 2001 "to improve sustainable development performance in the mining and metals industry" (ICMM 2016). Its members, currently 23 mining and metals companies as well as 34 national and regional mining associations and global commodity associations, are committed to contributing to the conservation of biodiversity (ICMM 2016).

In a joint workshop with IUCN held in 2003, ICMM agreed to explore the use of biodiversity offsets (ICMM 2010). Since then they have published two sets of case studies on integrating mining and biodiversity conservation (2004 and 2010). Furthermore, they issued a number of guidance documents, including a proposition paper (ICMM 2005a) and a briefing paper for the mining industry (ICMM 2005b) as well as "Good Practice Guidance for Mining and Biodiversity" (ICMM 2006).

ICMM members recognize the environmental values of protected areas and have pledged not to mine or explore in World Heritage Sites. Furthermore, they are committed to high standards of on-site mitigation. As of 2014, almost half of ICMM's members have a policy commitment on offsets. However, the actual involvement in offsets is likely to be higher (Globalbalance and The Biodiversity Consultancy 2014). Nevertheless, the application of offsets is largely limited to one or two sites per company.

> ICMM member companies recognize that mining activities will inevitably have an impact on biodiversity at a site level, even when high standards of on-site mitigation are employed. Biodiversity offsets potentially offer a mechanism of compensating for these impacts that may offer benefits to both biodiversity and business. Such offsets can channel resources into priority conservation activities, ideally near the site of impact, providing a higher conservation benefit and being more cost effective than on-site mitigation alone. (ICMM 2005b)

In 2012, ICMM and IUCN jointly commissioned an independent report on biodiversity offsets (ICMM IUCN 2012), which defines practical steps for implementing offsets, ensuring conservation outcomes and reducing business risks.

Similar to mining, the oil and gas industry operates in some of the world's most sensitive environments, e.g., deserts, wetlands, and tropical rainforests. Therefore, members of IPIECA, the global oil and gas industry association for environmental and social issues, and the International Association of Oil and Gas Producers (OGP) recognize the importance of operating sustainably in these sensitive environments.

Consequently, ICMM and IPIECA, together with the Equator Principles Association have united "to promote the use of scientifically valid, innovative and practical applications of the mitigation hierarchy to manage potential project related impacts on biodiversity and ecosystem services" (Equator Principles 2013). In 2013, they jointly launched a Cross Sector Biodiversity Initiative (CSBI), followed by a

Timeline Tool in 2014 and a cross-sector guide for implementing the mitigation hierarchy in 2015 (The Biodiversity Consultancy 2015).

The CSBI aims to explore and develop practical tools and share good practices for the effective application of the mitigation hierarchy and the International Finance Corporation Performance Standard 6 on biodiversity conservation. The Timeline Tool provides assistance on project planning in the extractive industries. The goal is to align the development and management of biodiversity impact management with financial timelines and milestones.

6.4.2 Case Study for Sectoral Biodiversity Offsets

In the scope of the screening of biodiversity offsets no case could be identified which could serve as an example to illustrate type 4 sectoral biodiversity offsets. An analysis even of the biodiversity case studies compiled by ICMM (2010) revealed that these were not primarily driven (and in several cases not at all) by sectoral requirements or guidance on offsetting, but could be characterized as corporate, enabled, conditional, local, or altruistic biodiversity offsets instead.

Therefore, as no appropriate evidence has been identified, no case study is presented for this offset type and it is not possible in the scope of this study to deliver additional insights specific to this type.

6.5 Type 5: Corporate Biodiversity Offsets

6.5.1 General Description

Corporate biodiversity offsets, together with enabled biodiversity offsets, are the most well-documented type of offsets. They represent the most direct exertion of the business case for biodiversity offsets. Their high visibility is clearly linked to a business' or corporation's motivations to implement offsets, in particular as a tool for reputation and risk management (cf. Sect. 4.2). This is particularly important for "global players" who operate in very different contexts worldwide. Examples include large mining corporations like Rio Tinto, Newmont, or De Beers.

With respect to offsets, legal and policy requirements as well as social and cultural conditions, e.g., environmental awareness and public scrutiny of the local population, may vary greatly within a single corporation with operations in a variety of locations. This hinders the transparency and transferability among different operations of a corporation and results in less standardization and higher expenditures. It also contains a risk to maintaining good public relations. Consumers no longer focus only on the local operations of a corporation that directly affect them, but also consider supply chain issues as well as the worldwide activities of corporations. Global players are thus increasingly under scrutiny by the public and campaigns by NGOs can greatly harm business activities.

For these reasons, in particular large corporations in highly criticized sectors with enormous impacts, e.g., mining, opt for a corporate strategy incorporating biodiversity offsets to showcase that they are addressing the environmental impacts of their activities. While environmental reporting has become a standard, biodiversity strategies and goals, as well as commitments towards no net loss or net gain of biodiversity, are becoming increasingly common:

> Company Biodiversity Commitment—company-wide commitments to "no net loss" and "net positive impact" to biodiversity are beginning to appear as are similar commitments for example around "zero harm", or "contributing to biodiversity protection in areas of operation". (ICMM 2010)

Figure 6.3 lists some corporate commitments that include the use of biodiversity offsets.

De Beers:

- De Beers Group's Approach towards mitigation, i.e. "anticipate and prevent", rather than "assess and repair"
- Application of the mitigation hierarchy, i.e. enhance, avoid, minimize, restore, compensate, offset

Rio Tinto:

- Rio Tinto Biodiversity Strategy targets "Net Positive Impact on Biodiversity"
- Piloting the biodiversity strategy on the ground, in particular the application of biodiversity offsets to compensate for unavoidable impacts as part of the mitigation hierarchy
- Pilot sites: Simandou in guinea, QMM in Madagascar, Rossing in Namibia, Palabora in South Africa and RTCA in Queensland, Australia

Newmont:

- Aligned with best international practice (including IFC PS 6) the goal is no net loss of biodiversity and net gain for areas of critical habitat
- Sites with key biodiversity values must have Biodiversity Action Plans that follow the Mitigation Hierarchy, i.e. avoid, minimize, restore/rehabilitate and offset
- Active offset programs: Akyem in Ghana, Boddington in Australia, and two offset programs in Nevada, US

Barrick Gold:

- Barrick Gold's Biodiversity Standard
- PS6-compliant biodiversity offset program for the Pueblo Viejo gold mine expansion in the Dominican Republic

Other examples:

- Shell
- Solid Energy New Zealand:
- Gold Reserves Inc.

Fig. 6.3 Examples of corporate commitments related to biodiversity offsets. Source: Author, building on ICMM (2010), BBOP (2009b)

However, the level of detail of these provisions varies greatly. They use common standards to a limited extent, primarily when corporations partner with NGOs and governments, e.g., with BBOP, BirdLife International or IUCN; that is, when corporate biodiversity offsets merge with enabled biodiversity offsets.

A particularity of corporate biodiversity offsets is that they are often related to large-scale development projects. These are not only highly visible and debated in public, but as a matter of their dimension have a huge impact on the corporation or joint venture which is planning or operating them. The problem for business is that success of the project hinges on a multitude of conditions, and even comparatively small influencing factors can hinder development or increase costs.

Thus, as has been noted above, corporate biodiversity offsets arise in a setting of business risks, which represents an ideal business case.

6.5.2 Case Study for Corporate Biodiversity Offsets: QIT Mining Madagascar QMM (Ilmenite Mining)[4] (Tables 6.20, 6.21, 6.22, 6.23, 6.24, and 6.25)

Table 6.20 General description for the case study for corporate biodiversity offsets: QIT Mining Madagascar QMM (ilmenite mining)

General description	
Location	Southeastern Madagascar, three separate deposits near the city of Tolagnaro, commonly known as Fort Dauphin
Development activity	The QMM project is a mineral sand mining operation extracting ilmenite and zirsill (zircon mixed with sillimanite) from coastal plain sand.
Environmental baseline and impact	Mining takes place in one of Madagascar's ecologically diverse areas, including a significant part of the littoral forest with many restricted-range and IUCN red list species. Mining will result in loss of 1665 ha of littoral forest (3.5% of the remaining 47,900 ha) through dredging, i.e., vegetation clearing and soil removal.

(continued)

[4] Information for the case study was derived from the following sources: Darbi et al. (2010), Temple et al. (2010), Rio Tinto and IUCN (2012), QMM (2007a–c), QMM (n.d.), QMM and Madagascar's National Office for the Environment (2001), Rio Tinto (2006) and Long (2011).

Table 6.20 (continued)

General description	
Impact size	1665 ha The total direct footprint is anticipated to be ca. 8000 ha over the mine's lifetime; however, the mine itself occupies ca. 50 ha at any one time, covering ca. 100 ha per year.
Offset size	6000 ha.
Goal of the offset	Ensure environmental (and social) net positive impact of the mining operations.
Loss–gain calculation	All forest and littoral forest loss are measured in Quality Hectares (QH), which express biodiversity values in terms of quantity and quality. 90 High priority terrestrial species (54 plants, 26 invertebrates, 10 vertebrates), measured using a novel metric developed for this analysis but conceptually related to Quality Hectares: Units of Global Distribution (UD); 1 UD is equivalent to 1% of a species' global population or 1% of its global distribution when population data are unavailable.
Type and scope of offset measures	The 230-hectare Mandena conservation zone has been established. It shelters 414 plant species, 21 species of mammals, 40 species of reptiles, and 44 species of birds.
Involved parties	1. Rio Tinto Group (the developer) 2. The Malagasy government (central government, local officials) 3. Lending organizations (World Bank and USAID) 4. Local businesses 5. NGOs (IUCN) 6. (Consultants)
Timeframe	1986–2005 operation and study phase, 2006–2009 construction phase, since 2009 production phase The QMM project is anticipated to span over 50 years, from operation to closure.

Source: Author

Table 6.21 Threshold criteria for the case study for corporate biodiversity offsets: QIT Mining Madagascar QMM (ilmenite mining)

Threshold criteria			
Intentionality	Yes	No	Comment
	X		QMM was declared the first "net positive" pilot project of the Rio Tinto Group (2006).
Additionality	Yes	No	Comment
	X		–

Source: Author

Table 6.22 Descriptive criteria (context) for the case study for corporate biodiversity offsets: QIT Mining Madagascar QMM (ilmenite mining)

Descriptive criteria: Context	
Domestic legislation in the home country	In Madagascar, no legal offsetting requirements are prevalent. However, legal references for biodiversity compensation issues can be found in the Malagasy Environmental Charter (Loi no. 90-033 du 21 décembre 1990, portant Charte de l'Environnement malagasy) and the Decree MECIE (Mise en Compatibilité des Investissements avec l'Environnement), which refer to the obligation to carry out an Environmental Impact Assessment for public and private investment projects. The Decree MECIE is accompanied by general and sectoral EIA guidelines (for tourism, roads, aquaculture, on- and off-shore oil, forests, textile, and mining). In its Madagascar Action Plan launched in 2006, the Malagasy Government has made a commitment to "cherish the environment." Given the significant position of the mining sector, one of the priority actions was identified as "develop[ing] a policy for mining companies and logging companies for biodiversity offsets and other mechanisms and incentives for environmental protection." So far, however, no evidence toward the implementation of this goal could be identified.
Competitive environment	The QMM project was the first and largest single industrial development project ever carried out in Madagascar. The construction phase generated an unprecedented level of economic activity in the region.
Cultural and social context	Madagascar consists of 18 ethnic groups and is one of the poorest countries in the world, with 68% of the population estimated to live below the international poverty line of US$1.25 per day. Due to its impoverished state, the majority of Madagascar's population depends on natural resources for their daily survival. The mining operations are located in a fragile and declining physical, biological and social environment in the remote Fort Dauphin region of southeastern Madagascar. Once the most viable French settlement in the country, it is now one of the least developed urban centers. The area is characterized by low school enrollment rates, limited access to health care, and high rates of unemployment. Physical infrastructure and social well-being continue to decline and deteriorate in one of the poorest areas in an already impoverished country. *"Sustainable development is not a programme independent from QMM, it is rather an approach underpinning the whole company's activities." (Ny Fanja Rakotomalala, President of QMM)*

Source: Author

Table 6.23 Core criteria for the case study for corporate biodiversity offsets: QIT Mining Madagascar QMM (ilmenite mining)

Core criteria

motivation and ultimate goal / influence and initiators	Pressure (compliance)	Incentive (cost-benefit)	*Altruism / responsibility*
Government	Case 1 (regulatory)	Case 3a (enabled: government)	
Financial institute	Case 2 (conditional)		
Sector	Case 4 (sectoral)		
Corporation	**Case 5 (corporate)**		
Local community	Case 6a (local: reputational risk)	Case 6b (local: license to operate)	
NGOs		Case 3b (enabled: new global norms)	
Without major influence			*Case 7 (altruism)*

comment	The mining project was initiated by the Rio Tinto Group, a UK-Australian based mining company in collaboration with the Malagasy government. In 1999 a regional planning process was initiated, involving a wide range of stakeholders. This includes local village leaders, central government, local officials, lending organizations (World Bank and USAID), local businesses and NGOs (in particular IUCN). A Regional Development Committee was put in place to steer the process. Rio Tinto is committed to achieving a Net Positive Impact (NPI) on biodiversity, a strategy launched at the 2004 IUCN World Conservation Congress. As a Rio Tinto company, QMM has made a strategic commitment to sustainable development. It has been chosen as a pilot site to test the tools designed to achieve and quantify a NPI on biodiversity. This case is clearly a compliance regime following Rio Tinto's corporate provisions. However, two aspects influence this position: first, the exploration phase was already ongoing for many years when the corporate commitment has been made and second, Rio Tinto promoted, encouraged and trialled QMM as a pilot for its commitment (thus including elements of an enabled offset).

	No influence (willful voluntary action)	Marginal influence (coercive offer)	Significant influence (vis compulsiva)	Strong influence (vis absoluta)	Comment
Choice or eligibility			X	X	The offset for QMM is clearly mandated by the Rio Tinto commitment towards "Net Positive Impact".

	Yes	No	Comment
Recompense or benefit			-

Source: Author

Table 6.24 Performance criteria for the case study for corporate biodiversity offsets: QIT Mining Madagascar QMM (ilmenite mining)

Performance Criteria			
Increased flexibility and cost efficiency	Yes	No	Comment
			n/a

Source: Author

Table 6.25 Quality criteria (Outcome and charitableness) for the case study for corporate biodiversity offsets: QIT Mining Madagascar QMM (ilmenite mining)

Quality criteria: Outcome and charitableness					
Counterfactual	Static (neutral)	Dynamic		Comment	
		Positive	Negative		
			X	Prior to the arrival of Rio Tinto QMM in the 1990s, the littoral forest was in rapid decline due to unsustainable forest exploitation (e.g., charcoal production and slash-and-burn agriculture) of an impoverished population. Biodiversity losses and gains were measured against a baseline scenario of no mining and an annual deforestation rate of 0.9% per year (the national average deforestation rate for all forest types can be extrapolated from c.1990 to c.2000). This baseline is likely to be conservative; the 1995–2005 annual deforestation rate was 3.89%.	
Outcome and charitableness relative to the counterfactual	Positive balance (Net Gain)	Neutral balance (No Net Loss	Minimum balance (reduced loss)	Not effective (loss)	Comment
	X				The project targets a net positive impact.

Source: Author

6.6 Type 6: Local/Consensual Biodiversity Offsets

6.6.1 General Description

Large-scale development projects that are imposed from the outside will inevitably lead to local conflicts and thus in economic losses for the developer. The concerns or support of local stakeholders can significantly affect the success or efficiency of a project. Therefore, capacity building and participation are common tools to engage local stakeholders. In this context, offsets are used by developers to respond to environmental concerns and to negotiate (social) license to operate.

This type of offsets does not follow a standardized approach, because it depends on the context that surrounds the location of the proposed development. Technical details, e.g., in terms of valuing the environmental baseline, the likely impacts and loss-gain calculations, may be transferred from different schemes, e.g., BBOP's approach or schemes developed under the offset legislation in Germany or Australia. By contrast, the governance of offsets hinges on its context sensitivity and the appropriate consideration of local culture and values.

Local biodiversity offsets can have two tendencies (or a mix of both):

1. **Case 6a local biodiversity offsets—reputational risk:** Reacting on perceived influences or conflicts during the planning or implementation stage of a project, a developer may choose to suggest an offset (which he would not otherwise be obliged to implement) in order to (re)gain acceptance with the local population, strengthen its reputation to facilitate local and regional collaboration and avoid delays in the timely implementation of the development.
2. **Case 6b local biodiversity offsets—license to operate:** However, ideally a developer will take a proactive approach, entering into a dialogue with local stakeholders on the proposed project or operation. This does not only help to develop legitimacy for the project, but also to establish a process to identify and address environmental (and social) impacts throughout the lifetime of the project.

Local knowledge can facilitate the application of the mitigation hierarchy and help to identify conservation priorities. These need to be put into relation to other uses of natural resources besides the proposed development, especially in the daily routine of the local population, e.g., agriculture, timber use, fisheries, in order to avoid indirect and cumulative impacts (see Fig. 6.4).

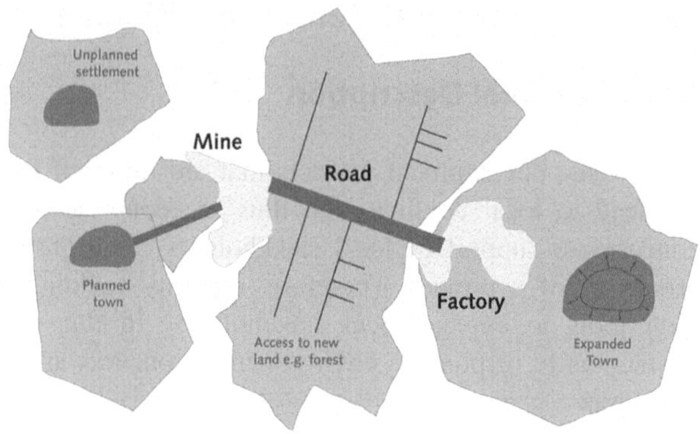

Fig. 6.4 Illustration of direct and indirect impacts of a mining project. Source: Ten Kate et al. (2011)

An example is the Community Reforestation Programme established by QIT Mining Madagascar in 2002 to reduce pressure by the local population on natural forest remnants. The aim is to plant 100 ha of fast-growing trees per year to supply local communities with wood for fuel and charcoal around the mining sites, supported also by a regional tree nursery.

This exemplifies that local biodiversity offsets usually need to go beyond the "no net loss of biodiversity and ecosystems" approach and are often accompanied by socioenvironmental measures as the aforementioned.

6.6.2 Case Study for Local Biodiversity Offsets: Network Rail's Thameslink Railway Project in the United Kingdom[5] (Tables 6.26, 6.27, 6.28, 6.29, 6.30, and 6.31)

Table 6.26 General description for the case study local biodiversity offsets: Network Rail's Thameslink railway project in the United Kingdom

General description	
Location	Streatham Common in South London, UK
Development activity	Thameslink Programme (TLP) is a major (4.6 billion British pounds) rail infrastructure enhancement project on the Thameslink service, one of the United Kingdom's busiest commuter routes, connecting Bedford and Brighton via central London. The objective is the transformation of the Thameslink route by providing more frequent, longer, and new trains as well as major improvements to three major stations in London (London Bridge, Farringdon, and Blackfriars).

(continued)

[5]Information based on Parsons Brinckerhoff (2013), Woodley and Baker (2014), Woodley (2015), Kemp (2014), Defra (2013), London Wildlife Trust (2013), Woodley et al. (2013), Baker et al. (2015), TLP (2012) and CIEEM (2013).

Table 6.26 (continued)

General description	
Environmental baseline and impact	Railway environment is varied and complex. The Thameslink railway line extends over long distances and thus cuts through a range of different urban and rural environments, covering a variety of vegetation and ecological features: urban depots with little surrounding vegetation cover, scrub-covered railway embankments in Greater London and woodland areas in the surrounding countryside. Key biodiversity features include green corridors, linking habitats, and migration routes for protected species. Despite onsite ecological enhancements, on many sites, an unavoidable loss of vegetation occurred, particularly at Selhurst. The railway-locked area of disused railway sidings that had become colonized by scrub, grassland, and trees had to be converted into railway sidings in order to store trains during the development of London Bridge station. The clearance affected trees, scrubs, areas of ruderal vegetation, and small patches of woodland along the Thameslink corridor, which may provide ecological linkages amongst habitat patches especially in urban areas.
Impact size	2 ha
Offset size	8 ha
Goal of the offset	The creation of new woodland in the nature reserve and public park of Streatham Common shall contribute to London Wildlife Trust's aim to restore sections of the Great North Wood. This is a dense forest that once covered huge parts of southern London. It provides a rich wildlife environment and recreational facilities for local residents and visitors.
Loss–gain calculation	Starting in 2012, loss–gain calculations were applied for TLP's second phase (KO2) in three steps: 1. Establishing the biodiversity baseline in terms of nature conservation value and primary ecological function of each site 2. Assessing the impact of habitat clearance on the baseline 3. Using the UK government's new habitat-based metric[a] for the assessment of the baseline condition, the calculation of biodiversity loss, and the determination of the offset demand The calculation is based on "biodiversity units," which are the product of the size, and the distinctiveness and condition of each habitat. As a result, the offset will deliver over 40 biodiversity units of woodland creation and enhancement at Streatham Common.

(continued)

Table 6.26 (continued)

General description	
Type and scope of offset measures	Following the mitigation hierarchy, ecological surveys were undertaken and any mitigation requirements were incorporated into the works. In TLP's first phase (KO1), the mitigation measures comprised: • The favoring of measures to avoid and mitigate impacts (including seasonal constraints), reducing the amount of vegetation cleared or relocating infrastructure installation, as well as on-site enhancements such as landscaping and planting schemes sympathetic to the biodiversity of the area. • The installation of a solar photovoltaic roof at Blackfriars station and a 700-sqm brown roof at Farringdon Station provide habitats for invertebrates and foraging opportunities for a number of birds. • Approx. 1500 trees have been planted at Woodland Trusts' Heartwood Forest to compensate for all permanent vegetation clearance works along the route. Prior to starting TLP's second phase (KO2), a feasibility study identified that offsite offset measures are required to reach the target of no net loss or net gain, because of planting restrictions on railway land (e.g., safety regulations, risk of disturbance from railway operations, and damage from vandalism or fly-tipping). Therefore, in the project's second stage (KO2) a biodiversity offset was designed to reach the net gain target: • A total of 8 ha of woodland have been planted and enhanced on Streatham Common and Brockwell Park in seven parcels, with the aim to increase species diversity. Native trees, shrubs, and other plants were used to complement existing tree lines and creating clusters of more wooded character.
Involved parties	4. Network Rail (the developer) 5. Parsons Brinckerhoff (consultants) 6. The London Wildlife Trust 7. Government: Lambeth Council, Defra (demonstration project), UK Environment Secretary
Timeframe	The Thameslink Programme runs in two key stages: Key Output 1 (KO1) ran from 2006 to 2011, including reconfiguration of Blackfriars and Farringdon station, construction of a new viaduct at Borough Market and expansion of Thameslink services outside London. Key Output 2 (KO2) started in 2012 and shall be completed in December 2018, including reconfiguration of London Bridge Station, remodeling of track layout, construction of depots and stabling site.

Source: Author

[a]In 2012, the Department for Environment, Food and Rural Affairs (Defra) launched a biodiversity unit calculation as part of a 2 year UK national pilot test on biodiversity offsetting (Woodley et al. 2013)

Table 6.27 Threshold criteria for the case study for local biodiversity offsets: Network Rail's Thameslink railway project in the United Kingdom

Threshold criteria			
Intentionality	Yes	No	Comment
	X		The Thameslink team made the voluntary decision to deliver a net gain target.
Additionality	Yes	No	Comment
	X		London Wildlife Trust had a vision to restore the Great North Wood of London; but this was just a vision without any source of funds (and no grant applications had been made). The Thameslink offset, by funding the woodland creation and enhancement at Streatham Common, was an additional outcome for nature conservation and became an initiator for London Wildlife Trust to obtain other funding elsewhere for other projects to restore the Great North Wood.

Source: Author

Table 6.28 Descriptive criteria (context) for the case study for local biodiversity offsets: Network Rail's Thameslink railway project in the United Kingdom

Descriptive criteria: context	
Domestic legislation in the home country	The project is a legally compliant development that entails biodiversity loss. Biodiversity offsetting is not required under UK legislation. Nevertheless the National Planning Policy Framework states that the planning system should contribute to and enhance the natural and local environment by minimizing impacts on biodiversity and providing net gains in biodiversity where possible. In addition, the UK government has provided extensive guidance on offsets, including the development of a biodiversity offsets metric, trialing a number of pilots and supporting additional offset projects. The Thameslink offset is promoted as a UK demonstration project.
Competitive environment	–
Cultural and social context	There have been concerns by the public over prior major rail infrastructure projects. The Thameslink project thus was planned and built into an atmosphere of high environmental awareness and scrutiny from the public. Railway embankments are not seldom subjects to vandalism. At the same time they provide habitats for wildlife. The developer needs to balance addressing these competing functions.

Source: Author

Table 6.29 Core criteria for the case study for local biodiversity offsets: Network Rail's Thameslink railway project in the United Kingdom

<table>
<tr><th colspan="6">Core criteria</th></tr>
<tr><th>motivation and ultimate goal / influence and initiators</th><th colspan="2">Pressure (compliance)</th><th colspan="2">Incentive (cost-benefit)</th><th>Altruism / responsibility</th></tr>
<tr><td>Government</td><td colspan="2">Case 1 (regulatory)</td><td colspan="3">Case 3a (enabled: government)</td></tr>
<tr><td>Financial institute</td><td colspan="4">Case 2 (conditional)</td><td></td></tr>
<tr><td>Sector</td><td colspan="4">Case 4 (sectoral)</td><td></td></tr>
<tr><td>Corporation</td><td colspan="2">Case 5 (corporate)</td><td colspan="2"></td><td></td></tr>
<tr><td>Local community</td><td colspan="2" rowspan="2">Case 6a (local: reputational risk)</td><td colspan="3">Case 6b (local: license to operate)</td></tr>
<tr><td>NGOs</td><td colspan="3">Case 3b (enabled: new global norms)</td></tr>
<tr><td>Without major influence</td><td colspan="2"></td><td colspan="2"></td><td>Case 7 (altruism)</td></tr>
<tr><td>comment</td><td colspan="5">The offset is a product of the influence of different stakeholders: To translate its commitment to sustainable development, the Thameslink Programme targets to enhance its habitat footprint to achieve a net biodiversity gain via the mitigation hierarchy. During the process the Thameslink team voluntarily revised its initial target of avoiding loss upwards to achieve a net gain as part of the new (2012) TLP Sustainable Development Policy, i.e. a commitment to have a positive effect on the natural environment and to deliver biodiversity benefits.
It should be noted that the decision to target a net gain in biodiversity was neither required for planning permission, nor was biodiversity offsetting undertaken to meet statutory requirements of protected species or habitats.
The project has also applied the Defra[34] biodiversity offset metric and consequently been named a complementary demonstration project by Defra.
The Thameslink team wanted to build local bonds and work closely and consult with another organization. Consequently, the project has been designed as a partnership among industry, conservation and local government to exceed its environmental targets and to generate long-term benefits for both the environment and the neighbouring communities, more precisely between Network Rail's Thameslink Programme, the London Wildlife Trust and the London Borough of Lambeth (Lambeth Council), supported by a consultant.</td></tr>
<tr><td rowspan="2">Choice or eligibility</td><td>No influence (willful
voluntary action)</td><td>Marginal influence (coercive offer)</td><td>Significant influence (vis compulsiva)</td><td>Strong influence (vis absoluta)</td><td>Comment</td></tr>
<tr><td>X</td><td></td><td></td><td></td><td>-</td></tr>
<tr><td rowspan="2">Recompense or benefit</td><td>Yes</td><td>No</td><td colspan="3">Comment</td></tr>
<tr><td></td><td></td><td colspan="3">-</td></tr>
</table>

[a] Department for Environment, Food and Rural Affairs, UK.

Source: Author

Table 6.30 Performance criteria for the case study for local biodiversity offsets: Network Rail's Thameslink railway project in the United Kingdom

Performance criteria			
Increased flexibility and cost efficiency	Yes	No	Comment
	X		Different locations for biodiversity offsets have been proposed and Streatham Common was chosen because it matched the criteria that Thameslink developed based on the biodiversity affected by works. It was also most cost efficient in terms of cost per biodiversity unit.

Source: Author

Table 6.31 Quality criteria (outcome and charitableness) for the case study for local biodiversity offsets: Network Rail's Thameslink railway project in the United Kingdom

Quality criteria: Outcome and charitableness					
Counterfactual	Static (neutral)	Dynamic		Comment	
		Positive	Negative		
				n/a	
Outcome and charitableness relative to the counterfactual	Positive balance (Net Gain)	Neutral balance (No Net Loss	Minimum balance (reduced loss)	Not effective (loss)	Comment
	X				To achieve its first project with a net gain biodiversity target, Network Rail follows a mitigation hierarchy of avoidance of biodiversity loss mitigation and, as a last resort, compensation.

Source: Author

6.7 Type 7: Altruistic Biodiversity Offsets

6.7.1 General Description

All of the biodiversity offset types described thus far have been driven or at least influenced by some pressures or incentives. While they would certainly qualify as different forms of voluntary offsets in that the project proponent makes a free and willful choice, the likely consequences of this choice, however, have a great influence and ultimately affect the choice.

By contrast, altruistic biodiversity offsets (at least theoretically) are exempt from any major influence and thus probably the most ideal representation of true voluntary biodiversity offsets, representing clear additionality.

As has been noted (see Sect. 3.2.1), altruism "is a motivational state with the ultimate goal of increasing another's welfare" (Batson 2014). In the case of biodiversity offsets, it is directed toward addressing biodiversity harm through nature conservation measures.

While the motivation for altruistic biodiversity offsets and altruistic behavior more generally is said to come from "within" the individual, not seldom a "moral requirement or responsibility", e.g., for the future of the planet and future generations, is said to be leading such behavior. Two crucial points can be derived for the case of altruistic biodiversity offsets:

1. Importance of individuals: Altruistic behavior is closely tied to the beliefs, values, and norms of individuals. Responsibility beyond compliance or striving for business benefit cannot be "felt" by corporations or abstract business entities, but only by individuals. This means that altruistic biodiversity offsets are usually driven by the commitment of single persons, e.g., the CEO of a firm.
2. Moral values shaped by (social) context: The moral values and norms that lead an individual towards responsibility or a requirement for action not only come from within this person, but are in fact the result of life-long socialization and the various influences of social and cultural context. This means that altruistic biodiversity offsets can emerge everywhere, but they are more likely in specific contexts than others.

6.7.2 Case Study for Altruistic Biodiversity Offsets: Antamina Mine in Peru[6] (Tables 6.32, 6.33, 6.34, 6.35, 6.36, and 6.37)

Table 6.32 General description for the case study for altruistic biodiversity offsets: Antamina mine in Peru

General description	
Location	Peruvian Andes, in the department of Ancash
Development activity	Open copper, zinc, and molybdenum mining and 302 km pipeline transporting the ore from the mine to a port facility in the coastal town of Huarmey
Environmental baseline and impact	At this high altitude (4300 m above sea level) grasslands and shrub ecosystems are dominating, forest parcels with the threatened Polylepis trees (vulnerable according to IUCN classification) exist in the region. These, however, have been heavily impacted over time by local communities, who use the area's resources for grazing and fuel wood.

(continued)

[6]Information based on BBOP (2009b), Lopez Arbeláez and Quintero Sagre (2015), Biodiversity Neutral Initiative (2005), ICMM (2010) and ten Kate et al. (2011). It is important to note that the information presented here refers to the design phase of the project. Several changes have occurred over time, including the extent of the forest restoration program and the implementing entities (personal communication with Jared Hardner, February 2016).

Table 6.32 (continued)

General description	
	The mine pit will result in loss of forest and grassland ecosystem, including draining of a 32-ha lake. 1 ha of vulnerable Polylepis forest will also be lost.
Impact size	1035 ha mine site (pit lake), 2221 ha direct impact area.
Offset size	Restoration of 874 ha of grassland and 1034 ha of aquatic habitat. 3000 ha of existing Polylepis forest protected, improving pasture management on more than 12,000 ha.
Goal of the offset	The main goal is the creation of Private Conservation Areas (ACPs) to protect and restore large and pristine Polylepis forests together with communities surrounding the mining area. Polylepis forests are a vanishing ecosystem urgently in need of conservation action. They are considered the most important ecological feature in the landscape from a biodiversity conservation perspective, providing habitats for several threatened bird species. The Polylepis program aims to contribute to the development of a conservation corridor between two protected areas, Huascarán National Park (a UNESCO World Heritage Area and Biosphere Reserve) and the Huayhuash Reserve Area.
Loss–gain calculation	Since the purpose of the Polylepis Program was not to offset residual loss from the mining development, no quantification of loss–gain calculation was undertaken. Furthermore, the program focuses on the threatened Polylepis forest habitat of which only one hectare is impacted by the mine. The scale of the offset could allow it to be counted as an out-of-kind offset for grassland and aquatic habitat impacts. The unit of currency is simply area of habitat in hectares and does not attempt to quantify ecosystem functions.
Type and scope of offset measures	To address the direct impacts of the mine, the closure plan for the mine foresees restoration of 874 ha of grassland and 1034 ha of aquatic habitat. The measures in the scope of the Polylepis Program, established via conservation agreements with local communities include: • Restoration of 322 ha of Polylepis forests • Protection of 3,000 ha of existing Polylepis forests • Improving pasture management (reducing grazing pressure on

(continued)

Table 6.32 (continued)

General description	
	Polylepis forests) over an area of more than 12,000 ha The total area of actual Polylepis forests to be restored in a distributed pattern across the Conchucos Valley is approximately 1,000 hectares in the first 5-year phase of the program. In addition, Antamina's Asociacion Ancash supports the management of the neighboring Huascaran National Park, which is an important freshwater source and harbors many endemic species, many of them threatened.
Involved parties	1. Compania Minera Antamina (the developer) 2. Asociación Ancash (a corporate foundation of Antamina, that carries out Antamina's voluntary projects) 3. Conservation International (CI, an independent conservation group) 4. The Mountain Institute (TMI, an independent conservation group) 5. Association of Andean Ecosystems (ECOAN, a Peruvian NGO that specializes in Polylepis restoration) 6. Peruvian Society of Environmental Law (SPDA, collaboration to achieve long-term legal protection of the forested areas)
Timeframe	Mine production started (operating life approx. 20 years) 2001, concept to restore and conserve Polylepis forests as a joint venture developed 2003, voluntary biodiversity conservation program to restore endangered Polylepis forests in the area around the mine started 2004.

Source: Author

Table 6.33 Threshold criteria for the case study for altruistic biodiversity offsets: Antamina mine in Peru

Threshold criteria			
Intentionality	Yes	No	Comment
		X	The Polylepis Program was not initially created as an offset, however, it was initiated with the intention of addressing business harm from the mining operation and balancing needs of the local population.
Additionality	Yes	No	Comment
	X		Polylepis restoration is not replacing any existing conservation program or government responsibility, it is implemented on private land to create a corridor between two protected areas.

Source: Author

Table 6.34 Descriptive criteria (context) for the case study for altruistic biodiversity offsets: Antamina mine in Peru

Descriptive criteria: Context	
Domestic legislation in the home country	No regulatory offset requirements were in place in Peru when the project started. Apart from that, according to Antamina, the mine meets or exceeds Peruvian regulatory requirements for environmental protection. The company follows the management requirements for environmental management set forth in its EIA. It also adheres to World Bank Operational Policy 4.04 (Natural Habitats), which includes requirements for biodiversity conservation.
Competitive environment	The mining complex is among the ten largest in the world and, at US$2.3 billion, represented the largest mining investment in the history of Peru until recently.
Cultural and social context	Most local communities are scattered throughout the region and are living in relative isolation and extreme poverty. As a result, the Polylepis forests in the region have been severely impacted by human intervention, leaving remnants of less than 10% of their natural extension. In particular, overgrazing by local communities is a huge problem. The project needed to take into consideration increased overgrazing in undisturbed grasslands adjacent to the mine. Therefore it was crucial to analyze and understand the complex legal, social, cultural, economic, governance, and ecological circumstances prevailing in the area, prior to designing and locating conservation activities and determining an optimum strategy for their effective implementation. This includes demonstrating tangible benefits to local communities in the short term, including nonmaterial cultural values of the communities. Therefore, the strategy of the Polylepis program builds on developing conservation agreements with local communities, thus attempting to balance economic and other incentives with local pride in stewardship. This way, the long-term sustainability of the conservation outcomes rests on the creation of a trust fund that will support the provision of socioeconomic benefits to the communities in exchange for their long-term commitment to conservation

Source: Author

Table 6.35 Core criteria for the case study for altruistic biodiversity offsets: Antamina mine in Peru

Core criteria

motivation and ultimate goal / influence and initiators	Pressure (compliance)	Incentive (cost-benefit)	*Altruism / responsibility*
Government	Case 1 (regulatory)	Case 3a (enabled: government)	
Financial institute	Case 2 (conditional)		X
Sector	Case 4 (sectoral)		X
Corporation	Case 5 (corporate)	X	X
Local community	Case 6a (local: reputational risk)	Case 6b (local: license to operate)	
NGOs		Case 3b (enabled: new global norms)	
Without major influence	X	X	***Case 7 (altruism)***

Comments	In 2003, Antamina management became interested in a voluntary conservation project in the region of the mine. From the various options for projects in accordance with technical advisors and conservation groups, a project to restore endangered Polylepis habitat was chosen, because it would provide an ideal blend of biodiversity conservation and local community benefits. The Polylepis program was not envisioned as an offset but rather as a voluntary initiative within the framework of the company's Corporate Social Responsibility strategy. The motivation for the Polylepis Program was to make a "positive contribution" to biodiversity conservation by uplifting local communities whilst simultaneously benefiting the restoration of landscape links and important habitat. It was considered crucial to focus on cooperation with local communities and to improve their quality of life, e.g. by helping to provide fuelwood from sustainable sources, introducing more fuel-efficient stoves, managing improved pastures and introducing improved breeds of cattle and sheep. The program was also partly triggered by Antamina's corporate policy. For the implementation a partnership with several NGOs has been built: Asociacion Ancash, CI, TMI and ECOAN. To help ensure financial sustainability over time, CI and TMI are promoting the creation of a trust fund that will provide benefits to the local communities in exchange for their continued commitment to protecting the restored areas as well as protecting other areas through the maintenance of fences and patrolling. The Peruvian Mining Society has recognized Antamina for environmental responsibility in 2006.

	No influence (willful voluntary action)	Marginal influence (coercive offer)	Significant influence (vis compulsiva)	Strong influence (vis absoluta)	Comment
Choice or eligibility					-

	Yes	No	Comment
Recompense or benefit			-

Source: Author

Table 6.36 Performance criteria for the case study for altruistic biodiversity offsets: Antamina mine in Peru

Performance criteria			
Increased flexibility and cost efficiency	Yes	No	Comment
			–

Source: Author

Table 6.37 Quality criteria (Outcome and charitableness) for the case study for altruistic biodiversity offsets: Antamina mine in Peru

Quality criteria: Outcome and charitableness					
Counterfactual	Static (neutral)	Dynamic		Comment	
		Positive	Negative		
				n/a	
Outcome and charitableness relative to the counterfactual	Positive balance (Net Gain)	Neutral balance (No net oss)	Minimum balance (reduced loss)	Not effective (loss)	Comment
	X				The project was not originally designed as an offset. Therefore, it did not target all affected ecosystems and will result in a long-term (postmine closure and restoration) increase in grassland and aquatic habitat, but a net loss of brush.The Polylepis Prgram was awarded the Sustainable Development Prize of the Ministry of Energy and Mines in 2006.

Source: Author

References

Albrecht J, Schumacher J, Wende W (2014) The German impact-mitigation regulation—a model for the EU's no-net-loss strategy and biodiversity offsets? Environ Policy Law 44(3):317–325

Asian Development Bank (2009) Safeguard policy statement. Asian Development Bank. Available at: http://www.adb.org/site/safeguards/policy-statement

Bainbridge Island Land Trust (2015) Powel Shoreline Restoration Project. Website. Available at http://www.bi-landtrust.org/default.asp?ID=66

Baker J, Woodley A, Wilkinson P (2015) Development with biodiversity net positive. PPT presentation. Available at https://www.iema.net/system/files/julia_baker_paul_wilkinson_amelia_woodley.compressed.pdf

Batson CD (2014) The altruism question. Toward a social-psychological answer. Psychology Press, New York

BBOP (2009a) Business, biodiversity offsets and BBOP: an overview. BBOP, Washington, DC. Available at http://www.forest-trends.org/biodiversityoffsetprogram/guidelines/overview.pdf

BBOP (2009b) Compensatory conservation case studies. BBOP, Washington, DC. Available at www.forest-trends.org/biodiversityoffsetprogram/guidelines/non-bbop-case-studies.pdf

BBOP (2010) Principles on biodiversity offsets supported by the BBOP Advisory Committee. Available at http://bbop.forest-trends.org/guidelines/principles.pdf

BBOP (2016) The Business and Biodiversity Offsets Programme (BBOP website). Available at http://bbop.forest-trends.org/

BBOP (n.d.) Biodiversity offsets and the Business and Biodiversity Offsets Programme (BBOP). A draft consultation paper for discussion and comment. Available at http://www.forest-trends.org/biodiversityoffsetprogram/documents/UNEP/BBOP-UNEP-CBD-COP-9-Inf-29.pdf

Biodiversity Neutral Initiative (2005) Environmental offset policies, principles, and methods: a review of selected legislative frameworks. Available at http://www.forest-trends.org/biodiversityoffsetprogram/library/new/Environmental%20Offset%20Legislative%20Frameworks%20DRAFT%20March%2030.pdf

British Columbia Ministry of Environment (2014) Environmental mitigation policy for British Columbia. Available at http://www.env.gov.bc.ca/emop/

Burgin S (2008) BioBanking: an environmental scientist's view of the role of biodiversity banking offsets in conservation. Biodiv Conserv 17(4):807–816

CIEEM (2013) Network Rail launch first net-positive biodiversity offset scheme. Available at http://www.cieem.net/news/161/network-rail-launch-first-net-positive-biodiversity-offset-scheme

City of Bainbridge Island (2009) BBOP pilot project case study. Bainbridge Island. Washington State, USA. Available at www.forest-trends.org/biodiversityoffsetprogram/guidelines/bainbridge-case-study.pdf

Commissariat Général Au Développement Durable. Direction De l'eau Et De La Biodiversité (2013) Lignes directrices nationales sur la séquence éviter, réduire et compenser les impacts sur les milieux naturels. Available at http://www.biodiversityoffsets.net/wp-content/uploads/2014/08/Lignes_directrices_ERC-France.pdf

Darbi M (2014) Doctrine ERC (Eviter-Reduire-Compenser)—application of the mitigation hierarchy in France. Biodiversity Offsets Blog. Available at http://www.biodiversityoffsets.net/doctrine-erc-eviter-reduire-compenser-application-mitigation-hierarchy-france/

Darbi M (2015a) Biodiversity law (including biodiversity offsets) under development in France. Biodiversity Offsets Blog. Available at http://www.biodiversityoffsets.net/biodiversity-law-under-development-in-france/

Darbi M (2015b) 2014 IUCN World Parks Congress launches dialogue on biodiversity offsets. Biodiversity Offsets Blog. Available at http://www.biodiversityoffsets.net/2014-iucn-world-parks-congress-launches-dialogue-on-biodiversity-offsets/

Darbi M (2015c) IUCN public consultation: IUCN seeks input on draft Biodiversity Offsets Policy. Biodiversity Offsets Blog. Available at http://www.biodiversityoffsets.net/iucn-public-consultation-iucn-seeks-input-on-draft-biodiversity-offsets-policy/

Darbi M, Ohlenburg H, Herberg A, Wende W (2010) Impact mitigation and biodiversity offsets-compensation approaches from around the world: a study on the application of Article 14 of the CBD (Convention on Biological Diversity). Naturschutz und Biologische Vielfalt 101. BfN-Schr.-Vertrieb im Landwirtschaftsverl.

Darbi M, Albrecht J, Schumacher J (2016) Die europäische No Net Loss-Initiative und Biodiversity Offsets. Einblicke in die aktuelle Diskussion zur Kompensation von Eingriffen in die Biodiversität aus deutscher Perspektive. In: Jahrbuch des Umwelt- und Technikrechts

Defra (2011) Biodiversity 2020: a strategy for England's wildlife and ecosystem services. Available at https://www.gov.uk/government/uploads/system/uploads/attachment_data/file/69446/pb13583-biodiversity-strategy-2020-111111.pdf

Defra (2013) Biodiversity offsetting in England Green paper. September 2013. Available at https://consult.defra.gov.uk/biodiversity/biodiversity_offsetting/supporting_documents/20130903Biodiversity%20offsetting%20green%20paper.pdf

Doswald N, Barcellos Harris M, Jones M, Pilla E, Mulder I (2012) Biodiversity offsets: voluntary and compliance regimes. A review of existing schemes, initiatives and guidance for financial institutions. UNEP-WCMC, Cambridge, UK. UNEP FI, Geneva, Switzerland

eftec, IEEP et al (2010) The use of market-based instruments for biodiversity protection—the case of habitat banking. Technical report. For European Commission DG Environment

Enetjärn A, Cole S, Kniivilä M, Hårklau SE, Hasselström L, Sigurdson T, Lindberg J (2015) Environmental compensation. Key conditions for increased and cost effective application

Equator Principles (2011) Biodiversity For Banks Program (B4B). Available at http://www.equator-principles.com/index.php/b4b

Equator Principles (2013) The Equator Principles. June 2013. A financial industry benchmark for determining, assessing and managing environmental and social risk in projects. Available at http://www.equator-principles.com/resources/equator_principles_III.pdf

European Investment Bank (2009) The EIB statement of environmental and social principles and standards, European Investment Bank. Available at: http://www.eib.org/attachments/strategies/eib_statement_esps_en.pdf

European Investment Bank (2015) Update: Nam Theun 2 Hydropower Project, Laos. Available at http://www.eib.org/infocentre/press/news/topical_briefs/2005-november-01/nam-theun-2-hydropower-project-laos.htm?lang=en

Globalbalance and The Biodiversity Consultancy (2014) Review of the International Council on Mining and Metals members' biodiversity performance management since 2003. Unpublished report to ICMM & IUCN

Herbert M (2015) German impact mitigation regulation—national, European, international. A comparison from the viewpoint of the Federal Agency for Nature Conservation. In: UVP Rep 29(3):149–151. Eingriffsregelung; Umweltverträglichkeitsprüfung; Strategische Umweltprüfung; Biologische Vielfalt

ICMM (2005a) Biodiversity offsets: a proposition paper from The International Council on Mining and Metals

ICMM (2005b) Biodiversity offsets—a briefing paper for the mining industry

ICMM (2006) Good practice guidance for mining and biodiversity

ICMM (2010) Mining and biodiversity. A collection of case studies—2010 edition

ICMM (2016) International Council on Mining and Metals. Homepage. Available at https://www.icmm.com/

ICMM IUCN (2012) Independent report on biodiversity offsets. Prepared by The Biodiversity Consultancy. Available at www.icmm.com/biodiversity-offsets

IEEP et al (2012) Policy options for an EU No Net Loss Initiative. A tender for the European Commission. Unpublished

IFC (2012) Performance standard 6. Biodiversity conservation and sustainable management of living natural resource. Available at http://www.ifc.org/wps/wcm/connect/bff0a28049a790d6b835faa8c6a8312a/PS6_English_2012.pdf?MOD=AJPERES. Accessed 24 Feb 2016

IFC (2015) Environmental and social performance standards and guidance notes. Available at http://www.ifc.org/wps/wcm/connect/topics_ext_content/ifc_external_corporate_site/ifc+sustainability/our+approach/risk+management/performance+standards/environmental+and+social+performance+standards+and+guidance+notes

International Advisory Group (IAG); Laos; Nam Thuen II Power Company Ltd.; EcoLao; Norplan (2005) Impact mitigation, restoration. The Nam Theun 2 Dam in Laos (vol 16), Fifth report of the International Advisory Group (IAG) on the World Banks handling of social and environmental issues. Available at http://www-wds.worldbank.org/external/default/WDSContentServer/WDSP/IB/2005/09/09/000011823_20050909112339/Rendered/PDF/E385v1615th1IAG0Report.pdf

Inter-American Development Bank (2006) Environment and safeguards compliance policy. Available at: http://idbdocs.iadb.org/wsdocs/getdocument.aspx?docnum=1481950

IUCN (2014) Biodiversity offsets technical study paper. IUCN, Gland, Switzerland, 65pp

Kemp D (2014) Thameslink upgrade uses offsetting to boost biodiversity. Available at http://www.cnplus.co.uk/innovation/sustainable-business/thameslink-upgrade-uses-offsetting-to-boost-biodiversity/8658417.article#.VjvSlNKrS1s

Key A, Darbi M (2015) Worldwide on-the-ground examples of Biodiversity Offsets—Koala Offset in South East Queensland. Biodiversity Offsets Blog. Available at http://www.biodiversityoffsets.net/worldwide-ground-examples-biodiversity-offsets-koala-offset-south-east-queensland/

Kormos R, Kormos C (2011) Towards a strategic national plan for biodiversity offsets for mining in the Republic of Guinea, West Africa with a focus on chimpanzees. Available at https://www.google.de/url?sa=t&rct=j&q=&esrc=s&source=web&cd=6&cad=rja&uact=8&ved=0CEwQFjAFahUKEwjWqtCk4PHIAhXD1RoKHTjPC-o&url=http%3A%2F%2Fwww.bankinformationcenter.org%2Fwp-content%2Fuploads%2F2013%2F07%2FGuineaBiodiversityOffsetKormosKormos2011.pdf&usg=AFQjCNGej-JDKlGWpvEmrp-vQjNPPlyFew&sig2=tZ6JFlRjWhTDyk4kwIsjzw

London Wildlife Trust (2013) Environment secretary launches London's first biodiversity offset scheme. Available at http://www.wildlondon.org.uk/news/2013/12/17/environment-secretary-launches-london%E2%80%99s-first-biodiversity-offset-scheme

Long R (2011) Tracking development. A collection of QIT Madagascar Minerals' (QMM) socio-economic contributions. Rio Tinto/QMM, December 2011. Available at http://www.riotintomadagascar.com/pdf/tracking.pdf

Lopez Arbeláez DM, Quintero Sagre JD (2015) Compensaciones de biodiversidad: experiencias en Latinoamérica y aplicación en el contexto colombiano. Gestión y Ambiente 18(1):159

Mercados de Medio Ambiente (2013) La "Ley de Evaluación Ambiental 21/2013" ve la luz. Available at http://www.mercadosdemedioambiente.com/actualidad/la-ley-de-evaluaci-n-ambiental-21-2013-ve-la-luz/#gdQ08gNgf9BuAUv0

Mercados de Medio Ambiente (2014) La nueva "Ley de Evaluación Ambiental", de un vistazo. Available at http://www.mercadosdemedioambiente.com/opinion/la-nueva-ley-de-evaluaci-n-ambiental-de-un-vistazo/#gdQ09uAkn4e8QdGs

Ministère de l'Écologie, du Développement durable, des Transports et du Logement (2012) DOCTRINE relative à la séquence éviter, réduire et compenser les impacts sur le milieu naturel. Available at http://www.biodiversityoffsets.net/wp-content/uploads/2014/08/doctrine-ERC-France.pdf

Naturvårdsverket (2015) Kompensation vid förlust av naturvärden. En prövningsvägledning. Available at http://www.naturvardsverket.se/upload/stod-i-miljoarbetet/remisser-och-yttranden/remisser-2015/Provningsvagledning-remissversion-20151005.pdf

OECD (2013) Scaling-up finance mechanisms for biodiversity, OECD Publishing. Available at https://doi.org/10.1787/9789264193833-en

Parsons Brinckerhoff (2013) Parsons Brinckerhoff helps deliver London's first net positive biodiversity offset scheme for Network Rail. Available at https://www.pbworld.com/pdfs/regional/uk_europe/thameslink%20biodiversity%20offset%20announcement.pdf

Pilgrim JD, Ekstrom JMM (2014) Technical conditions for positive outcomes from biodiversity offsets. An input paper for the IUCN Technical Study Group on Biodiversity Offsets. IUCN, Gland, Switzerland, 46pp

QMM (ed) (2007a) The QMM—Rio Tinto biodiversity programme. Available at http://www.riotintomadagascar.com/siteware/pdfs/biodiversity-prog-large.pdf

QMM (ed) (2007b) Fact Sheet—QIT Madagascar minerals—environment—Biodiversity. Available at http://www.riotintomadagascar.com/siteware/pdfs/environ.pdf

QMM (ed) (2007c) The QMM biodiversity programme, Madagascar. Available at http://www.riotintomadagascar.com/siteware/pdfs/biodiversity.pdf

QMM (ed) (n.d.) Community Relations strategy and plan. Available at http://www.riotintomadagascar.com/siteware/pdfs/community-relations-plans.pdf

QMM; Madagascar's National Office for the Environment (ed) (2001) Ilmenite project—social and environmental impact assessment. Environmental management plan. Available at http://www.riotintomadagascar.com/siteware/pdfs/env_mg_pln.pdf

Rio Tinto (ed) (2006) Madagascar to pilot biodiversity indicators. Available at http://www.riotintomadagascar.com/siteware/pdfs/npi.pdf

Rio Tinto and IUCN (2012) Rio Tinto QMM on track to a net positive impact on biodiversity. Available at http://www.riotintomadagascar.com/pdf/NPI.pdf

Serusclat J-P (2004) Nam Theun 2 Hydroelectric Project Lao PDR. Available at https://www.google.de/url?sa=t&rct=j&q=&esrc=s&source=web&cd=3&cad=rja&uact=8&ved=0ahUKEwj20arc_qvJAhWFGCwKHV1uBgAQFggxMAI&url=http%3A%2F%2Fsiteresources.worldbank.org%2FINTLAOPRD%2FResources%2F293582-1092106399982%2F492430-1092106479653%2F492431-1095757484636%2Ftechnical_safeguards.ppt&usg=AFQjCNEQ_56T5nVMG_cs87Y6-W53AXXpYw&sig2=1846xtwxGquUVBeMe2utPA&bvm=bv.108194040,d.bGg

Temple H, Edmonds B, Butcher B, Treweek J (2010) Biodiversity offsets: testing a possible method for measuring losses and gains at Bardon Hill Quarry, UK. Practice 70:11–14

ten Kate K (2007) Biodiversity offsets and conservation finance. PPT presentation. Available at https://www.google.de/url?sa=t&rct=j&q=&esrc=s&source=web&cd=10&cad=rja&uact=8&ved=0CG4QFjAJahUKEwi9r5zu3ufIAhXBPxoKHYC0CI0&url=https%3A%2F%2Fwww.bfn.de%2Ffileadmin%2FMDB%2Fdocuments%2Fina%2Fvortraege%2F14_ten_Kate.pdf&usg=AFQjCNHNmpmZ8YtV-NFwXhkQk1v_zy58uQ&sig2=ggwsHlI0U4VjH4GWwP1Elg&bvm=bv.106130839,d.d2s

ten Kate K, von Hase A, Boucher J, Cassin J, Victurine R (2011) Opportunities for environmental funds in compensation and offset schemes. RedLAC capacity building project for environmental funds—Rio de Janeiro: RedLAC, 2011. Available at https://www.google.de/url?sa=t&rct=j&q=&esrc=s&source=web&cd=5&cad=rja&uact=8&ved=0CEAQFjAEahUKEwj5id_P9-nIAhXI2BoKHWeiBhU&url=http%3A%2F%2Ftoolkit.conservationfinance.org%2Fsites%2Fdefault%2Ffiles%2Fdocuments%2Fredlac-capacity-building%2F5-opportunities-environmental-funds-compensation-and-offset-schemes.pdf&usg=AFQjCNGcxk5akDMfV_11U2syzDuHxj6vsg&sig2=BG9IsPZnohSD_ww6wlDhrw&bvm=bv.106379543,d.d2s

The Biodiversity Consultancy (2015) A cross-sector guide for implementing the Mitigation Hierarchy. Available at http://www.icmm.com/document/9460

The World Bank (2010) Convenient solutions to an inconvenient truth: ecosystem-based approaches to climate change. World Bank Series

The World Bank (2012) Safeguard policies. Available at http://go.worldbank.org/WTA1ODE7T0

TLP (2012) Thameslink programme sustainable development policy. Available at www.networkrail.co.uk/aspx/6345.aspx

Tucker G, Allen B, Conway M, Dickie I, Hart K, Rayment M, Schulp C, van Teeffelen A (2013) Policy options for an EU No Net Loss Initiative. Report to the European Commission. Institute for European Environmental Policy, London

UNEP Finance Initiative (2016) Positive impact finance. A common vision for the financing of the sustainable development goals (Sdgs) Available at http://www.unepfi.org/wordpress/wp-content/uploads/2017/06/POSITIVE-IMPACT-MANIFESTO-JUNE-17.pdf

von Hase A (2014) Working towards NNL of Biodiversity and Beyond: Strongman Mine—a case study. Available at http://www.forest-trends.org/documents/strongman_2014

von Hase A, Cooke A, Andrianarimisa A, Andriamparany R, Mass V, Mitchell R, ten Kate K (2014) Working towards NNL of biodiversity and beyond: Ambatovy, Madagascar—a case study. Forest Trends. Available at http://www.forest-trends.org/documents/index.php?pubID=4813

Wende W, Darbi M, Stein C (2015) Annex 4 evidence of the costs of offsetting in Germany. Case study provided by Prof. Wolfgang Wende, Marianne Darbi and Christian Stein (Technische Universität Dresden). In: The Institute for European Environmental Policy (IEEP), Biotope, eftec and ICF (ed 2015): supporting the elaboration of the impact assessment for a future EU Initiative on No Net Loss of Biodiversity and Ecosystem Services. 2nd Draft Interim/Task 2 Report. Unpublished draft

Woodley A (2015) Greening the Thameslink Route. Available at https://www.zsl.org/sites/default/files/media/2015-10/Greening%20the%20Thameslink%20route.pdf

Woodley A, Baker J (2014) Thameslink programme—delivering network rails first net positive biodiversity offset. Available at http://www.cieem.net/data/files/Resource_Library/Conferences/2014_Spring_Bio_Offsetting/13_Julia_Baker.pdf

Woodley A, Baker J, Britton C (2013) Thameslink Programme: delivering biodiversity benefits. CIRIA briefing. March 2013

Working Group on No Net Loss of Ecosystems and their Services (2013) Scope and objectives of the No Net Loss Initiative. Available at http://ec.europa.eu/environment/nature/biodiversity/nnl/pdf/Subgroup_NNL_Scope_Objectives.pdf

Discussion of the Development of a Typology of Biodiversity Offsets

7

Abstract

The applied research methodology was fit for purpose under the given research premises, and it can provide a model for future research. Nevertheless, one major limitation of the typology is that in practice mostly a combination of influences and motivations occurs, i.e., the individual types are represented in their purest form only to a limited extent in reality. Thus, intersections of these types are commonly encountered, especially for large-scale projects. In fact, the different cases or approaches are not mutually exclusive; rather, they can even be of enhancing and reinforcing nature. For example, an existing regulatory frame will not prevent, but most likely foster, additional engagement of developers in offsets.

Despite these observed overlaps, differences were identified between the types of biodiversity offsets in terms of magnitude, location, and particularities. For example, a general gap in offset implementation was observed for most Asian countries. In addition, offset cases can differ with regard to the scale of the development impacts, the sectors/types of developments addressed, and the governance of their implementation. Throughout the screening of biodiversity offset cases, mainly large-scale development projects have been identified. Furthermore, a clear dominance for certain sectors was observed, most notably mining. Differences between the types also occur with regard to top down vs. bottom up organized processes. Whereas regulatory, conditional, sectoral, and corporate biodiversity offsets are by definition top down, induced by regulators, lenders, (sectoral) business associations and corporations, local and altruistic biodiversity offsets build on a collaborative bottom up process with local stakeholders.

M. Darbi, *Biodiversity Offsets Between Regulation and Voluntary Commitment*,
https://doi.org/10.1007/978-3-030-25594-7_7

Keywords

Mixed types · Limits of typification · Mitigation hierarchy · Vis absoluta · Vis compulsive · Coercive offer · Willful voluntary action · Biodiversity offsets types · Biodiversity offsets case studies

7.1 Discussion and Critique of Methodology

7.1.1 Appropriateness of the Research Methodology

The present study builds on an expert and Internet based research methodology, using netnography and Science 2.0. In terms of appropriateness of this research methodology a number of questions can be asked:

1. Did the expert and Internet based research approach prove to be applicable and deliver appropriate results?
2. What are the main strengths and constraints?
3. Do alternative approaches exist to reach the envisaged goal?

Appropriateness of the Results

The methodology has proven to be practicable throughout the research process. No major problems were encountered and the foreseen results and information could be retrieved. This refers in particular to exploring the whole range of different voluntary biodiversity offsets with the ultimate goal of drafting a typology. With certain limitations the built types deliver appropriate results and contribute to a better understanding of the various offset contexts.

Strengths and Constraints

As with any methodology, the chosen research approach has a number of advantages, but also a few disadvantages. Both strengths and constraints are displayed in Table 7.1.

Most importantly, the use of online-based tools is especially suitable when dealing with experts: online-based research is more efficient and intuitive; relations are established naturally and quickly on the basis of common interest (act–react–react–. . .). Also, it can be observed that online networks and forums are more actively shaped and have a higher response rate than questionnaires, email queries, and other modes of inquiry.

Thus, information gains arise mainly through networking. This direct contact and exchange with experts enables firsthand information accessed from the home desk which otherwise would require extensive field trips, interviews, etc.

In practical terms, this is the core advantage of the chosen strategy. Information is easily accessible with low costs and allows for global coverage with no travel required.

Table 7.1 Strengths and constraints of the applied research methodology

Strengths	Constraints
• Firsthand information through direct expert contact • Relations to experts established naturally and quickly on the basis of common interest • Online networks and forums are more actively shaped and have a higher response rate (than questionnaires, email queries and others) • Easily accessible with low costs • Global coverage, no travel required • Feasible in a relatively short period of time • Supports learning processes and can be adjusted over time (compared to field trips) • Helps to explore and track the newest topical literature of various forms • Good means of documentation (early and online, use of blog as Content Management System)	• A lot of communication efforts needed • Necessity for continuity • In a global setting inequalities of Internet access • Lacking representativeness and underrepresentation of small-scale projects (hardly documented online)

Source: Author

At the same time it is also a major constraint of the chosen approach: It is clear that an Internet-based research is bound to the specific characteristics of this medium, most notably large inequalities of access to the Internet and information quality (especially for smaller projects that are usually less well documented online) in a global setting.

For this matter, the chosen approach can by no means ensure representativeness. It is however most suitable in explorative cases as the present study.

Alternative Approaches

The most efficient alternative way to envisage a similar scope with reasonable efforts, covering different sectors and locations as well as offset implementation schemes worldwide, would be an analysis of secondary and gray literature. This, however, has already been incorporated into the present study.

The alternative would be to reduce the scope to a specific location, sector etc. and build on field trips. While this is in principle promising, it would be most appropriate not in replacing the current study, but instead building on its results and further detailing them.

Summarizing, given the explorative character of the research in a narrowly limited research field, that however geographically extends over a multitude of places worldwide, the research methodology has proved suitable to tackle the described research problem. It has provided means to effectively and efficiently answer the research questions and build a typology of biodiversity offsets.

The applied research methodology was fit for purpose under the given research premises. Thus, it would need to be adjusted to be transferred to other research contexts. Nevertheless, it can provide a model upon which to draw for developing an adequate research methodology for the specific case.

7.1.2 Mixed Types: The Limits of Typification to Represent Real World Examples

As with any typology, this typology of biodiversity offsets is an attempt to theoretically categorize and thus to better understand the practice of offsets and the particular differences that arise from different context conditions. As diverse as these may be, however, a number of common tendencies or major drivers for offset development have been identified and translated to the present seven types.

The analysis of offset cases from around the world (see Appendix 1: Screening of Biodiversity Offset cases worldwide) and in particular the illustrative case studies for the built types of biodiversity offsets (see Chap. 6) have shown that these are ideal types that are represented in their purest form only to a limited extent in reality. Rather, intersections of these types are most commonly encountered in practice. In fact, the larger a development project, the more likely it is that the project is actually a conglomerate of several or all of the mentioned types. The interplay and mutual reinforcement is a major characteristic and can add to the success of the development project.

Nevertheless, one major limitation is inherent to the built typology: when in practice mostly a combination of influences and motivations is encountered, it is difficult to judge which is the predominant driver or initiator. This judgment obviously can only be made normatively. Therefore, it must be kept in mind when analyzing or drawing upon the typology that its reliability is bound to normative choices and hence it cannot be assured to be reproduced in the same way, i.e., the results might be different when undertaken by someone else or in a different setting.

7.1.3 Difficulties of Comparability and Clear Classification of Biodiversity Offsets

Offset, Compensatory Conservation, and Additional Conservation Outcomes: What Is an Offset and What Not?

As has been noted (see Sect. 1.2), biodiversity offsets are now used to refer to a wide array of compensation approaches worldwide (Fig. 7.1). Consequently, differences exist with regard to the terminology, e.g., biodiversity offsets vs. environmental compensation vs. ecological restoration vs. impact mitigation etc.

These differences not only refer to the terminology, but also to different measures and outcomes associated with them in practice. Taking this ambiguity into consideration, for example ten Kate et al. (2011) differentiate between "compensation" and "offset" as is shown in Fig. 7.2, whereby only offsets would meet the BBOP standard (cf. Fig. 4.5) and qualify to reach the goal of no net loss or net gain. Consequently, the mitigation hierarchy can be modified to take into account a more refined range of compensatory actions, e.g., including additional conservation actions, as shown in Fig. 7.3.

Building on these observations, this leads to the crucial question of what is or what qualifies as an offset. Despite an elaborate discussion and development of the

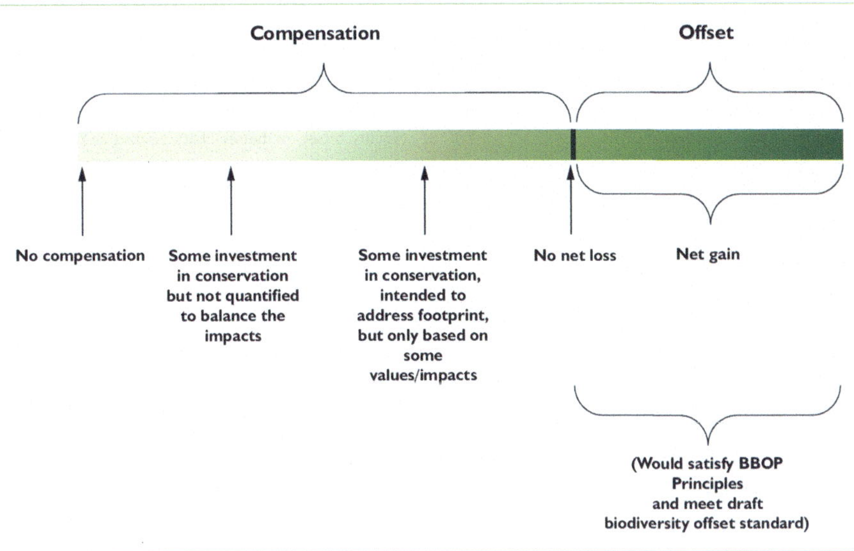

Fig. 7.1 Offset versus compensation (Source: ten Kate et al. 2011)

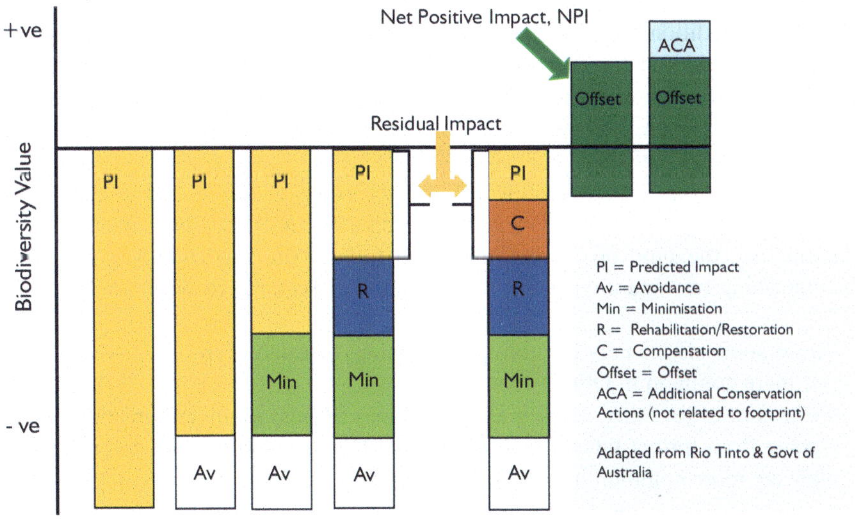

Fig. 7.2 Suggestion of a refined mitigation hierarchy (Source: ten Kate et al. 2011)

definition (cf. Sect. 1.2) and quality standards (cf. Sect. 4.3) for biodiversity offsets, some fuzziness remains in practice as to what counts as an offset. Ten Kate et al. (2011) have attempted to solve this problem with a checklist as presented in Fig. 7.3.

The present study has used a broad understanding of biodiversity offsets as defined by the involved stakeholders (i.e., if it was presented as an offset). In this

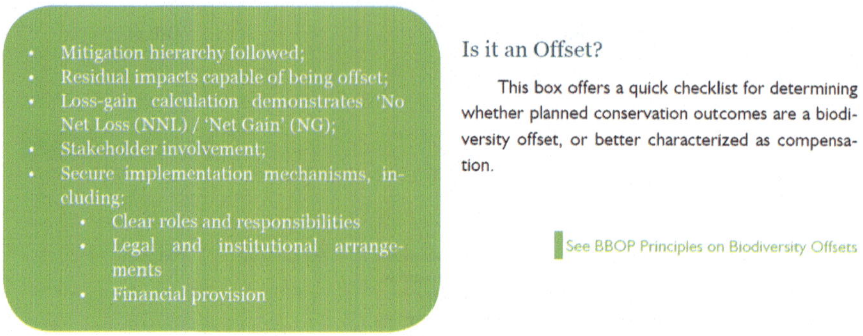

Fig. 7.3 Is it an offset? A checklist (Source: ten Kate et al. 2011)

regard, the mentioned checklist for biodiversity offsets presents an ideal case that is not always applicable in practice. Sometimes, measures are christened and promoted as biodiversity offsets, even though following these criteria they would qualify as other forms of conservation action. In other cases, such as the presented case study on the Antamina mine in Peru (see Sect. 6.7.2) an action that would qualify as an offset was not initially intended as one.

For the exploration of evidence on biodiversity offset cases worldwide this presents an additional difficulty and a limitation of the study, that could not be satisfactorily addressed in this study. Rather, it was bound to the goal of giving a broad general overview of biodiversity offset types in a worldwide context, instead of an in-depth presentation of a small number of case studies that would exactly fit under these specifications.

Moreover, even if conservation actions according to the checklist qualify as offsets, there may still be large variation among them. Most notably, offsets can be classified by the type of conservation used to generate biodiversity gains, thus distinguishing two major types as has been noted in Sect. 4.3 (ICMM IUCN 2012):

- Restoration offsets entail restoring, enhancing, or establishing biodiversity, and are more common in OECD countries.
- Protection offsets (also known as averted loss offsets) involve protecting biodiversity from further threats such as grazing, fire, overfishing, and deforestation; they are more common in non-OECD countries.

The Issue of Scale and Global Differences in Offset Schemes

The discussion about biodiversity offsets covers two important aspects: first, how to assure the quality of an offset and second, how to implement an offset (leaving aside more general ethical considerations of framing with regard to the replaceability of nature and the appropriateness of offsets). While from a quality or outcome oriented point of view offsets need to be judged by similar standards, an implementation point of view considers the governance of offset schemes (this has been the focus of this study) and thus necessarily refers to different contexts and conditions.

In fact, throughout this study it has become very clear that context is very important. In particular, huge differences exist globally between offsets in the developed and in the developing world. Therefore, the question arises, whether we can ultimately compare and judge by the same standards a large mine in Madagascar or a huge hydropower project in Laos and a gas station in the United Kingdom. There are three reasons why this is challenging:

1. *Scale*: The scale of a development project and the associated impacts and offsets matters a lot. Not only do larger projects usually bear a greater magnitude of cumulative and indirect effects, their mere size requires huge efforts of cooperation and collaboration of multiple stakeholders. At best, this can be a source of control to prevent unforeseen outcomes; at worst, it can absorb much of the capacity, necessary for implementing conservation measures, for communication means. Whereas small-scale development shall fit to the local context above all, large developments tend to focus more on strategic issues. The latter applies particularly to developing countries, where development projects with huge adverse environmental effects (e.g., from hydropower) are related to overarching goals of promoting renewable energy and combatting climate change etc.
2. *Naturalness*: The impact of a proposed development needs to be judged in relation to the natural conditions into which it is implanted. Obviously, offset schemes in culturally shaped landscapes as in most parts of Western Europe (in particular in the Netherlands) build on completely different prerequisites than offsets in large pristine areas. In large parts of the developing world (almost) untransformed nature exists, but is usually under threat of many emerging influences, both from the local population and external developers.
3. *Human environment*: Development projects cannot be seen isolated from their socioeconomic context. While this holds to be true for any development context, it is more than elsewhere shaping the implementation of offsets in developing countries: "Islands of biodiversity cannot survive in a sea of poverty." (IAG 2005) Therefore, appropriate mechanisms need to be put in place to ensure the offset can be effective while not penalizing the local communities that live in and depend on this biodiversity.

Another example of global differences is evident in the discussion about the relation of biodiversity offsets and protected areas and the associated ultimate question whether the two are mutually exclusive or intertwined. This topic has been explored in detail at the 2014 World Parks Congress in Sydney where IUCN launched an in-depth debate related to "no-go" decisions, permanence, and additionality and ultimately appropriateness and potential implementation of offsets in protected areas (Darbi 2015).

Generally, in terms of no net loss, neither development nor biodiversity offsets should be implemented in protected areas. Otherwise, additionality in the strictest sense could not be reached if the offset merely replaces other responsibilities (in particular conservation goals and duties governments have committed to). However, as many examples from the developing world show (e.g., the mentioned Nam

Theun 2 Hydropower Facility in Laos, see Sect. 6.2.2), offsets are not seldom used to secure or improve protected areas.

Also the issue of averted loss offsets (conservation) in relation to protected areas is overriding in the developing world (as compared to restoration offsets). As ICMM and IUCN (2012) note, "recognized biodiversity priority sites such as Key Biodiversity Areas and some unfunded protected areas [. . .] may offer potential, mainly in non-OECD-countries."

7.2 Discussion of the Research Hypotheses

This study is based on five hypotheses, which are briefly discussed in Fig. 7.4.

Voluntariness as a Gradual Continuum: Blurring the Edges?

While in fact the voluntariness of a specific characteristic or criterion can be described as a gradual continuum, in practice usually a superposition of multiple characteristics is encountered. The degree of voluntariness may vary from one characteristic to another, thus requiring a normative decision as to which characteristic is the preferred criterion to judge the voluntariness of an action. As this decision cannot be taken at a generalized level, the degree of voluntariness varies greatly from one case to another (or even within the same case!) depending on the outcome of this decision.

Notwithstanding this general difficulty to localize biodiversity offset cases along a linear gradual continuum between mandatoriness at one edge and voluntariness at the other, it is important to distinguish and describe different types of voluntary biodiversity offsets to better understand the specific features and circumstances of these particular types. The limitations of such a typology lie in quantitatively comparing these types, i.e., value judgments such as "case A is less voluntary than case B" are tempting, but remain mostly inappropriate and do not fit with practice.

Nevertheless, following two attempts to build a hierarchical sequence from mandatory to voluntary are shown as illustrated in Figs. 7.5 and 7.6:

1. The first builds on the criterion of *choice or eligibility of options*. In the scope of this study this has been defined to include the following four categories regarding the question of whether there is an influence/change to the eligibility of options for a rational person (ranging from mandatory to voluntary): strong (vis absoluta), significant (vis compulsiva), marginal (coercive offer), and none (willful voluntary action).
2. The second builds on the likely consequences of the different offset types in terms of the possibility of sanctions. Here, three categories are distinguished (from mandatory to voluntary): formal sanctions, informal sanctions, and no sanctions.

As a result of the present work it becomes clear that such a sequence can indeed be constructed normatively and can be supported with individual cases, but that for the practical and theoretical debate the specifics and characteristics of the individual

1. **Differences exist regarding the voluntariness of Biodiversity Offsets.**

The described types and illustrative case studies show the variety of different voluntary biodiversity offsets (see Chapter 6). However, there are also similarities as well as intersections, in particular with regard to mixed types (see chapter 7.1.2).

2. **Biodiversity Offsets cannot be adequately explained as a dichotomy of mandatory vs. voluntary offsets.**

As has been assumed, a dichotomy of a regulatory and a voluntary approach can be judged an oversimplification. This is confirmed both in the literature at the general level and in practice for the specific case of biodiversity offsets.

In the literature, a distinction between three poles is being made in the environmental sector, namely 1. Regulation, 2. Market forces and 3. Voluntary approaches. Furthermore, both voluntary approaches and market forces have not proven to replace traditonal environmental regulation as has been widely assumed some decades ago, nor does private action make public action dispensable. Instead a "level playing field [...] of effective global rules" (Baker 2001) includes elements of regulation, market incentives, social norms and responsibility.

It is a distinct feature of our current era that these are increasingly intertwined as can be exemplarily shown at the new concept of "voluntary compliance".

3. **The voluntariness of biodiversity offsets can be described as a gradual continuum.**

Contrary to the other four, this hypothesis has been refuted in the scope of this study. Therefore, it is discussed in greater detail below.

4. **A typology of biodiversity offsets (and different types) can be build to analyse and illustrate the space between the two poles of this continuum.**

The typology, consisting of seven types (including four subtypes) is presented in Chapter 6 and discussed in Chapter 7 (see especially Section 7.1.2 Mixed types – the limits of typification to represent real world examples).

5. **(These types help to analyse and understand the different outcomes of biodiversity offsets).**

This hypothesis has not been the explicit focus of this study. Nevertheless, it has been mentioned with regard to the potential of the present work for future research. As argued in Chapter 9, the ultimate goal of the presented typology would be to evaluate the outcome of the different types of biodiversity offsets with regard to differences in effectiveness and efficiency.

Fig. 7.4 Hypotheses in the scope of this study (Source: Author)

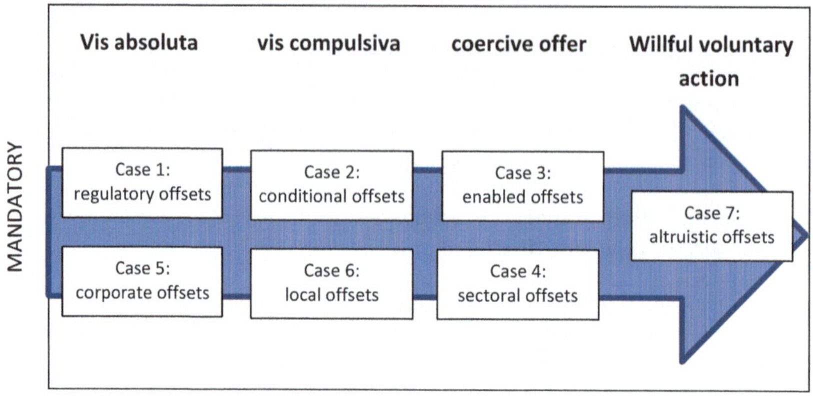

Fig. 7.5 Four-step gradual continuum of voluntariness of biodiversity offsets building on the choice or eligibility of options (Source: Author)

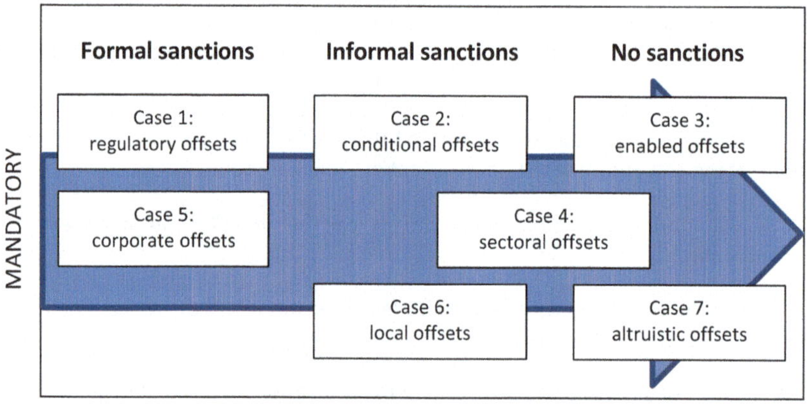

Fig. 7.6 Three-step gradual continuum of voluntariness of biodiversity offsets building on the possibility of sanctions (Source: Author)

types (including the degree of institutionalization) are of greater importance than the possible (and indeed not very reliable) sequence.

Accordingly, the third hypothesis, "The voluntariness of biodiversity offsets can be described as a gradual continuum," is refuted as a result of this study. It is true that voluntariness can indeed be understood as a gradual phenomenon, but this cannot be measured on a scale. In addition, it has been shown for the specific subject of biodiversity offsets that voluntariness cannot be fixed to a single criterion, but rather is the result of an intersection of several criteria. From this intersection of qualitative criteria no universal ranking or sequence can be derived.

7.3 Discussion of the Single Types

In the following, the seven developed types of biodiversity offsets are analyzed one by one, building on the practical evidence from the screening of biodiversity offset cases (see Sect. 5.5 and Appendix 1). This includes three aspects:

1. *Magnitude*: a quantitative representation of the number of cases per type in tabular format, comprising (a) a list of cases that are primarily attributed to this type and (b) a list of cases that are secondarily attributed to this type (i.e., in addition to being primarily attributed to another type)[1]
2. *Location*: a geographical representation of the cases [(a) as solid dots and (b) as circles] on a world map
3. *Particularities*: a qualitative analysis of the particularities and crucial aspects that were observed in the evidence base

7.3.1 Magnitude, Location, and Particularities of Type 1 Regulatory Biodiversity Offsets

Magnitude of Type 1 Regulatory Biodiversity Offsets

As can be seen from Table 7.2, from the identified total of 72 cases, 27 have been primarily attributed to type 1 regulatory biodiversity offsets. For this type no secondary attribution has been noted because it is assumed that in cases where several drivers coexist, compliance with regulatory or government provisions is most likely to be the ultimate goal.[2]

The number of cases attributed to this type represents more than one-third of all cases and is by far the largest category by type. Nevertheless, in absolute terms, the number of cases of regulatory biodiversity offsets (type 1) is clearly underrepresented in this study. This is due to the fact that the long-standing experiences with US Wetland Mitigation and German Impact Mitigation are only exemplarily included (with three and one cases, respectively). While no clear numbers on the total number or area of offsets that have been implemented under the two schemes occur, it can be estimated that at least several hundreds of habitat banks exist for each of them (Madsen et al. 2011; Böhme et al. 2005).

Furthermore, as Darbi et al. (2010) find, Brazil has two major offsetting schemes in place, the "forest set-aside offsets" after the Forest Code and "project developers' offsets." These might also be underrepresented in the present study; however, the

[1]For the majority of cases a clear primary attribution to one type was possible; however, a few cases remain that are split between two equally important drivers. Secondary attribution of cases can refer to different types (one or more).

[2]Notwithstanding, as has been observed in Sect. 5.4, for example the pressure that a corporation exerts on the local or project level (type 5) may in some instances more directly and more strictly affect the specific case than the regulative pressure in type 1.

Table 7.2 Type 1 regulatory biodiversity offsets: primary and secondary attribution of identified cases

No.	Name	Country	Continent	Driver (influence) for this type[a]
Primary attribution				
8	Anglo Coal Isibonelo colliery (open cast coal mining)	South Africa	Africa	Government
10	Mount Royal Golf Estate (expansion of golf course with residential and commercial development)	South Africa	Africa	Government
11	Pulp United Pulp Mill (bleached chemi thermos mechanical pulp mill)	South Africa	Africa	Government
22	Koala Offset in South East Queensland	Australia	Australia and Oceania	EPBC Act
24	Mc Arthur River Mine Open Cut Project	Australia	Australia and Oceania	Northern Territory Government
26	Basslink Undersea Power Cable	Australia	Australia and Oceania	Victoria's Native Vegetation Management
27	Raventhorpe Nickel Project	Australia	Australia and Oceania	EPA
28	Karuah bypass offset	Australia	Australia and Oceania	NSW Biobanking scheme
29	Cumberland Plain Woodland of Western Sydney (clearing of natural vegetation)	Australia	Australia and Oceania	NSW Biobanking scheme
31	Open cut extension at the Warkworth Coal Mine (Rio Tinto Coal Australia)	Australia	Australia and Oceania	EPBC and TSC Act
33	Burgberg Zschaitz habitat bank (offset for windpower development)	Germany	Europe	German IMR
34	Contournement Nîmes Montpellier—A new high speed rail bypass	France	Europe	EU BHD, national law: Environmental Code
36	Bardon Hill Quarry	United Kingdom	Europe	Natural England
45	Offsetting the ecological impact of the A65 Langon-Pau motorway	France	Europe	EU BHD, national law: Environmental Code
46	Apennine Wind Farms	Italy	Europe	EU BHD, national law
48	Ume Delta River Railway	Sweden	Europe	EU BHD, national law

(continued)

Table 7.2 (continued)

No.	Name	Country	Continent	Driver (influence) for this type[a]
49	Kennecott Utah Copper Mine (expansion of the tailings of a copper mine)	United States	North and Central America	US Wetland Mitigation
57	Tygart Valley River Stream and Wetland Bank	United States	North and Central America	US Wetland Mitigation
58	South Fork Hoppers Creek Mitigation	United States	North and Central America	US Wetland Mitigation
61	NGTL pipeline proposals in Horne River	Canada	North and Central America	NEB
62	Total Joslyn oil sands mine: offsets for species at risk	Canada	North and Central America	Canadian Environmental Assessment Act, NEB
63	Enbridge Northern Gateway pipelines	Canada	North and Central America	Canadian Environmental Assessment Act, NEB
67	Gasoducto Gasbol (pipeline)	Bolivia, Brazil	South America	SNUC Act
72	Gramalote Mining Project	Colombia	South America	Regulation on no net loss
Secondary attribution				
–	–	–	–	–

[a]Abbreviations: *EPBC* Environment Protection and Biodiversity Conservation Act 1999, *EPA* Environmental Protection Act, *NSW* New South Wales, *TSC* Threatened Species Conservation Act, *EU BHD* Birds and Habitats Directives of the European Union, *SNUC* National System of Conservation Units Act, *NEB* National Energy Board Act
Source: Author

evidence base and number of cases that have been implemented in practice is unclear.[3]

Location of Type 1 Regulatory Biodiversity Offsets

With a view to the map (Fig. 7.7) three major clusters can be identified: (1) North America, (2) Europe, and (3) Australia. This result corresponds to expectations, because in particular the United States and Canadian as well as Australian offset schemes are well known and documented in the scientific literature (e.g., Madsen et al. 2011; Bayon et al. 2012; Poulton 2015; Miller et al. 2015).

[3]This may partly be due to language restrictions in the scope of this study, which did not include sources in Portuguese.

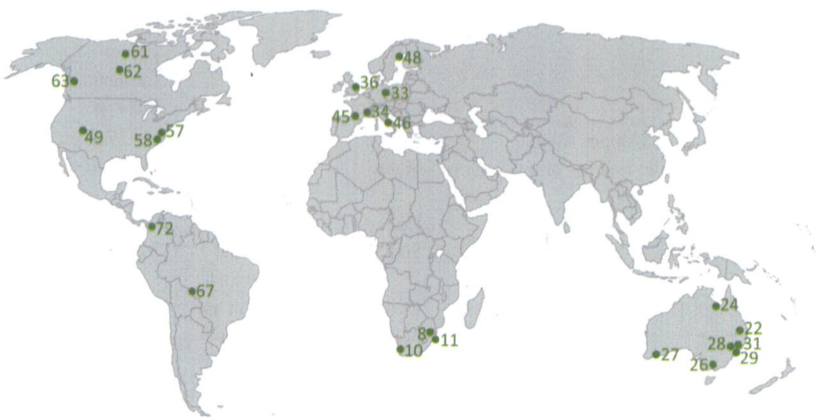

Africa
8. Anglo Coal Isibonelo colliery
10. Mount Royal Golf Estate
11. Pulp United Pulp Mill

Asia
-

Australia
22. Koala Offset in South East Queensland
24. Mc Arthur River Mine
26. Basslink Undersea Power Cable
27. Raventhorpe Nickel Project
28. Karuah bypass offset
29. Cumberland Plain Woodland / Western Sydney
31. Warkworth Coal Mine

Europe
33. Burgberg Zschaitz habitat bank
34. CNM high speed rail bypass
36. Bardon Hill Quarry
45. A65 Langon-Pau motorway
46. Apennine Wind Farms
48. Ume Delta River Railway

North & Central America
49. Kennecott Utah Copper Mine
57. Tygart Valley River Stream and Wetland Bank
58. South Fork Hoppers Creek Mitigation
61. NGTL pipeline proposals in Horne River area
62. Total Joslyn oil-sands mine
63. Enbridge Northern Gate-way pipelines

South America
67. Gasoducto Gasbol
72. Gramalote Mining Project

Fig. 7.7 Type 1 regulatory biodiversity offsets: map of attributed cases from the worldwide screening (Source: Author)

Moreover, the six cases that have been gathered for Europe focus on the implementation of the EU Birds and Habitats Directive as translated into different national contexts, with the exemption of the Burgberg Zschaitz habitat bank in Germany (no. 33 in the screening). This is in line with the observation that EU legislation currently creates limited demand for compensation except for damage to Natura 2000 sites (see Sect. 4.1.3).

Further cases are mapped for South Africa, which despite Provincial Guidelines on Biodiversity Offsets (Department of Environmental Affairs and Development Planning 2007) has no mandatory system in terms of law, but rather the individual offsets are mandated through Environmental Impact Assessment regulation (Manuel 2013).

In addition, the two cases that have been mapped for South America point to the rise in interest of offsetting policies in Latin America (cf. Villarroya et al. 2014; Lopez and Quintero 2015). Alongside Brazil, which has a longer offsetting tradition (Darbi et al. 2010), Colombia has only recently set up an offsetting scheme (Lopez and Quintero 2015). While Villarroya et al. (2014) find that most Latin American countries "enable the use of offsets only Brazil, Colombia, Mexico and Peru explicitly require their implementation," no sufficient evidence could be found supporting broad implementation of these offset requirements in practice.[4]

Particularities of Type 1 Regulatory Biodiversity Offsets

In a global context, when considering the obligation for restoration of impacts on biological diversity and natural resources, impact mitigation regulations mostly focus on Environmental Impact Assessments (EIA) as the main implementation tool (Darbi et al. 2010). An example for this is South Africa, as has been noted above. However, EIAs (and the corresponding mitigation and compensation measures) are only conducted for major engineering projects and programs. This is a major limitation.[5] Consequently, a large number of impacts are not covered by mandatory compensation approaches, due to the absence of impact mitigation for small-scale projects and traditional land use practices etc. (Darbi et al. 2010).

As has been noted, offset requirements usually result from provisions fixed in different laws (e.g., related to the environment, forests, fisheries, environmental impact assessment etc.). However, several offset cases in South Africa show that regulatory biodiversity offsets do not necessarily stem from an offset-requiring legislation or framework, but may be conditioned by governmental bodies.

For example the Department of Environmental Affairs and Development Planning (2007) mentions that the case studies comprised in the Provincial Guideline on Biodiversity Offsets of the Western Cape Province "provide an indication of the possibilities of an offset system," yet "should be interpreted as having taken place before the development of this guideline." This means that neither are they pilots nor do they fall under the new regulative framework. Nevertheless, they had been requested by the authorities, e.g., the authorization of the Anglo Coal mine near Kriel in Mpumalanga (no. 8 in the screening) was bound to the condition for wetland offsite mitigation (Isibonelo Offsite Wetlands Rehabilitation). Similarly, the extension of the Mount Royal Golf and Country Estate (no. 10 in the screening) was approved, requiring that a conservation area be established and incorporated as a Contractual Nature Reserve into Cape Nature's Stewardship Programme (Department of Environmental Affairs and Development Planning 2007). These examples highlight the power of permitting authorities even in cases of nonexistent offset regulation.

In the case of the extension of the Bardon Hill Quarry in Leicestershire in the United Kingdom (no. 36 in the screening), offsets were proposed to address concerns raised by Natural England[6] in its planning consultation response to provide "an

[4]Again, language restrictions apply and it would thus be a topic for further research.

[5]An exception is the German Impact Mitigation Regulation, which follows an area-wide approach as has been described in Sect. 4.1.2.

[6]The nondepartmental public body of the UK government responsible for ensuring that England's natural environment, including its land, flora and fauna, freshwater and marine environments, geology and soils, are protected and improved.

agreed programme of mitigation and compensation that will be delivered as part of this proposed development" (Aggregate Industries UK Ltd. 2010).

There are also examples for weaker forms of legislation that however give authorities the opportunity to request offsets from developers. For example in Sweden offsets are only strictly mandated for impacts in protected areas, but authorities may justify mandating offsets even outside protected areas if there is a common public interest (personal communication with Asa Granberg).

The former represent examples of government pressure in the absence of a regulatory offset framework and highlights the crucial role of government bodies in regulatory biodiversity offsets.

Doswald et al. (2012) note that companies performing regulatory biodiversity offsets with a "no net loss" objective sometimes choose to go for "net gain," thus adding a voluntary offset to the mandatory offset, which however are both operated under a regulatory biodiversity offsets scheme, e.g., in the case of the Kennecott Utah Copper Mine (no. 49 in the screening).

7.3.2 Magnitude, Location, and Particularities of Type 2: Conditional Biodiversity Offsets

Magnitude of Type 2 Conditional Biodiversity Offsets

Conditional biodiversity offsets, i.e., offsets mandated by the requirements of lenders, are observed to be the main driver in 12 cases (primary attribution) and another 4 cases are at least partially influenced by them. This relates to one-sixth (16.7%) or more than one-fifth (22.2%), respectively.

This number is considerably low compared to the other types and also lower than expected—given that cases under this type are major development projects (often with public involvement) in particular in the energy and extractives sector and thus are assumed to be documented online to a large extent.

Among the compiled cases a clear dominance of the World Bank and its private sector arm, the International Finance Corporation (IFC) can be observed. Thus, as has been expected, IFC Performance Standard 6 (see Sect. 6.2.1) is the major driver for biodiversity offsets under type 2 conditional biodiversity offsets.

However, IFC PS6 which was originally adopted in 2006 and updated in 2012, building on lessons from first implementation experience (ten Kate et al. 2011), has only been in place a few years. Consequently, the implementation is still in its early stages as first demonstration projects demonstrate, e.g., Rio Tinto's Oyu Tolgoi copper and gold mine in Mongolia (no. 18 in the screening) is "the first project to publicly disclose full documentation that conforms with the revised (2012) PS6 requirements" (ICMM IUCN 2012). Thus, an increase in the number of conditional offset cases is expected. However, their number will remain limited by the number of large-scale development projects under the scope of World Bank, IFC, and other lender financing (Table 7.3).

Table 7.3 Type 2 conditional biodiversity offsets: primary and secondary attribution of identified cases

No.	Name	Country	Continent	Driver (influence) for this type[a]
Primary attribution				
4	Piloting aggregated offsets and mining in Liberia	Liberia	Africa	World Bank
7	Bujagali Energy Limited (Hydropower plant and transmission line)	Uganda	Africa	World Bank, African Development Bank
9	Petroleum Development and Pipeline Project	Chad and Cameroon	Africa	World Bank, IFC
17	Nam Theun 2 Hydropower Project	Laos	Asia	World Bank, Asian Development Bank
18	Oyu Tolgoi LLC (copper/gold mine)	Mongolia	Asia	IFC
54	Via San Lorenzo—Olanchito (road infrastructure)	Honduras	North and Central America	World Bank
55	Hidroelectrica Reventazon (hydroelectric power station)	Costa Rica	North and Central America	IDB
60	Pueblo Viejo gold mine expansion	Dominican Republic	North and Central America	IFC
65	Hidroelectrica Yacyreta (hydro-electric power station)	Argentina and Paraguay	South America	World Bank, IDB
66	Pasto Mocoa in the Amazon (Mocoa-Saanciso road)	Colombia	South America	IDB
68	Brisas Gold and Copper Project (Open pit mine, Gold Reserves Inc)	Venezuela	South America	IFC, Equator principles
72	Gramalote Mining Project	Colombia	South America	IFC
Secondary attribution				
1	Ambatovy BBOP Pilot Project Case Study (Nickel mine)	Madagascar	Africa	n/a
3	Simandou (open-pit iron-ore mine)	Guinea	Africa	IFC
5	BBOP Pilot Case Study Akyem Gold Mining Project	Ghana	Africa	n/a
67	Gasoducto Gasbol (pipeline)	Bolivia, Brazil	South America	IDB

[a]Abbreviations: *IFC* International Finance Corporation, *IDB* Inter-American Development Bank
Source: Author

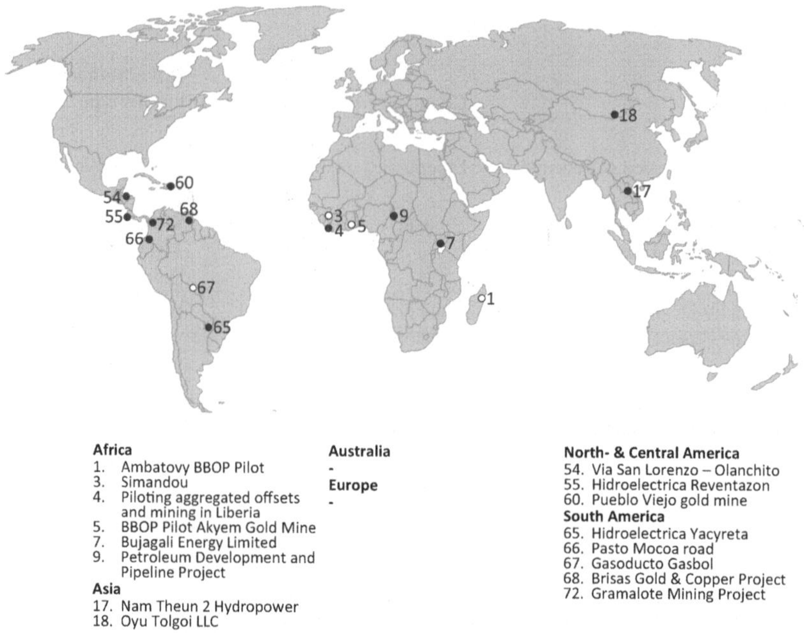

Fig. 7.8 Type 2 conditional biodiversity offsets: map of attributed cases from the worldwide screening (Primary attributions are presented as solid dots and secondary attributions as circles) (Source: Author)

Location of Type 2 Conditional Biodiversity Offsets

As can be seen from the map, there is a strong focus on Latin American and African countries. This can be explained partly by the investment focus of the lenders (in particular World Bank and IFC) on developing and threshold countries.

Despite this observation, Asian countries are clearly underrepresented in the screening. It remains unclear whether this is due to methodological and procedural limitations, e.g., in terms of lacking documentation in general and no documentation online in particular, or language barriers, or whether a real implementation deficit of IFC PS6 exists in the Asian context. Further research would be needed to explore this question (Fig. 7.8).

Particularities of Type 2 Conditional Biodiversity Offsets

Lending organizations, with the World Bank and the International Finance Corporation and the Equator Banks at the forefront, have formulated clear requirements and detailed guidance on quality standards for environmental and social performance and management in development projects. This includes rather specific requirements on biodiversity offsets as is outlaid in IFC PS6 (see Sect. 6.2.1). However, as the screening of biodiversity offsets revealed, practical evidence is often lacking an appropriate and traceable level of detail.

Given the huge scope, acreage, and nature of development projects in combination with the context in which the development is implemented, offsets may not always be used to their full potential. This is particularly the case in the context of prevailing economic and social considerations, but also overarching environmental considerations at very general level as in the case of the presented Nam Theun 2 Hydropower project (no. 17 in the screening, see Sect. 6.2.2), where the environmental benefit of the project was said to contribute to tackling climate change by fostering the exploration of renewable energy. The problem with this argumentation is, however, that this environmental benefit at global level goes along with an enormous transformation and loss of ecosystems and habitats at site level and severe disturbances of the water regime of the dammed river.

Despite an accompanying due diligence process by lenders through the establishment of steering committees or an International Advisory Group to the World Bank, which are commissioned for reporting, the proper evaluation and follow-up on the offsets appears to remain limited and few concrete observations can be retrieved from the reports. This is basically due to the fact that projects involve sometimes competing priorities and biodiversity offsets form just one small part of the overall project management.

On the contrary, other examples can be cited, e.g., the abovementioned Oyu Tolgoi mine in Mongolia (no. 18 in the screening), where a thorough application of the provisions set out in IFC PS6 has been carried out with the help of a specialist consultancy.[7]

Seeing the only recent adoption of the IFC Performance Standards as well as their translation into the Equator Principles on the one hand and the long project timeframes and lifetimes on the other hand, it is most likely that the implementation of conditional biodiversity offsets in future will mature in practice building on best practice experience.

7.3.3 Magnitude, Location, and Particularities of Type 3: Enabled Biodiversity Offsets

Magnitude of Type 3 Enabled Biodiversity Offsets

In terms of numbers of cases, enabled biodiversity offsets with one quarter (25%) represent the second largest group of primarily attributed cases after type 1 regulatory biodiversity offsets. In addition, for a large number of cases (13) offset enabling frameworks are acknowledged as additional motivations (i.e., seconday attributions). Together, primary and secondary attribution of enabled biodiversity offsets with more than 40% represent the largest number of offset cases (31) observed in the screening.

[7]UK-based boutique consultancy "The Biodiversity Consultancy," http://thebiodiversityconsultancy.com/

This number appears slightly too high in comparison to the other types, due to the comprehensive inclusion of evidence from the United Kingdom offset pilots (no. 37-44 in the screening), Nevertheless, enabled biodiversity offsets are among the major drivers of offset implementation in practice, as the number of cases attributed to type 3b (enabled biodiversity offsets driven by NGOs) justifies (21) (Table 7.4).

Location of Type 3 Enabled Biodiversity Offsets

With regard to the global distribution of enabled biodiversity offsets no clear pattern can be observed. This is especially true for enabled biodiversity offsets driven by nature conservation NGOs such as the International Union for the Conservation of Nature, Conservation International or in a wider sense the Business and Biodiversity Offsets Program (type 3b new global norms). These appear to be acting on a broad global scale in a variety of contexts that go beyond their original regional or national context, e.g., the US based NGO "The Nature Conservancy" has extended its "development by design" scheme, e.g., to projects in Colombia and Mongolia (ICMM 2010).

For enabled biodiversity offsets driven by governments (type 3a) the situation on the ground is currently dominated by the two recent changes in Europe, with the United Kingdom offsetting pilot phase and the enforcement of offset implementation in France (see Sect. 4.1.3). Nevertheless, government enabled offset frameworks can also be envisaged in different contexts as the example of an offsetting scheme being trialed for petroleum and natural gas development in Uzbekistan (no. 21 in the screening) shows (Fig. 7.9).

Particularities of Type 3 Enabled Biodiversity Offsets

As has been noted, two broad categories of enabled biodiversity offsets have been observed: type 3a driven by government and type 3b driven by NGOs. For both of these categories, as is obvious from the BBOP pilots but also from the controversy on the UK pilots, enabled biodiversity offsets are based on a comparably weak institutional arrangement. Thus, unlike regulatory regimes, the continuity of offset enabling frameworks cannot be safeguarded in perpetuity. Thus, as has been noted earlier, government enabled offset schemes (type 3a) are often of temporary nature and represent an exploration or trial phase leading toward the ultimate goal of a regulatory offset scheme fixed by law. This can for example be observed in the case of the UK pilots (no. 37–44 in the screening), but also in the case of the Operation Cossure habitat banking pilot in southern France (no. 35) which aims to underpin the existing legal offsetting requirement in France.

By contrast, organizations of civil society such as BBOP, IUCN or The Nature Conservancy aim at setting new global standards for biodiversity offsets (type 3b), thus establishing a baseline against which future offsets shall be evaluated. A number of pilots often serve for illustrative purposes and to establish best practice evidence and guidance.

However, as these pilots in every case rely on a specific set of stakeholders (and committed individuals, their persistence depends on a variety of context factors and

Table 7.4 Type 3 enabled biodiversity offsets: primary and secondary attribution of identified cases

No.	Name	Country	Continent	Driver (influence) for this type[a]
Primary attribution				
1	Ambatovy BBOP Pilot Project Case Study (Nickel mine)	Madagascar	Africa	BBOP
5	BBOP Pilot Case Study Akyem Gold Mining Project	Ghana	Africa	BBOP
20	PEARL GTL PROJECT, QATAR BBOP Pilot (Gas To Liquid)	Qatar	Asia	BBOP
21	Offsets for Petroleum and Natural Gas Development in the Ustyurt	Uzbekistan	Asia	Uzbek government, UNDP
23	Strongman Coal Mine, Solid Energy New Zealand BBOP Pilot (coal mine)	New Zealand	Australia and Oceania	BBOP
30	IndometCoal Project (metallurgical coal resource, BHP Billiton)	Indonesia	Australia and Oceania	FFI
35	Operation Cossure	France	Europe	French Ministry of Ecology
37	South Devon UK pilot	United Kingdom	Europe	UK government
38	The North Devon UNESCO Biosphere Reserve UK pilot	United Kingdom	Europe	UK government
39	Essex UK pilot (infrastructure development, housing scheme)	United Kingdom	Europe	UK government
40	Doncaster UK pilot	United Kingdom	Europe	UK government
41	Development Project in Dudley, West Midlands	United Kingdom	Europe	UK government
42	Project Pine-wood (ARUP)	United Kingdom	Europe	BBOP, UK government
43	Nottinghamshire UK pilot	United Kingdom	Europe	UK government
44	Warwickshire, Coventry and Solihull UK pilot	United Kingdom	Europe	UK government
52	Case study of the BBOP pilot biodiversity offset work on Bainbridge Island (real estate and infrastructure)	United States	North and Central America	BBOP, local government
53	Jonah Natural Gas Field	United States	North and Central America	TNC
71	Amaila Fall Hydro Project	Guyana	South America	CI

(continued)

Table 7.4 (continued)

No.	Name	Country	Continent	Driver (influence) for this type[a]
Secondary attribution				
2	QIT Mining Madagascar QMM (ilmenite mining)	Madagascar	Africa	IUCN
6	Potgietersrust platinums limited BBOP Pilot (Platinum mine)	South Africa	Africa	BBOP
12	De Beers Marine Kleinzee and Alexander Bay Sea Areas (off-shore diamond mining	South Africa	Africa	WWF
14	Rossing Uranium Mine (Rio Tinto)	Namibia	Africa	FFI
15	Palabora	South Africa	Africa	FFI
19	Kumtor Gold Mine (open pit gold mine)	Kyrgyzstan	Asia	FFI
24	Mc Arthur River Mine Open Cut Project	Australia	Australia and Oceania	CI
32	Thameslink Railway offsets	United Kingdom	Europe	Local government: Defra
64	Antamina Copper and Zinc Mine	Peru	South America	
66	Pasto Mocoa in the Amazon (Mocoa-Saanciso road)	Colombia	South America	WWF
67	Gasoducto Gasbol (pipeline)	Bolivia, Brazil	South America	WCS
70	Cerrejón Coal (coal mining)	Colombia	South America	CI
72	Gramalote Mining Project	Colombia	South America	BBOP

[a]Abbreviations: *BBOP* Business and Biodiversity Offsets Program, *FFI* Fauna & Flora International, *TNC* The Nature conservancy, *IUCN* International Union for Conservation of Nature, *WWF* World Wide Fund For Nature, *CI* Conservation International, *WCS* Wildlife Conservation Society
Source: Author

thus cannot be ensured, unlike the offset enabling framework in itself. As an example the offset pilot portfolio of the Business and Biodiversity Offsets Program can be cited. While in 2009, six pilots have been presented, in 2015 only two remain and are updated. This may in some cases suggest that the offset commitment may have been abandoned or not further pursued (e.g., Shell Qatar Gas-To-Liquid, no. 20 in the screening), in other cases it simply denotes undocking from the enabling framework (e.g., Bainbridge Island, no. 52 in the screening).

Another significant part of enabled biodiversity offsets consists of partnerships between NGOs and single companies, mostly at the local level, e.g., in the case of Cerréjon Coal Mine (no. 70 in the screening) where the project proponent engages with Conservation International Colombia through a cooperative agreement. While

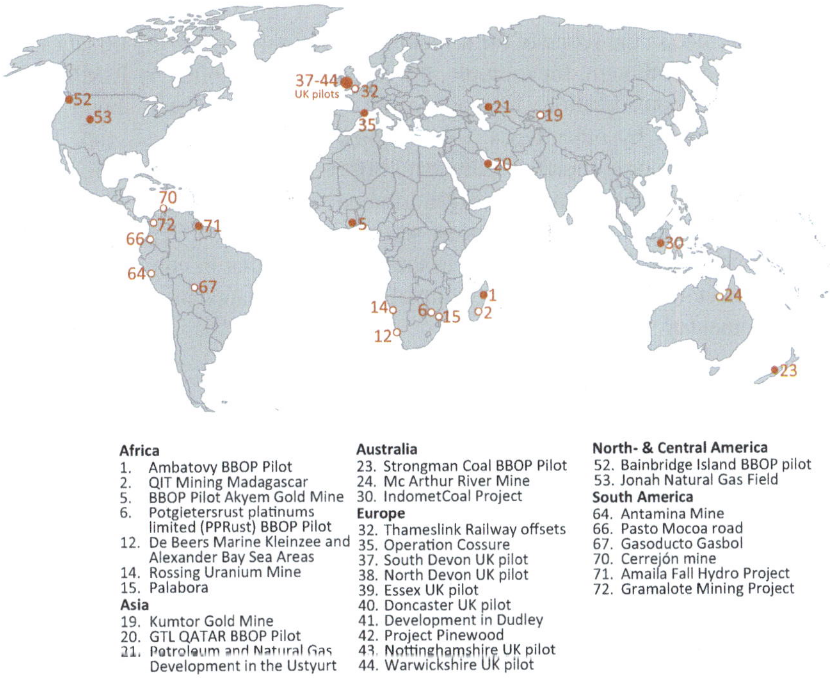

Fig. 7.9 Type 3 enabled biodiversity offsets: map of attributed cases from the worldwide screening (Primary attributions are presented as solid dots and secondary attributions as circles.) (Source: Author)

the advantage of this project level enabled biodiversity offsets is certainly the focus on the single business and its particularities, in the long term the offset enabling framework is most likely to vanish once this bilateral agreement comes to an end.

Nevertheless, other examples of long-term partnerships are tied at corporate rather than project level, e.g., the FFH–Rio Tinto partnership which translates into a number of offset cases implemented in different locations worldwide (no. 14, 15 in the screening) (Marsh n.d.).

This sort of long-term bilateral cooperation of corporations and NGOs highlights the role of NGOs not only to ensure consideration of environmental and social aspects at the local level, but increasingly strategically in establishing and promoting global standards and best practice for biodiversity offsets.

7.3.4 Magnitude, Location, and Particularities of Type 4 Sectoral Biodiversity Offsets

Magnitude of Type 4 Sectoral Biodiversity Offsets

Contrary to the expectations that had been suggested by a large variety of sectoral activities in particular in the extractives industry (i.e., mining as represented through the ICMM and oil and gas as represented through IPIECA), very little evidence could be identified in the scope of the screening of sectoral biodiversity offset cases. This occurred even though, for example, the case studies compiled by ICMM have been taken into consideration. In practice, all of these have been found to be primarily motivated by other types, e.g., enabled biodiversity offsets (no. 1 and 30 in the screening) or corporate biodiversity offsets (no. 12 in the screening).

Thus, no cases have been primarily attributed to type 4 sectoral biodiversity offsets and only seven cases have been secondarily attributed, representing less than 10% of the total cases. This leads to the assumption that in comparison to the other types, sectoral biodiversity offsets are of minor importance in terms of driving offset implementation (Table 7.5).

Location of Type 4 Sectoral Biodiversity Offsets
As has been noted, the number of sectoral offset cases as a result of the screening appears insignificant. Thus, on the basis of this small number only very restricted conclusions can be drawn regarding the geographical distribution of this offset type. A view on the map reveals that evidence could be found for North and South America, Africa and Southeastern Asia, thus being relatively evenly spread, and no pattern could be identified. Nevertheless, given the concentration on mining and oil and gas, it is likely to conclude that to date sectoral biodiversity offsets are bound to the location of extractive industry (Fig. 7.10).

Particularities of Type 4 Sectoral Biodiversity Offsets
As can be concluded from the screening of biodiversity offset cases, "sectoral biodiversity offsets" appear to present one of the weakest institutional arrangements. In fact, despite the role of sectors in promoting best practice, such as the commitment by ICMM members not to mine or explore in World Heritage sites (ICMM 2010), these usually remain accompanying motivations besides others drivers (i.e., secondary attributions)—most notably either pressure by lending institutions, government provisions, or corporate policies (e.g., Rio Tinto's Net Positive Impact commitment). One reason for this may also be that sectoral agreements or commitments to no net loss or the use of biodiversity offsets are a relatively recent development.

Despite the marginal role of sectoral biodiversity offsets as a distinct type, the role of different sectors and its related representations and lobby groups should not be underestimated. While not being the major driving force for any particular offset examples, they greatly contribute to the discussion and advancement of biodiversity offset practices. This has various forms, e.g., by providing guidance and exchange among their members (in particular the example of ICMM has to be highlighted) they encourage the creation of corporate biodiversity offset standards or commitments as well as the exploration of biodiversity offset pilots. ICMM has also supported research in this field and commissioned the documentation of a number of case studies (ICMM 2016).

To conclude, the different sectors and their particularities as well as their (historically grown) relation with the public form an important part of the institutional context of biodiversity offset emergence and implementation.

Table 7.5 Type 4 sectoral biodiversity offsets: primary and secondary attribution of identified cases

No.	Name	Country	Continent	Driver (influence) for this type[a]	Primary driver[b]
Primary attribution					
–	–	–	–	–	–
Secondary attribution					
1	Ambatovy BBOP Pilot Project Case Study (Nickel mine)	Madagascar	Africa	ICMM	Type 3b enabled biodiversity offsets (BBOP)
12	De Beers Marine Kleinzee and Alexander Bay Sea Areas (off-shore diamond mining	South Africa	Africa	ICMM	Type 5 corporate biodiversity offsets (De Beers)
30	IndometCoal Project (metallurgical coal resource, BHP Billiton)	Indonesia	Australia and Oceania	ICMM	Type 3b enabled biodiversity offsets (FFI)
50	Suncor Biodiversity Offset project in Canada (oil sands extraction)	Canada	North and Central America	COSIA	Type 7 altruistic biodiversity offsets
51	Resolution Copper Mining	United States	North and Central America	ICMM	Type 6b local biodiversity offsets
64	Antamina Copper and Zinc Mine	Peru	South America	ICMM	Type 6b local biodiversity offsets
69	AngloGold Ashanti Brasil Mineracao (gold mine)	Brazil	South America	ICMM	Type 7 altruistic biodiversity offsets

[a]Abbreviations: *ICMM* International Council on Mining and Metals, *COSIA* Canada's Oil Sands Innovation Alliance
[b]Abbreviations: *BBOP* Business and Biodiversity Offsets Programme, *FFI* Fauna and Flora International
Source: Author

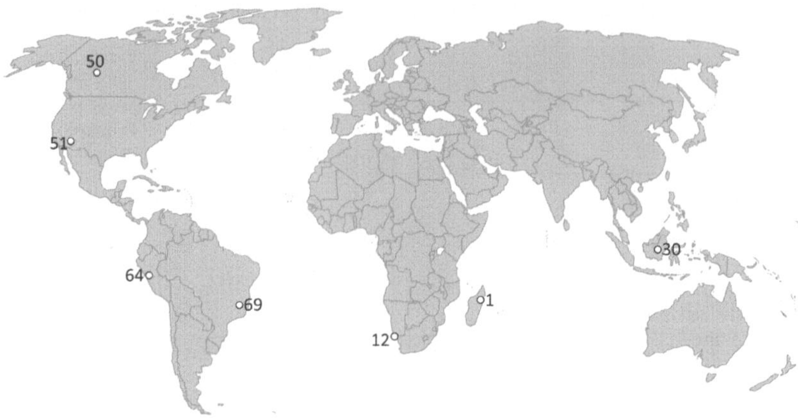

Fig. 7.10 Type 4 sectoral biodiversity offsets: map of attributed cases from the worldwide screening (Primary attributions are presented as solid dots and secondary attributions as circles) (Source: Author)

7.3.5 Magnitude, Location, and Particularities of Type 5: Corporate Biodiversity Offsets

Magnitude of Type 5 Corporate Biodiversity Offsets

Besides regulatory and enabled biodiversity offsets, corporate biodiversity offsets present the strongest driver for biodiversity offsets. It is thus not surprising that a large number of examples could be identified worldwide, making it one of the most significant biodiversity offset types.

More than one-fifth of all biodiversity offset cases (20.8%) are characterized as corporate biodiversity offsets in this study and almost every third (29.2%) case identified in the worldwide biodiversity offset screening was to some extent driven by corporate commitments toward biodiversity offsets.

This is also indicative of the current trend of engaging business in nature conservation and in particular the increasing role of global players.

Even though comparably high, the numbers clearly do not support an exposed or dominating position of business as expressed through corporate biodiversity offsets, in comparison to the other drivers, most notably government and NGOs, as could have been assumed in the light of the growing promotion of the "business case" for biodiversity offsets (Table 7.6).

Table 7.6 Type 5 corporate biodiversity offsets: primary and secondary attribution of identified cases

No.	Name	Country	Continent	Driver (influence) for this type[a]
Primary attribution				
2	QIT Mining Madagascar QMM (ilmenite mining)	Madagascar	Africa	Rio Tinto
3	Simandou (open-pit iron-ore mine)	Guinea	Africa	Rio Tinto
5	BBOP Pilot Case Study Akyem Gold Mining Project	Ghana	Africa	Newmont
12	De Beers Marine Kleinzee and Alexander Bay Sea Areas (off-shore diamond mining)	South Africa	Africa	De Beers
13	Namdeb Diamond Corporation (diamond mine, De Beers)	Namibia	Africa	De Beers
14	Rossing Uranium Mine (Rio Tinto)	Namibia	Africa	Rio Tinto
15	Palabora	South Africa	Africa	Rio Tinto
16	Shell Foundation offset to support Smithsonian Institution's Monito-ring and Assessment of Biodiversity Program on Gamba Complex	Gabon	Africa	Shell
18	Oyu Tolgoi LLC (copper/gold mine)	Mongolia	Asia	Rio Tinto
19	Kumtor Gold Mine (open pit gold mine)	Kyrgyzstan	Asia	Centerra
20	PEARL GTL PROJECT, QATAR BBOP Pilot (Gas To Liquid)	Qatar	Asia	Shell
25	Newmont Boddington mine (Gold mine)	Australia	Australia and Oceania	Newmont
49	Kennecott Utah Copper Mine (expansion of the tailings of a copper mine)	United States	North and Central America	Rio Tinto
59	Newmont Long Canyon Project (Gold mine)	United States	North and Central America	Newmont
68	Brisas Gold and Copper Project (Open pit mine, Gold Reserves Inc.)	Venezuela	South America	Gold Reserves Inc.
Secondary attribution				
23	Strongman Coal Mine, Solid Energy New Zealand BBOP Pilot (coal mine)	New Zealand	Australia and Oceania	Solid Energy New Zealand
27	Raventhorpe Nickel Project	Australia	Australia and Oceania	BHP Biliton
31	Open cut extension at the Warkworth Coal Mine (Rio Tinto Coal Australia)	Australia	Australia and Oceania	Rio Tinto

(continued)

Table 7.6 (continued)

No.	Name	Country	Continent	Driver (influence) for this type[a]
50	Suncor Biodiversity Offset project in Canada (oil sands extraction)	Canada	North and Central America	Suncor
51	Resolution Copper Mining	United States	North and Central America	Rio Tinto
60	Pueblo Viejo gold mine expansion	Dominican Republic	North and Central America	IFC

[a]Abbreviations: *IFC* International Finance Corporation
Source: Author

Location of Type 5 Corporate Biodiversity Offsets

Corporate biodiversity offsets are observed in various locations worldwide. However, a concentration of eight cases on the African continent can be observed as opposed to none in Europe.

These differences can be explained mainly by looking at the type of development activities triggering corporate biodiversity offsets commitments. Clearly, mining projects are dominating (18), accompanied by a few examples from the oil and gas sector (3).

This restriction to extractive industry and the focus on large corporations (Rio Tinto alone accounts for eight of the above listed corporate biodiversity offsets) explain the location of corporate biodiversity offsets in some of the global mining hotspots, in particular in sub-Saharan Africa (Fig. 7.11).

Particularities of Type 5 Corporate Biodiversity Offsets

Some companies have included a commitment to have a "net positive impact" (or similar) on biodiversity in their environmental or biodiversity strategies, e.g., the Rio Tinto Group (ICMM 2010). This way, they are setting a requirement (similar to legal or policy requirements) to implement offsets at their operations (Doswald et al. 2012).

The business risks that these large corporations face, e.g., withdrawal of license to operate, project delays, consumer boycotts and reputational damage of a brand, clearly justify or mandate the business case for biodiversity offsets, i.e., corporate biodiversity offsets are usually the result of a meticulous weighing of costs and benefits. However, this also shows a potential weakness of this type: if a commitment toward biodiversity offsets is conditioned by envisaged benefits by the company, there is always a risk to abandon this commitment if no longer relevant to the envisaged benefits or if these can be better reached by other means. In addition, biodiversity offsets are mostly not positively framed in public discourses and it might even negatively affect the reputation of a company if it promotes biodiversity

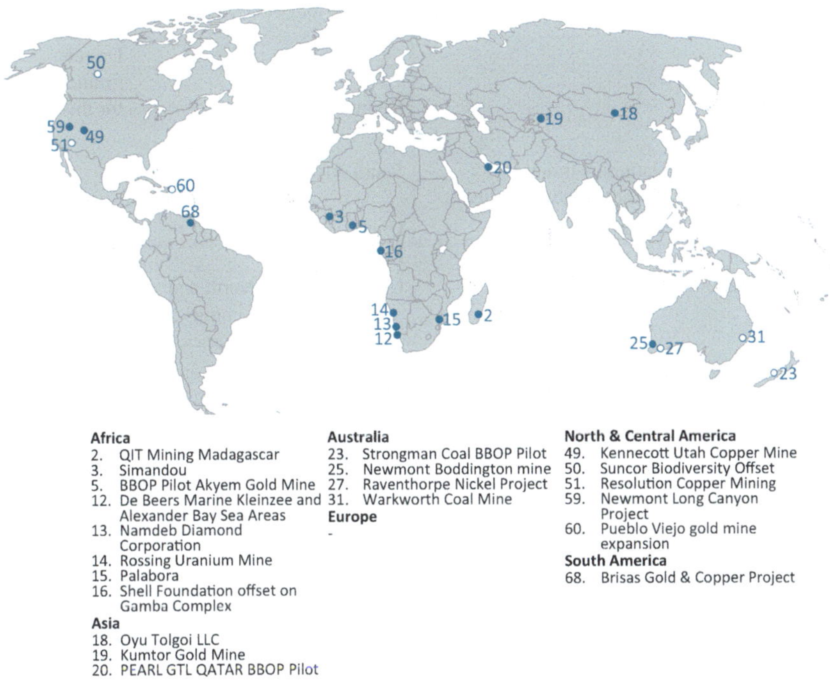

Fig. 7.11 Type 5 corporate biodiversity offsets: map of attributed cases from the worldwide screening (Primary attributions are presented as solid dots and secondary attributions as circles) (Source: Author)

offsets. Furthermore, even if not totally abandoned, biodiversity offsets bear the risk of existing only "on paper" (cf. Quétier et al. 2014: No net loss of biodiversity or paper offsets?).

Thus, the major problem associated with corporate biodiversity offsets is that they rely on a corporate commitment that is not externally mandated and thus cannot be externally tracked or enforced.

7.3.6 Magnitude, Location, and Particularities of Type 6: Local Biodiversity Offsets

Magnitude of Type 6 Local Biodiversity Offsets

Six cases have been identified in the screening of biodiversity offset cases that appear to have been driven by local conditions and stakeholders. This relatively small number, representing only 8.3% of all cases, extends to more than one-fifth (15 cases) when taking into consideration also secondary attribution of cases. From this it can be concluded that local biodiversity offsets are seldom the major offset driver; however, they can underpin and coexist alongside other major drivers such as corporate commitments and enabled offset schemes and may be interwoven with

Table 7.7 Type 6 local biodiversity offsets: primary and secondary attribution of identified cases

No.	Name	Country	Continent	Driver (influence) for this type
Primary attribution				
6	Potgietersrust platinums limited (PPRust) BBOP Pilot (Platinum mine)	South Africa	Africa	n/a
32	Thameslink Railway offsets	United Kingdom	Europe	n/a
51	Resolution Copper Mining	United States	North and Central America	n/a
52	Case study of the BBOP pilot biodiversity offset work on Bainbridge Island (real estate and infrastructure)	United States	North and Central America	n/a
56	Establishment of the Braulio Carrillo National Park for the construction of a highway	Costa Rica	North and Central America	n/a
70	Cerrejón Coal (coal mining)	Colombia	South America	n/a
Secondary attribution				
1	Ambatovy BBOP Pilot Project Case Study (Nickel mine)	Madagascar	Africa	n/a
2	QIT Mining Madagascar QMM (ilmenite mining)	Madagascar	Africa	n/a
3	Simandou (open-pit iron-ore mine)	Guinea	Africa	n/a
5	BBOP Pilot Case Study Akyem Gold Mining Project	Ghana	Africa	n/a
16	Shell Foundation offset to support Smithsonian Institution's Monito-ring and Assessment of Biodiversity Program on Gamba Complex	Gabon	Africa	n/a
20	Pearl Gtl Project, Qatar BBOP Pilot (Gas To Liquid)	Qatar	Asia	n/a
23	Strongman Coal Mine, Solid Energy New Zealand BBOP Pilot (coal mine)	New Zealand	Australia and Oceania	n/a
33	Burgberg Zschaitz habitat bank (offset for windpower development)	Germany	Europe	n/a
35	Operation Cossure	France	Europe	n/a

Source: Author

these (Table 7.7). For example, considerations of stakeholder participation, equity, and inclusion of traditional knowledge are prominently rooted in the BBOP principles on biodiversity offsets (see Fig. 4.5).

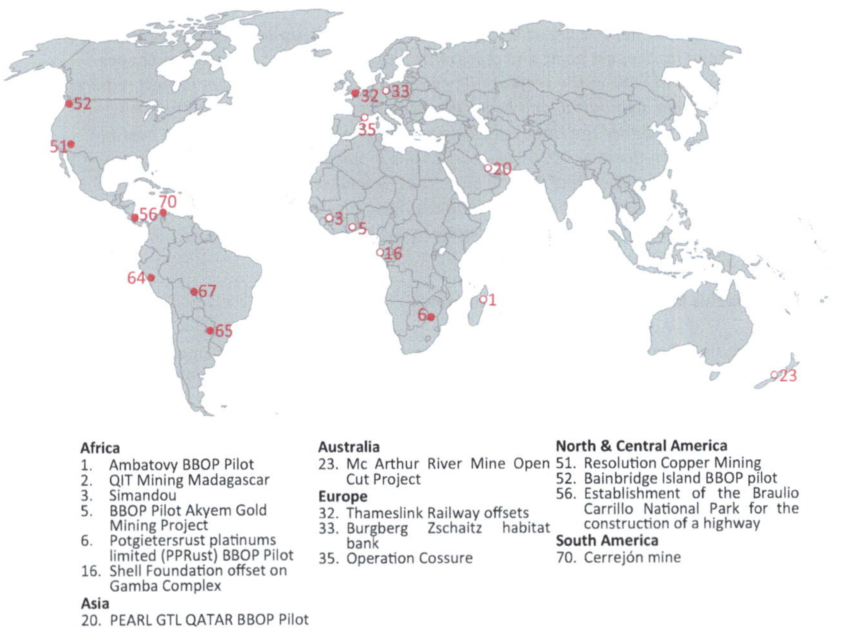

Fig. 7.12 Type 6 local biodiversity offsets: map of attributed cases from the worldwide screening (Primary attributions are presented as solid dots and secondary attributions as circles) (Source: Author)

Location of Type 6 Local Biodiversity Offsets

The location of cases that are either primarily or secondarily attributed to the type of local biodiversity offsets shows no clear distribution pattern. While slightly more cases have been compiled for the African continent (5), evidence can also be found for Europe, North and Latin America, Asia and Oceania. Due to the nature of this type in responding to the specific local conditions and stakeholder expectations it is assumed that local biodiversity offsets may appear anywhere in a global context. However, further research would be needed to explore if certain regions or local contexts provide stronger drivers for the implementation of biodiversity offsets, e.g., protected areas and or threatened areas, indigenous communities, level of environmental awareness and mobilization, etc. (Fig. 7.12).

Particularities of Type 6 Local Biodiversity Offsets

As has been observed in the screening of biodiversity offset cases, local biodiversity offsets can be very different—ranging from very abstract ambitions to consider the integrity of the natural environment and the local communities, based on consultation and participation of local stakeholders, to very specific motivations that are crucial for the project success. While most of the cases attributed to local biodiversity offsets fall into the first category, the case of Resolution Copper Mining in

In 1995, a large copper deposit was discovered near Superior, Arizona. The deposit lies nearly 7,000 feet deep at its deepest point and represents one of the largest undeveloped copper resources in the world. Still in the prefeasibility stage, the Resolution Copper Project requires extensive engineering, geological, and hydrological studies as well as a comprehensive environmental review under the National Environmental Policy Act (NEPA) before possibly reaching production in the mid- 2020s. As part of this project, Resolution Copper Company is trying to exchange eight parcels of private land (that Resolution has purchased) totaling 5,344 acres for a parcel of Tonto National Forest land overlying and adjacent to the project totaling 2,422 acres. This land exchange is under consideration by the U.S. Congress.

1. **Why is this land exchange being done now?**

   Resolution Copper will use block cave mining to extract ore from its unpatented mining claims. This mining will cause subsidence and make the area inaccessible. Because that area will not be accessible once mining has begun, Resolution Copper would prefer to own the area that will be affected. As an offset, Resolution Copper will place lands with a high ecological value back into the public trust.

   Further Resolution believes the ore continues beneath an area that Resolution does not hold mining claim too. Resolution needs access to about 760 acres – including the Oak Flat campground – to effectively study the size and quality of the whole copper ore body.

Fig. 7.13 Land exchange for the resolution copper mine in Arizona, USA as an example for a local offset (Source: Resolution Copper Company 2013)

Arizona, USA operated by Rio Tinto (no. 51 in the screening) is an example for the second as is illustrated in Fig. 7.13.

In this case the developer suggested an offset in terms of a land exchange for a publicly owned area, the Tonto National Forest, that is likely to become inaccessible due to the mining operations, i.e., the offset shall counterbalance the disturbed access to nature and can be considered a socio-environmental compensation. This example shows that local biodiversity offsets besides ecological considerations often also focus on social aspects such as equity, access to land, traditional land use, and indigenous communities.

Another particularity of local biodiversity offsets is that they are usually shaped by a diverse set of drivers or actors and their various interactions. While their common focus is on a specific location, i.e., a piece of land or a region, their motivations often significantly vary (if not completely contrasting). Thus, the major obstacle of local biodiversity offsets is in bringing together and mediating between these different motivations, e.g., local communities, governmental bodies at various levels, NGOs, business associations, individual citizens, etc. In this regard, local biodiversity offsets often build on a "partnership approach" (Baker et al. 2015) that places stakeholder participation and contribution at the core of developing an offset.

Due to the diverse set of actors and their motivations, local biodiversity offsets have the highest overlap with other types of biodiversity offsets, in particular with conditional, enabled and corporate biodiversity offsets. An example for the latter is the QMM ilmenite mine in Madagascar (no. 2 in the screening) which has been presented as a case study for corporate biodiversity offsets (see Sect. 6.5.2), but as mentioned above (see Sect. 6.6.1) with its Community Reforestation Programme has a strong local offset component.

7.3.7 Magnitude, Location, and Particularities of Type 7: Altruistic Biodiversity Offsets

Magnitude of Type 7 Altruistic Biodiversity Offsets

Type 7 altruistic biodiversity offsets are most difficult to evaluate and even to underpin with evidence. By its very nature, altruistic behavior would not explicitly seek public benefit. Thus, the screening of biodiversity offset cases does not provide much evidence on this case and relies on the four cases that are said to be undertaken voluntarily in the strict sense, i.e., for mainly altruistic reasons, by the project developers. The four cases that have been primarily attributed to this type represent little more than 5% of all offset cases from the screening. Three further cases that are noted to at least partially being encouraged by altruistic motivations (see table 7.8). In total, this represents less than 10% of all cases, which leads to the conclusion that altruistic biodiversity offsets are a fairly weak type compared to the other types.

Table 7.8 Type 7 altruistic biodiversity offsets: primary and secondary attribution of identified cases

No.	Name	Country	Continent	Driver (influence) for this type
Primary attribution				
47	Compensation Outline Open-Pit Mining in Sweden—Mertainen	Sweden	Europe	n/a
50	Suncor Biodiversity Offset project in Canada (oil sands extraction)	Canada	North and Central America	n/a
64	Antamina copper and Zinc Mine	Peru	South America	n/a
69	AngloGold Ashanti Brasil Mineracao (gold mine)	Brazil	South America	n/a
Secondary attribution				
52	Case study of the BBOP pilot offset on Bainbridge Island (real estate and infrastructure)	United States	North and Central America	n/a
53	Jonah Natural Gas Field	United States	North and Central America	n/a
70	Cerrejón Coal (coal mining)	Colombia	South America	n/a

Location of Type 7 Altruistic Biodiversity Offsets

Similar to the small number of cases observed for sectoral biodiversity offsets, for altruistic biodiversity offsets the conclusions on the geographical distribution also remain limited. As can be seen, all cases but one are located in the Americas, with one in Europe. As an attempt to explain this, it might be assumed that altruistic biodiversity offsets are more likely to emerge in America and Europe where voluntary environmental approaches have a longer tradition than elsewhere (Fig. 7.14).

Particularities of Type 7 Altruistic Biodiversity Offsets

While it certainly cannot be considered purely theoretical, the practical examples that have been explored for type 7 altruistic biodiversity offsets leave room for interpretation: are they driven by the perceived responsibility of business or just the result of clever emotional marketing? The case of AngloGold Ashanti's gold mine in the Brazilian state of Minas Gerais (no. 69 in the screening) illustrates this dilemma (see Fig. 7.15).

The rare examples that have been encountered and grouped into this type are not ideal representations. Rather, they are a form of mixed types, which however are characterized by a strong responsibility motivation. In fact, altruistic biodiversity offsets are most likely to intersect with other types of biodiversity offsets in practice, in particular local and corporate biodiversity offsets. This is due to the fact that altruistic and egoistic motivations are not mutually exclusive (and altruistic action or commitment toward biodiversity offsets are at least indirectly shaped by the social

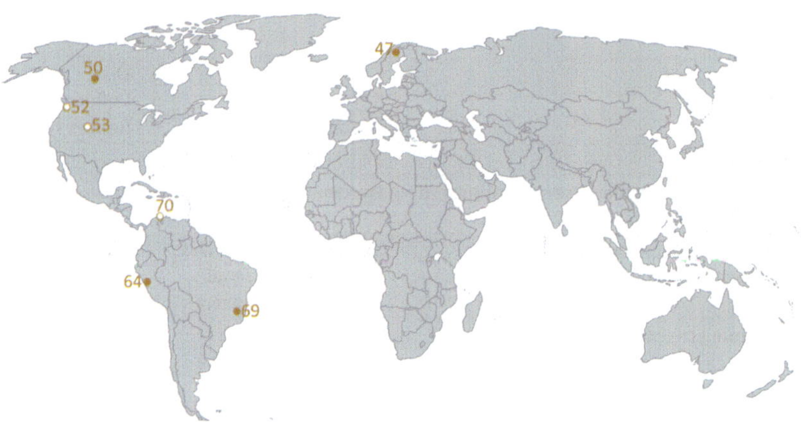

Africa
-
Asia
-

Australia
-
Europe
47. Openpit Mining Mertainen

North & Central America
50. Suncor Biodiversity Offset
52. Bainbridge Island BBOP pilot
53. Jonah Natural Gas Field
South America
64. Antamina Mine
69. AngloGold Ashanti Brasil Mineracao
70. Cerrejón mine

Fig. 7.14 Type 7 altruistic biodiversity offsets: map of attributed cases from the worldwide screening (Primary attributions are presented as solid dots and secondary attributions as circles) (Source: Author)

Following is an excerpt from the foreword of Biodiversidade da Mata Samuel de Paula by José Carlos Carvalho, Secretary of the Environment and Sustainable Development of the State of Minas Gerais:

"In the midst of so much concern and disbelief in the future, what reason could there possibly be for AngloGold Ashanti launching the book Biodiversidade da Mata Samuel de Paula at this particular time?

We thought of the best one of all: hope. That is to say, even in the midst of doom and gloom about the future, complicated by the unbridled ambition of many, there is still a place among us for companies and entrepreneurs who, in addition to seeking profit, can find a way in the intricate world of capitalism to build something for those who come after us so that they, too, may enjoy the blessings of the earth.

That objective is what led AngloGold Ashanti to set up, next to the beautiful capital of Minas Gerais, a Private Natural Heritage Reserve (RPPN), which it created and has maintained since 2000. Furthermore, it has supported studies, helping to understand the phenomena that take place within it, with the intention of compiling rich reference material for both specialists in the ecotones of the Atlantic Forest and Cerrado and students in general.

Through this initiative, the company has added its name to the list of mining companies in Minas Gerais that have left behind a past of predatory exploitation and disregard for the community well-being and joined a group of companies that are focused on complying with the constitutional requirement of the social function of economic ventures, aware that protecting the environment is more than just a modern marketing strategy: it is a commitment to humanity."

Fig. 7.15 Altruistic behavior or clever emotional marketing? The case of AngloGold Ashanti's gold mine in the Brazilian state of Minas Gerais (Source: ICMM 2010)

context in which they arise) and the dedication for the good of others ultimately also adds to one's own benefit.

As the example of the Antamina mine and its Polylepis Program (no. 64 in the screening, see the detailed case study in Sect. 6.7.2) shows, there is also an overlap between altruistic biodiversity offsets and other conservation actions more generally. Antamina's Polylepis Program was not initially created as an offset. Indeed, it is characteristic for altruistic biodiversity offsets that this type has seamless transitions toward other forms of voluntary action or initiatives.

Furthermore, altruistic biodiversity offsets represent also a temporal phenomenon. Similar to government enabled biodiversity offsets, e.g., in the scope of pilots (type 3a government enabled biodiversity offsets) that often forego the establishment of a regulatory offset scheme (type 1), altruistic biodiversity offsets can be interpreted as pilots preceding the development of corporate commitments or policies of offsets (type 5).

7.4 Discussion of the Typology

7.4.1 Differences Between the Types

As has been shown in Sects. 7.3.1–7.3.7, differences exist between the built types of biodiversity offsets in terms of magnitude, location, and particularities (see Fig. 7.16, cf. Appendix 1: Screening of Biodiversity Offset cases worldwide for a larger map).

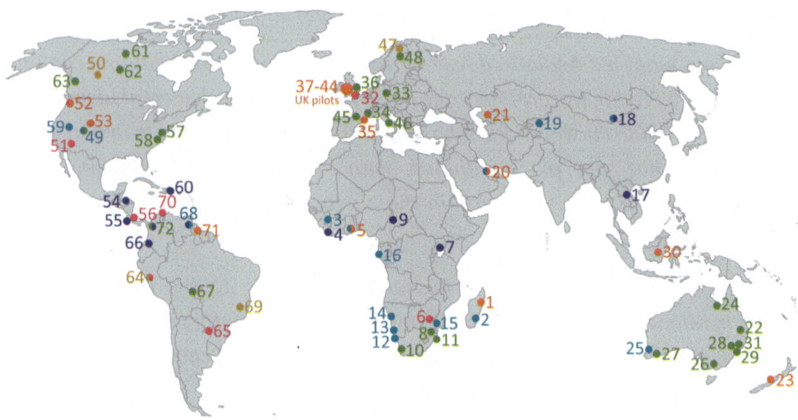

Color scheme: green: type 1 regulatory biodiversity offsets, dark blue: type 2 conditional biodiversity offsets, orange: type 3 enabled biodiversity offsets, turquoise: type 5 corporate biodiversity offsets, red: type 6 local biodiversity offsets, ochre: type 7 altruistic biodiversity offsets

Africa
1. Ambatovy BBOP Pilot
2. QIT Mining Madagascar QMM
3. Simandou
4. Piloting aggregated offsets and mining in Liberia
5. BBOP Pilot Akyem Gold Mining Project
6. Potgietersrust platinums limited (PPRust) BBOP Pilot
7. Bujagali Energy Limited
8. Anglo Coal Isibonelo colliery
9. Petroleum Development and Pipeline Project
10. Mount Royal Golf Estate
11. Pulp United Pulp Mill
12. De Beers Marine Kleinzee / Alexander Bay Sea Areas
13. Namdeb Diamond Corporation
14. Rossing Uranium Mine
15. Palabora
16. Shell Foundation offset on Gamba Complex

Asia
17. Nam Theun 2 Hydropower
18. Oyu Tolgoi LLC
19. Kumtor Gold Mine
20. PEARL GTL QATAR BBOP Pilot
21. Petroleum and Natural Gas Development in the Ustyurt

Australia
22. Koala Offset in South East Queensland
23. Strongman Coal BBOP Pilot
24. Mc Arthur River Mine
25. Newmont Boddington mine
26. Basslink Undersea Power Cable
27. Raventhorpe Nickel Project
28. Karuah bypass offset
29. Cumberland Plain Woodland / Western Sydney
30. IndometCoal Project
31. Warkworth Coal Mine

Europe
32. Thameslink Railway offsets
33. Burgberg Zschaitz habitat bank
34. CNM high speed rail bypass
35. Operation Cossure
36. Bardon Hill Quarry
37. South Devon UK pilot
38. North Devon UK pilot
39. Essex UK pilot
40. Doncaster UK pilot
41. Development in Dudley
42. Project Pinewood
43. Nottinghamshire UK pilot
44. Warwickshire UK pilot
45. A65 Langon-Pau motorway
46. Apennine Wind Farms
47. Openpit Mining Mertainen
48. Ume Delta River Railway

North & Central America
49. Kennecott Utah Copper Mine
50. Suncor Biodiversity Offset
51. Resolution Copper Mining
52. Bainbridge Island BBOP pilot
53. Jonah Natural Gas Field
54. Via San Lorenzo – Olanchito
55. Hidroelectrica Reventazon
56. Braulio Carrillo National Park / construction of a highway
57. Tygart Valley River Stream and Wetland Bank
58. South Fork Hoppers Creek Mitigation
59. Newmont Long Canyon Project
60. Pueblo Viejo gold mine
61. NGTL pipeline proposals in Horne River area
62. Total Joslyn oil-sands mine
63. Enbridge Northern Gateway pipelines

South America
64. Antamina Mine
65. Hidroelectrica Yacyreta
66. Pasto Mocoa road
67. Gasoducto Gasbol
68. Brisas Gold & Copper Project
69. AngloGold Ashanti Brasil Mineracao
70. Cerrejón mine
71. Amaila Fall Hydro Project
72. Gramalote Mining Project

Fig. 7.16 Screening of biodiversity offset cases worldwide (nonrepresentative and not exclusive) (Source: Author)

For example a general gap in offset implementation has been observed for most Asian countries, which may partly be due to a lack of documentation. However, compensation schemes are known to be in place or under development in several countries, including China (Eco-compensation), South Korea, India, Mongolia, and Japan (Darbi et al. 2010). Nevertheless, these have not been found to be backed by a sufficient body of evidence in this study.

In addition, offset cases can differ for example with regard to the following aspects:

- The scale of the development impacts
- The sectors/types of developments addressed
- The governance of the implementation of offsets

Throughout the screening of biodiversity offset cases mainly large-scale development projects have been identified. This was noted in particular for type 2 conditional biodiversity offsets and type 5 corporate biodiversity offsets. While theoretically it would be possible to extend the scope of these types to also include smaller scale projects and impacts in the future, this is not likely to happen in the short term.

By contrast, type 1 regulatory biodiversity offsets, type 3 enabled biodiversity offsets and to some extent also type 6 local biodiversity offsets, have been observed to cover both large-scale development projects, e.g., open cast mining, and small-scale projects, e.g., housing or single wind turbines (cf. Fig. 7.17).

While different types of development projects have been found to be addressed by the offset cases identified in the screening, a clear dominance for certain sectors was observed, most notably mining.

Figure 7.18 displays the number of offset cases per sector/development type split up according to the built types of biodiversity offsets. It can be seen that regulatory biodiversity offsets cover the broadest spectrum of development projects, whereas corporate biodiversity offsets to date remain limited to mining, oil, and gas.

Furthermore, it is interesting to note that all four hydropower projects covered in this study have been assigned to type 2 conditional biodiversity offsets. Thus, despite the small and not representative number of biodiversity offset cases

Type / Scale	Type 1: Regulatory biodiversity offsets	Type 2: conditional biodiversity offsets	Type 3: Enabled biodiversity offsets	Type 4: Sectoral biodiversity offsets	Type 5: Corporate biodiversity offsets	Type 6 Local biodiversity offsets	Type 7: Altruistic biodiversity offsets
Large scale	X	x	x	x	x	x	x
Small scale	X	-	x	-	-	x	-

Fig. 7.17 Scale of development projects associated to the different types (Source: Author)

Type / Sectors	Type 1: Regulatory biodiversity offsets	Type 2: conditional biodiversity offsets	Type 3: Enabled biodiversity offsets	Type 4: Sectoral biodiversity offsets	Type 5: Corporate biodiversity offsets	Type 6: Local biodiversity offsets	Type 7: Altruistic biodiversity offsets	TOTAL
Mining	**7** (8, 24, 27, 31, 36, 49, 72)	**5** (4, 18, 60, 68,72)	**4** (1, 5, 23, 30)	-	**12** (2, 5, 12, 13, 14, 15, 18, 19, 25, 49, 59, 68)	**3** (6, 51, 70)	**3** (47, 64, 69)	**34**
Oil and gas	**4** (61, 62, 63, 67)	**1** (9)	**3** (20, 21, 53)	-	**2** (16, 20)	-	**1** (50)	**11**
Infrastructure	**4** (28, 34, 45, 48)	**2** (54, 66)	-	-	-	**1** (32)	-	**7**
Housing/real estate	**3** (10, 22, 29)	-	**9** (37-44, 52)	-	-	**2** (52,56)	-	**14**
Hydropower	-	**4** (7, 17, 55, 65)	**1** (71)	-	-	-	-	**5**
Energy	**3** (26, 33, 46)	-	-	-	-	-	-	**3**
Other industry	**1** (1)	-	-	-	-	-	-	**1**
n/a	**2** (57, 58)	-	-	-	-	-	-	**2**

Fig. 7.18 Sectors/types of development projects associated to the different types (Source: Author)

compiled, correlations between the built types of biodiversity offsets and different sectors/types of developments can be observed.

The implementation of biodiversity offsets can follow either one of two governance approaches. Consequently, differences between the types occur with regard to top down vs. bottom up organized processes as illustrated in Fig. 7.19.

Whereas regulatory, conditional, sectoral, and corporate biodiversity offsets by their definition are top down—induced by regulators, lenders, (sectoral) business associations and corporations—local and altruistic biodiversity offsets build on a collaborative bottom up process with local stakeholders.

Type 3, enabled biodiversity offsets, represents a hybrid form in that it is encouraged or incentivized by government or global norms and standards (e.g., by BBOP), but they also largely built on collaborative processes at local scale.

7.4.2 Similarities and Overlap Between the Types

In terms of similarities it is obvious that the development of biodiversity offsets builds on a diversified but restricted set of actors that have multiple overlapping responsibilities. This can be shown with the example of the Business and Biodiversity Offsets Programme, which has become the global leader in this field. It builds on

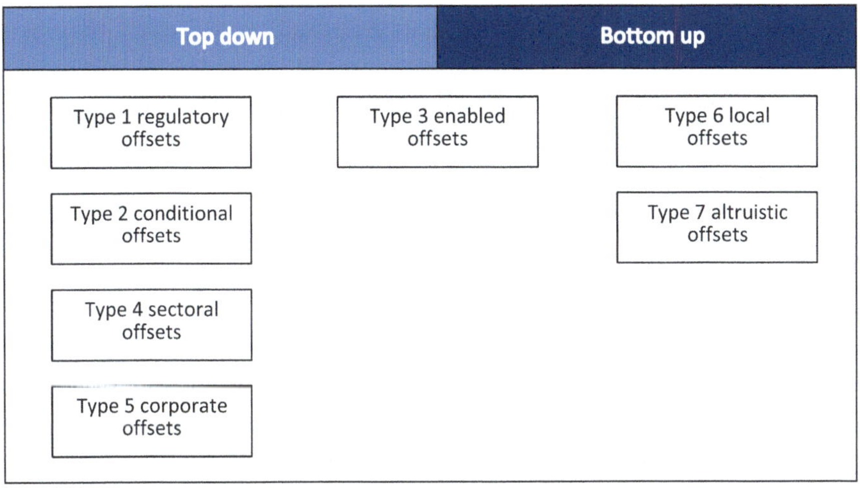

Fig. 7.19 Bottom up versus top down implementation of biodiversity offsets according to the different types (Source: Author)

membership and collaboration with international organizations and NGOs for nature conservation (e.g., UNDP, IUCN, Birdlife International, WWF, TNC), business (e.g., TOTAL, Ambatovy, New Britain Palm Oil Ltd., CEMEX), financial institutes (e.g., IFC, GEF, EBRD), government agencies (e.g., Ministry of Ecology, Energy, Sustainable Development, and Spatial Planning, France, Ministry of Environment and Tourism, Government of Namibia, Ministry of Infrastructure and the Environment, The Netherlands) as well as consultants, researchers and individuals.[8] Many of these have different functions and are engaged in offset development in different spheres, e.g., the no net loss initiative of the European Commission and the related working group, the International Council for Mining and Metals (ICMM) of the extractives sector, IPIECA of the oil and gas sector and the Business and Biodiversity Initiative of IUCN.

This shows that both a personal and technical overlap can be observed, i.e., the set of actors is highly interwoven with a number of key players.

In this setting of high interdependence and personal relations it is difficult to distinguish the driving forces behind an offset and to isolate the roles of the single actors.[9]

An example for this is when firms voluntarily choose to go beyond what is required by law, i.e., cases exist were an offset is driven by an existing regulatory requirement in a first instance, but later accompanied by additional offsetting

[8]Cf. Appendix 4: List of members of the Advisory Group of the Business and Biodiversity Offsets Programme.

[9]This also includes a problem of definition, i.e., what counts as a corporation—Rio Tinto as global business or also Ambatovy as a huge local project based joint venture?

Rio Tinto's Kennecott Utah Copper mine

"In the mid-1990s, Rio Tinto Kennecott Utah Copper mine, North America's largest copper mine, needed additional storage capacity for 'tailings'. The company purchased an area of degraded saltpans and industrial land containing designated wetland habitat. To offset their impact on the wetlands (required by US law), Kennecott purchased water shares and 2,500 acres (1,011 hectares) of degraded lands for creation of a shorebird and waterfowl refuge. A wetland plan was designed, establishing Kennecott's obligations for construction, operation, maintenance and monitoring. After the initial successes, Kennecott went beyond its obligation by purchasing additional land and water to expand the site to more than 3,600 acres (1,460 hectares) with the added benefit of mitigating for impacts from other projects affecting wetlands in the same watershed. After completion, 'The Kennecott Inland Sea Shorebird Reserve' now shows a 1,000-fold increase in bird use." (TEEB 2010)

Fig. 7.20 Additional voluntary biodiversity offsets: the case of Rio Tinto's Kennecott Utah Copper mine (Source: TEEB 2011 and Rodricks 2010)

activities. An example for this is the Kennecott Utah Copper mine operated by Rio Tinto (no. 49 in the screening, see Fig. 7.20).

As has been said, the built typology represents ideal types and in practice an overlap of these is likely to occur. Typical intersections that have been encountered in the course of this study include:

- Type 3 enabled biodiversity offsets and type 5 corporate biodiversity offsets
- Type 3 enabled biodiversity offsets and type 6 local biodiversity offsets
- Type 2 conditional biodiversity offsets and type 5 corporate biodiversity offsets

However, the greatest overlap can be seen with regard to the two subtypes under type 3 enabled biodiversity offsets and type 6 local biodiversity offsets. Case 3a and 3b are distinctive in terms of stakeholders or initiators who foster biodiversity offsets (i.e., governments vs. NGOs), but also similar in terms of implementation mechanisms such as standards and guidance.

By contrast, types 6a and 6b could hardly be distinguished in practice. As has been found, in a collaboration at local level the focus is not primarily on the distinction of pressures vs. incentives, but rather on an integrative/cooperative approach.

What can be learned from this is that the different cases or approaches are not mutually exclusive—on the contrary, they can even be of enhancing and reinforcing nature. This means that an existing regulatory frame will not prevent, but most likely foster additional engagement of developers in offsets.

Furthermore, in practice the pressure (or even conflict) of different intersecting drivers resulting from a complex setting of actors can often be addressed by similar means, i.e., the driving forces or pressures do greatly vary, but the response can be almost the same.

7.4.3 Limitations of the Typology

The suggested typology has not taken into consideration a number of aspects. The two most important ones are temporal aspects and the distinction between demand and supply side for biodiversity offsets.

Temporal Aspects of Biodiversity Offsets

With regard to temporal aspects it is important to note that offsetting schemes usually are dynamic systems that undergo changes. This is in particular true for voluntary compensation schemes. This can be illustrated at the example of the Malua BioBank in Sabah, Malaysia (no. 73 in the screening). Initiated as a credit-based voluntary certification system set up as a cooperation between the Forestry Department, Sabah (public authority) and New Forests (a private timberland investment management company with a focus on sustainable forestry and emerging environmental markets), the biobank now implements biodiversity offsets. To date all activities are on voluntary basis, but establishing a mandatory regulatory framework is being discussed. A general recommendation of a workshop held in 2010 was to go toward making offsets and no net loss mandatory with a voluntary trial period (personal communication with Fred Kugan and Darius Sarshar 2010).

Another Case is that of the Akyem gold mine in Ghana operated by Newmont Mining Corporation (no. 5 in the screening). The mining company had set up a plan to implement voluntary biodiversity offsets. Through the inclusion of this plan in the Environmental Impact Assessment and the approval procedure this initial voluntary engagement was converted into a legal permit requirement (personal communication with Nana Adunsiah 2010).

Generally, in addition to the two cited examples, it is not unusual for new offset regulations to emerge or existing ones are enforced. In such cases voluntary biodiversity offsets are likely to be (anticipatory) "pre-compliance measures" (personal communication with Nathaniel Carroll 2010).

Demand and Supply Side for Biodiversity Offsets

As previously mentioned there is often a discrepancy between a project proponent who is responsible for the negative environmental impacts of his activities and a third party who implements biodiversity offsets (compensation services provider). According to the "polluter pays principle," generally the project proponent is liable for the damages caused by the project, and has therefore to put in place appropriate compensation measures.

Nevertheless, the development and implementation of offsets can be carried out either on a project-by-project basis, i.e., by the project proponent providing actions

Table 7.9 Examples of implementation options for regulatory and voluntary offsets

	Examples	
Type of offsets	Regulatory	Voluntary
Private conservation banks	Clean Water Act Compensatory Mitigation ("wetland banking", US); Corporation of the Society of the Missionaries of the Sacred Heart BioBank (New South Wales, Australia); Endangered Species Program Conservation Banks (US); Environmental Offsets Policy (Western Australia)	CDC Biodiversité (Bouches-du-Rhône, France); the Environment Bank (UK), eg Thames River Conservation Credits Bank
Government conservation banks	BushBroker (Victoria, Australia); biobanking (New South Wales, Australia)	Unlikely as a medium for voluntary offsets
Public–private conservation banks		Environment Agency-Associated British Ports (UK); Malua BioBank (Sabah, Malaysia)
Contracts with private organizations	Clean Water Act Compensatory Mitigation ("wetland banking", US)	Possible with private consultancy firms and NGOs
Partnerships or contracts with nonprofit organizations (eg with NGOs)	Some Australian mining companies are considering partnerships with existing NGOs to deliver their offset commitments	Rio Tinto QMM and Missouri Botanical Gardens (Madagascar); Walmart (USA: "Acres for America")
DIY offsets by developers	BushBroker (Victoria, Australia); Environmental Offsets Policy (Western Australia); Fish Habitat Compensation Banks (Canada)	Ambatovy Sherritt and Wildlife Conservation Society (Madagascar)
In lieu fees	Clean Water Act Compensatory Mitigation ("wetland banking", US); Environmental Offsets Policy (Western Australia)	

Source: adapted from ICMM IUCN (2012)

and measures, or delegated to a third party (Escorcio Bezerra 2007). The latter includes a variety of different approaches as is exemplarily shown in Table 7.9, e.g., mitigation banks (see Sect. 4.1.1, US Wetland Mitigation Banking), conservation banks, in-lieu fee arrangements, auction, and brokering schemes (Darbi et al. 2010). Banking frameworks are those schemes in which entrepreneurs, by developing their own offset initiatives, can earn credits and then recapture their investment by selling these credits to project developers with offset obligations (Escorcio Bezerra 2007) (Table 7.9).

This shift of responsibility from the project proponent to a third party is particularly important for voluntary biodiversity offsets. A clear distinction between the responsibility and motivation of the project proponent on the one side (demand side) and the compensation services provider on the other side (supply side) must be made.

For compensation services providers, biodiversity offsets represent a market opportunity. Thus, an offset always needs to be judged in relation to the developer who uses offsets to counterbalance the negative externalities of his business activities and not in relation to the offset provider for whom the delivery of offsets presents his usual business model.

In practice, however, this distinction is sometimes difficult—especially with regard to habitat banking, where in a first instance only the offset and the offset provider exist and only later are paired with a developer, i.e., technically a habitat bank is a business model which only turns into an offset once a specific area and conservation action is assigned to a specific development with the aim to counterbalance the negative residual impacts of this development.

References

Aggregate Industries UK Ltd (2010) Proposed Bardon Hill Quarry Extension. Detailed mitigation and compensation implementation plan

Baker J, Woodley A, Wilkinson P (2015) Development with biodiversity net positive. PPT presentation. https://www.iema.net/system/files/julia_baker_paul_wilkinson_amelia_woodley.compressed.pdf

Bayon R, Carroll N, Fox J (2012) Conservation and biodiversity banking: a guide to setting up and running biodiversity credit trading systems. Earthscan, London

Böhme C, Bruns E, Bunzel A, Herberg A, Köppel J (2005) Flächen- und Maßnahmenpools in Deutschland. Bundesamt für Naturschutz (ed.) Schriftenreihe "Naturschutz und Biologische Vielfalt" Heft 6, Bonn, Bad Godesberg

Darbi M (2015) Biodiversity Law (including biodiversity offsets) under development in France. Biodiversity Offsets Blog. http://www.biodiversityoffsets.net/biodiversity-law-under-development-in-france/

Darbi M, Ohlenburg H, Herberg A, Wende W (2010) Impact mitigation and biodiversity offsets-compensation approaches from around the world: a study on the application of Article 14 of the CBD (Convention on Biological Diversity). BfN-Schr-Vertrieb im Landwirtschaftsverl

Department of Environmental Affairs and Development Planning (2007) Provincial guideline on biodiversity offsets. Republic of South Africa, Provincial Government of the Western Cape, Department of Environmental Affairs & Development Planning, Cape Town

Doswald N, Barcellos Harris M, Jones M, Pilla E, Mulder I (2012) Biodiversity offsets: voluntary and compliance regimes. A review of existing schemes, initiatives and guidance for financial institutions. UNEP-WCMC/UNEP FI, Cambridge, UK/Geneva, Switzerland

Escorcio Bezerra LG (2007) Biodiversity offsets in national (Brazil) and regional (EU) mandatory arrangements: towards an international regime? http://www.forest-trends.org/biodiversityoffsetprogram/library/new/Dissertation%20Biodiversity%20Offsets%20LGB%20IUCN%20BBOP.doc

ICMM (2016) International Council on Mining and Metals. Homepage. https://www.icmm.com/

ICMM (2010) Mining and biodiversity. A collection of case studies—2010 edition

ICMM IUCN (2012) Independent report on biodiversity offsets. Prepared by The Biodiversity Consultancy. www.icmm.com/biodiversity-offsets

International Advisory Group (IAG), Laos, Nam Thuen II Power Company Ltd., EcoLao, Norplan (2005) Impact mitigation, restoration. The Nam Theun 2 Dam in Laos (Vol. 16), Fifth report of the International Advisory Group (IAG) on the World Banks handling of social and environmental issues. Available at: http://www-wds.worldbank.org/external/default/WDSContentServer/WDSP/IB/2005/09/09/000011823_20050909112339/Rendered/PDF/E385v1615th1IAG0Report.pdf
Lopez Arbeláez DM, Quintero Sagre JD (2015) Compensaciones de biodiversidad: experiencias en Latinoamérica y aplicación en el contexto colombiano. Gestión y Ambiente 18(1):159
Madsen B, Carroll N, Kandy D, Bennett G (2011) Update: state of biodiversity markets. Forest Trends, Washington, DC
Manuel J (2013) Overview of the South African framework for biodiversity offsets. Presentation in a joint workshop of the Business and Biodiversity Offsets Programme and the European Union No Net Loss Working on 22 May 2013 as part of BBOP's Community of Practice and hosted by the European Commission in Brussels, Belgium. http://bbop.forest-trends.org/documents/files/jmanuelppt.pdf
Marsh D (n.d.) Rio Tinto partnership. http://www.fauna-flora.org/initiatives/rio-tinto-2/
Miller KL, Trezise JA, Kraus S, Dripps K, Evans MC, Gibbons P et al (2015) The development of the Australian environmental offsets policy: from theory to practice. Environ Conserv 42 (04):306–314
Poulton D (2015) Key issues in biodiversity offset law and policy: a comparison of six jurisdictions. Report for Ontario Nature
Quétier F, Regnery B, Levrel H (2014) No net loss of biodiversity or paper offsets? A critical review of the French no net loss policy. Environ Sci Pol 38:120–131
Resolution Copper Company (2013) Frequently asked questions
Rodricks S (2010) Mainly based on Department of Environment Climate Change and Water 2009 & Mamouney et al. 2009; TEEBcase: Biodiversity Banking and Offset Scheme of NSW, Australia (2010). Available at: TEEBweb.org
TEEB—The Economics of Ecosystems and Biodiversity for Local and Regional Policy Makers (2011) (Temple HJ, Anstee S, Ekstrom J, Pilgrim JD, Rabenantoandro J, Ramanamanjato J-B, Randriatafika F, Vincelette M) Forecasting the path towards a Net Positive Impact on biodiversity for Rio Tinto QMM. Rio Tinto-IUCN Technical Series No. 2
ten Kate K, von Hase A, Boucher J, Cassin J, Victurine R (2011) Opportunities for environmental funds in compensation and offset schemes. RedLAC capacity building project for environmental funds. RedLAC, Rio de Janeiro. https://www.google.de/url?sa=t&rct=j&q=&esrc=s&source=web&cd=5&cad=rja&uact=8&ved=0CEAQFjAEahUKEwj5id_P9-nIAhXI2BoKHWeiBhU&url=http%3A%2F%2Ftoolkit.conservationfinance.org%2Fsites%2Fdefault%2Ffiles%2Fdocuments%2Fredlac-capacity-building%2F5-opportunities-environmental-funds-compensation-and-offset-schemes.pdf&usg=AFQjCNGcxk5akDMfV_11U2syzDuHxj6vsg&sig2=BG9IsPZnohSD_ww6wlDhrw&bvm=bv.106379543,d.d2s
Villarroya A, Barros AC, Kiesecker J (2014) Policy development for environmental licensing and biodiversity offsets in Latin America. PLoS One 9:e107144

Putting the Developed Typology of Biodiversity Offsets into Context: Some Conclusions

8

Abstract

The results of this study are neither exhaustive nor representative. Nevertheless, global tendencies become visible, allowing for some generalized conclusions and recommendations for different locations and contexts, as well as for different stakeholders. For example, diversified roles and responsibilities of government bodies beyond the conventional understanding in terms of regulator and enforcing authority were observed under the different types. This also offers an enormous potential for new forms of private–public partnerships, etc.

Moreover, four trends were identified for the current situation of offset implementation worldwide: (1) dominance of regulatory biodiversity offsets and acceleration of offset-related legislation, (2) increasing cooperative approaches and growing role of NGOs and civil society groups encouraging developers to undertake biodiversity offsets, (3) biodiversity offsets as part of business operations, and (4) emerging lender requirements and standards.

Despite these global tendencies, the typology highlights the context dependency of biodiversity offsets, e.g., regarding differences between offset implementation in the developed and developing world, pristine vs. transformed landscapes, but also different cultural and legal traditions as well as different forms of governance associated with them. Thus, the typology helps to identify and focus on the major characteristics and challenges in a given context.

While building on biodiversity offsets as a practical case, the developed typology could potentially be applied to a variety of environmental (and most likely also social) aspects beyond legal compliance. Thus, the typology could add to the discussion about voluntary environmental approaches and corporate responsibility both in research and in practice.

M. Darbi, *Biodiversity Offsets Between Regulation and Voluntary Commitment*,
https://doi.org/10.1007/978-3-030-25594-7_8

Keywords

Business and biodiversity · Biodiversity offsets types · Biodiversity offsets context · No net loss · Net gain · Altruistic biodiversity offsets · Voluntary biodiversity offsets · Corporate biodiversity offsets · Voluntary environmental approaches

8.1 Conclusions from the Built Typology

It has to be noted that there is certainly a large number of more or less well-documented biodiversity offset cases, which are not captured in the present study. However, the limited scope of this study did not allow for the exploration and incorporation of a broader empirical evidence base. Thus, as has been stated (cf. Chap. 2), the results are neither exhaustive nor representative.

Nevertheless, global tendencies become visible, allowing for some generalized conclusions that will be discussed below and can provide the starting point for further research. Different conclusions can be drawn and recommendations are given which types could be preferred in a global context (see Sect. 8.1.2), for different locations and contexts (see Sect. 8.1.3) as well as for different stakeholders (see Sect. 8.1.1).

8.1.1 Stakeholder Orientation: Which Conclusions Can Be Drawn About the Roles and Responsibilities of Different Actors?

Biodiversity offsets are complex and shaped by the interplay of a number of actors with distinct roles. A short overview of these is presented below.

Government

Government and its various bodies and representatives were, are and most likely will be the most important actors related to biodiversity offsets. The functions, however, are diverse: from regulator and enforcing authority, to enabling pilot schemes and providing incentives, to providing baseline data and strategic nature conservation goals.

As has been shown throughout the different types, government involvement is often crucial to offsets. However, the type and scope of government involvement can be very different according to the specific circumstances, i.e., government involvement is not restricted to the role of a regulator, but also facilitator, mediator, etc.

Even in the absence of or preempting legal offsetting provisions, government can actively shape and influence the use of offsets: first through the power of permitting authorities who may bind approval to offset conditions and second through actively facilitating pilots, commissioning of trial projects, scientific advice, and elaboration of (independent) guidance and baseline data.

As various examples worldwide have shown (e.g., Germany, Australia, and Brazil), the different levels of administrative competencies have a major influence on biodiversity offsets, especially for large-scale projects and in government-mandated or -enabled schemes. Thus, for government it is crucial to consider and weigh the differences between the federal (and a possible framework law) and provincial level or federal states' competencies early on. It is recommended to provide comprehensive guidance at the federal level, on which provincial guidelines and more specific provisions can build upon.

Financial Institutes
Financial institutes increasingly take nearly the same role as government. They have a high degree of power when offsets are bound as a condition for project financing. Financial institutes, with the World Bank and the International Finance Corporation at the forefront), have seen the potential in managing the risk (especially with respect to consumers and the public) of financing large-scale development projects (often in the developing world).

Conservation Organizations, NGOs, and Networks
In the light of the limited effectiveness of traditional conservation actions to halt global biodiversity loss, nature conservation experts are exploring ways of engaging a new set of actors (in particular business) as well as the broader public. Thus, Natural Capital is the new paradigm for many nature conservation organizations and increasingly discussed. This discussion centers around the economic valuation of nature and biodiversity and the role of businesses in addressing the harm arising from their activities (i.e., via biodiversity offsets and others).

However, conservationists are split between sceptics (e.g., FERN, Friends of the Earth) and those who are actively working in partnership with businesses to collectively achieve the best outcome out of biodiversity offsets (e.g., The Nature Consevancy and Birdlife International).

Furthermore, various groups and actors of civil society can actively shape and advance offsets. As has been highlighted throughout this study, a growing number of organizations have provided guidance and best practice on biodiversity offsets. For best results it is crucial to join forces (not "reinvent the wheel") and ensure that standards (e.g., from financial organizations, sectoral associations, and NGOs) are mutually reinforced.

Businesses
Business is very much in focus now ("Business and biodiversity," the natural capital discussion, etc.), but what appears to be their own motivation is often driven by a combination of other influences (e.g., government, financial institutes, or the public).

Biodiversity offsets, along with impact assessment and impact mitigation are not the prior interest of business, but necessary by-products that arise from pushing development forward. Thus, businesses are usually not trained in biodiversity offsets and for most companies, it does not make sense to build capacity and develop competencies in house to respond to this necessity. Thus, they should seek assistance from trained professionals. Similar to the field of Environmental Impact Assessment, there are now numerous consultancies that have a sound knowledge of biodiversity offsetting, the application of different methodologies and governance regimes. These range from small boutique consultancies (e.g., the United Kingdom based "The Biodiversity Consultancy") to global consultancies (e.g., PricewaterhouseCoopers).

For businesses, it pays off to invest in the experience and expertise of specialized consulting firms to accompany or carry out all operations. For biodiversity offsets, this has the additional advantage that the objective view of an expert can add to the unbiased development of an offset.

Consultants

Seeing the complexity of biodiversity offsets, the role of consultants as experts in implementation is crucial. However, their role is seldom appropriately acknowledged, with rare exceptions, e.g., IFC PS6 paragraph 10:

> When a client is considering the development of an offset as part of the mitigation strategy, external experts with knowledge in offset design and implementation must be involved. (IFC 2012)

Nevertheless, many consultants are increasingly shaping further development and the improvement of biodiversity offsets, e.g., in the scope of the Business and Biodiversity Offsets Programme, both with a stronger global (e.g., the British "The Biodiversity Consultancy" and the French "Biotope") and national focus (e.g., Swedish "Ecogain").

Consumers and Society as a Whole

Given the exposure of business activities (mainly fostered through globalization and rapid share of information via new media), the public is increasingly attentive toward the environmental harm that these create. As has been shown with the type of local biodiversity offsets (see Sect. 6.6), local communities as well as society, in general, are influencing and sometimes even driving biodiversity offsets. One constraint to this is that biodiversity offsets are not yet relevant to consumers. This bears a huge potential for the future.

Society as a whole, as well as the institutions and individuals it consists of, has huge power and influence. Especially globalization and the use of the internet have opened up new pathways to the involvement of a broad public into various fields. A complex task such as biodiversity offsets can benefit significantly from this corrective power of the masses—not only the targeted actions of NGOs, but also, for example, grassroots efforts and individuals can help to provide accurate and topical baseline data and to ensure compliance and monitoring.

For long-term success, it is viable to build strong partnerships, e.g., with local communities and authorities, nature conservation organizations, and other NGOs, as well as to explore synergies with local firms and developers.

Summarizing, Fig. 8.1 specifies the roles and responsibilities of the abovementioned stakeholders, differentiated according to the built types of biodiversity offsets.

Actors / types	Governments	NGOs	Financial institutes	Local communities	Businesses
Type 1: Regulatory biodiversity offsets	**Regulator, enforcing authority**	Partner with government on standards and implementation, verify outcome, monitoring	Enforcement of the legal framework as part of the lending conditions	Monitor enforcement, report failure	Compliance with governmental requirements
Type 2: conditional biodiversity offsets	Cooperation and mediation at national/local level, Foster and reinforce context sensitive implementation	Partner with business to help them meet lending conditions (e.g. IFC PS6)	**Anchor offset requirements as part of the lending conditions (e.g. IFC PS6)**	Be involved to strengthen context sensitive implementation	Compliance with lending conditions
Type 3: Enabled biodiversity offsets	**Enabling pilot schemes or incentives, guidance**	**Set (global or national) standards for offsets, launch offset programs, enter in partnership with business**	Encourage use of offset enabling schemes in line with lending conditions	Be involved to strengthen context sensitive implementation	Partner with governmental bodies or NGOs
Type 4: Sectoral biodiversity offsets	Cooperation and mediation at national/local level, Foster and reinforce context sensitive implementation	Contribute to setting sectoral standards (knowledge and know how)	Create standards for sectors, e.g. World Bank offset pilot scheme in mining in Liberia	Be involved to strengthen context sensitive implementation	Initiate/take part in sectoral offset initiatives, partner with other businesses to set sectoral standards
Type 5: Corporate biodiversity offsets	Cooperation and mediation at national/local level, Foster and reinforce context sensitive implementation	Foster and underpin corporate commitments by strategic partnerships and know how/training	May trigger a changed business behavior preempting lending conditions	Be involved to strengthen context sensitive implementation	**Establish corporate biodiversity offset policies and no net loss or net gain targets**
Type 6: Local biodiversity offsets	Structuring governance processes (bottom up)	Give local stakeholders a voice	Work together with local stakeholders to facilitate and verify compliance with the lending conditions	**Drive and influence offset implementation, strengthen context sensitive implementation**	Engage with local stakeholders, facilitate bottom up participation and offset development
Type 7: Altruistic biodiversity offsets	-	Contribute to a proactive natue conservation context and social norms	-	Contribute to a proactive natue conservation context and social norms	**Embrace corporate responsibility without recompense**
Cross-cutting issues	Provide baseline data, set strategic nature conservation goals, landscape planning	Experts in conservation, help to gain credibility and legitimacy, assist in complex methodological issues, baseline data, monitoring	Enforcement of the mitigation hierarchy, outbalancing develop	Provide local knowledge, help to gain credibility and social license, monitoring	Balance different business priorities (offsets being one), ensure compliance, best practice, stakeholder orientation

Fig. 8.1 Role, challenges, and crucial points of different stakeholders with regard to the different types (It is highlighted in bold where actors play the role as major drivers for a specific type, e.g., financial institutes are the major driver for type 2 conditional biodiversity offsets). Source: Author

TRENDS:

- More **governments** introducing or exploring policy on biodiversity offsets;
- More **companies** undertaking offsets voluntarily for business reasons;
- More **banks** and investors requiring biodiversity offsets as a condition for access to credit or investment;
- More **NGOs** and civil society groups encouraging developers to undertake biodiversity offsets; and
- **BBOP** set up to develop, share and encourage the use of best practice.

Fig. 8.2 Trends in biodiversity offsets implementation. Source: ten Kate et al. (2011)

8.1.2 The Way Ahead: Which of the Offset Types Are Promising and Why?

The analysis of the evidence from the screening of biodiversity offset cases together with the characterization of the types and underpinned by the state of the art presented in the literature draws a picture of the current situation of offset implementation worldwide and reveals tendencies for future development.

In general, four trends can be observed (cf. Fig. 8.2) with major emphasis on the following types:

- Dominance of regulatory biodiversity offsets and acceleration of offset-related legislation.
- Increasingly cooperative approaches and the growing role of NGOs and civil society groups encouraging developers to undertake biodiversity offsets.
- Biodiversity offsets as part of business operations.
- Emerging lender requirements and standards.

Most notably, a dominance of offset schemes required or enabled by government has been observed. The increasingly diversified roles and responsibilities of governmental bodies in offset implementation have been already discussed in Sect. 8.1.1.

However, despite the expanded involvement of government beyond regulation, many offsets in the screening have been driven by regulatory pressures (type 1). This

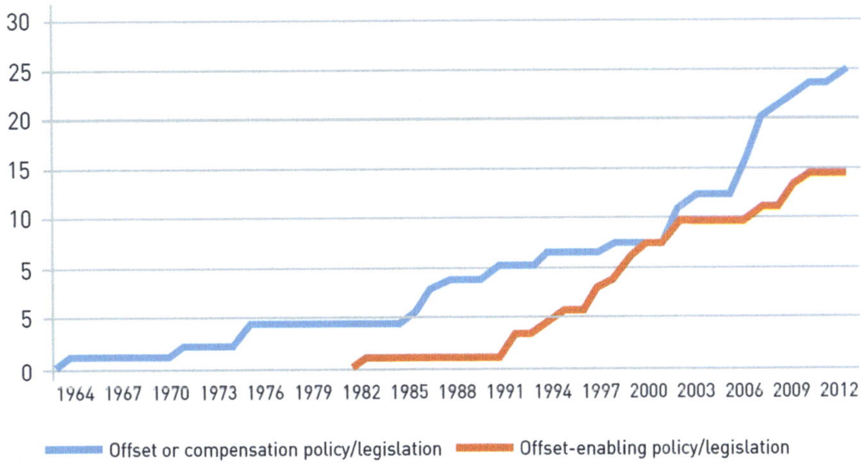

Fig. 8.3 Rise in government policies, guidance, and legislation that require/enable biodiversity offsets. Source: ICMM IUCN (2012)

is in line with findings by PricewaterhouseCoopers LLP (2010) that "(t)he main business drivers for addressing biodiversity are reputational risk and regulatory compliance."

Furthermore, ICMM IUCN (2012) and FFI (n.d.) also highlight regulation as one of the main drivers of business involvement in biodiversity offsets.

Similarly, IEEP et al. (2012) claim that offsetting and habitat banking initiatives "require appropriate regulations to be effective. Experience from around the world has shown that it is possible to effectively regulate without excessive transaction costs" (Fig. 8.3).

The evidence from this study clearly underpins regulatory biodiversity offsets as the dominant offset implementation scheme. This means that offsetting cannot be considered merely as a neoliberal market-based instrument applied by businesses, but instead constitutes a tool in environmental policy to underpin and reach nature conservation goals as several authors have argued (cf. Albrecht et al. 2014; Herbert 2015; Darbi et al. 2016; eftec, IEEP et al. 2010).

Nevertheless, as has been observed with regard to the location and geographical distribution of regulatory biodiversity offsets, their use to date remains limited and fails to comprehensively address major impacts in a worldwide context, e.g., leaving Asian and African countries largely out.

The recent rise in biodiversity offset policies in Latin America (Lopez Arbeláez and Quintero Sagre 2015; Villarroya et al. 2014) hints to a paradigm shift in terms of a wider dissemination of offset regulations beyond the traditional locations (i.e., North America, Europe, and Australia).

However, it is unknown whether this trend will continue. Thus, in the short- to medium-term regulatory biodiversity offsets alone cannot provide the solution to cope with major environmental impacts at a global level and consequently to halt biodiversity loss.

Besides regulatory biodiversity offsets, cooperative approaches are increasingly observed, with a growing role of NGOs. This hints to the promotion of enabled biodiversity offsets (type 3). Furthermore, these are found in comparably large numbers and are relatively evenly spread, which makes them particularly promising in a global context.

> Experience suggests that, in addition to working with local and indigenous communities, the involvement of conservation NGOs is a critical component in generating credible BDO initiatives. This is not merely recognizing the obvious: mining companies are better at mining, whilst NGOs and other stakeholders have an edge on conservation. Without the involvement of legitimate NGOs, most BDO concepts may not gain credibility and would not be able to contribute to a social license. The involvement of international NGOs will also help to develop practical approaches to complex methodological issues surrounding BDOs. NGOs can assist in assessing and validating baselines and benchmarks, selecting appropriate "offset currency" and indicators (hectares, trees or frogs?), identifying eligible components in view of the project specific context (planting trees, capacity building or trading-up to higher biodiversity priorities?) and use of multipliers (two trees planted for each tree removed?). (Nazari and Proebstel 2009)

The main focus of the increased promotion of biodiversity offsets from the mid-2000s has been on the role of business. Biodiversity offsets have been presented as the response to business risks and opportunities from which corporate commitments toward no net loss or net gain have arisen.

> Pressure is increasing from governments, financial institutions and civil society for businesses to take responsibility for their impacts on biodiversity. From a corporate perspective, this equates to considerable regulatory, financial and reputational risk. However, a survey of over 1,000 business executives showed that nearly 60% viewed biodiversity as more of an opportunity for their companies than as a risk. Some estimates suggest that sustainability-related global business opportunities in natural resources (including energy, forestry, food and agriculture, water and metals) could reach between US$ 2-6 trillion annually by 2050 at 2008 prices. (FFI n.d.)

As has been found in this study, the increasing global promotion of the business case for biodiversity offsets is not backed by a correspondingly high number of corporate offset cases. In fact, corporate biodiversity offsets to date are restricted to certain sectors (in particular mining) and a relatively small number of a few global players (e.g., Rio Tinto). However, these may serve as role models, and may inspire further expansion of corporate biodiversity offsets.

Financial institutes have a large power. However, they are restricted to a limited number of large-scale development projects in particular in developing and threshold countries. Despite their limited number, they are anticipated to have a crucial (and growing) role in the future, because the developments they are associated with usually have an enormous footprint with large negative residual impacts. Those development projects are often highly exposed in the public and may, therefore, serve as an (either positive or negative) prototype or role model for smaller projects in the region.

> (IFC PS6) is now perhaps the largest single driver of corporate biodiversity management. With broad uptake and support form financial institutions, industry, governments and civil society, it is rapidly gaining recognition as global best practice guidance. (The Biodiversity Consultancy 2015)

To summarize, all of the mentioned types have certain strengths and restrictions, which leads to the conclusion that no one alone should be preferred in a global context to counterbalance the negative externalities of business operations.

> For voluntary offsets and those driven by financial lending requirements, significant consensus has emerged around high-level principles, but the details of what these mean in practice remain open to question. This is partly because different situations require different approaches, so there is no single "right" way of conducting offsets. (ICMM IUCN 2012)

Most likely a combination of enabled and corporate biodiversity offsets is preferable, i.e., building on corporate commitment and a favorable offset enabling framework provided by governments and increasingly by NGOs such as BBOP, IUCN, Flora & Fauna International, and BirdLife International.

While these in many cases have been seen to be restricted to specific sectors (the dominance of the extractive industry has been mentioned several times), regulatory biodiversity offsets that account for a larger scale of development activities, e.g., from other industries, infrastructure, and housing, will also make an important contribution, especially with regard to incorporating smaller-scale impacts.

On the role of conditional biodiversity offsets and lenders' requirements to implement biodiversity offsets (with IFC PS6 at the forefront), there appears to be more expectations in the literature (e.g., FFI n.d.; ten Kate et al. 2011; TBC 2014; ICMM IUCN 2012) than evidence has been found in this study to confirm these broad expectations. To date, it is too early to judge whether conditional biodiversity offsets and IFC PS6 will only translate to a small number of World Bank and IFC flagship developments, or whether they can broadly influence lenders' requirements even for a large array of development projects at different scales. It is even imaginable that they could educate businesses in the long term toward accepting these standards as common practice.

Moreover, local biodiversity offsets, even though less significant in quantitative terms, have been found to be an accompanying motivation in about twenty percent of all offset cases encountered. Beyond the mere numbers, a context-sensitive approach toward offsets building on collaborations with local stakeholders can be used as a means to maximize satisfaction of stakeholders and support for the offset, and thus also to enhance the persistence of the offset.

Sectoral biodiversity offsets have been found to be insignificant in absolute numbers. Their restriction to the extractives is a further major concern. Nevertheless, a chance lies in sectoral biodiversity offsets to encourage other sectors to explore the way biodiversity offsets could be induced in their operations and their supply chains and finally to identify and define standards and best practices.

Lastly, altruistic biodiversity offsets represent the purest understanding of biodiversity offsets in the narrowest sense of the word. This makes them questionable at least and mostly inoperational as a concept in practice. Altruistic biodiversity offsets are a unicorn which we might imagine in great detail, however, in practice evidence rather proves to these motivations are observed only in combination with one of the other types.

8.1.3 Global Variety: What Are the Implications for Different Contexts Worldwide?

While the global tendencies and recommendations have been discussed in the preceding section, the importance of context and the context dependency have been stressed throughout this study. Therefore, some brief conclusions on different geographical regions will be given to complement the global view.

Developed versus Developing World

First, it can be noted that differences exist between offset implementation in the developed and developing world. This refers not only to the distinction between pristine vs. transformed landscapes, but also to different cultural and legal traditions as well as different forms of governance associated with them.

Consequently, regulatory biodiversity offsets exist mainly in developed countries and have highly matured in several of them (e.g., the United States, Germany, Australia, and the European Union). They may, therefore, serve as role models. For example, the transfer of lessons learnt from the German Impact Mitigation Regulation to the Japanese context is being explored (Darbi and Wende 2013).

Another example is the habitat compensation under the EU Birds and Habitats Directive. As has been noted, these EU level offset requirements have been translated into several national contexts, e.g., in Sweden and France (see Sect. 4.1.3), however, not comprehensively and further expansion should be undertaken.

Enforcement of offset regulation can be observed, for example, in France and Canada.

Thus, regulatory biodiversity offsets still bear a large potential in the short term—in particular for the developed world, and maybe in the long term even beyond this, as the recent development of offset policies in Latin American countries indicates.

As regards the developing world, corporate and offset requirements by financial institutes, supported by NGOs, local communities, and civil society groups, can respond to the need for biodiversity offsets most quickly today and in the near future. While this is already largely present in South American and sub-Saharan countries, further potential for expansion exists in Asian and northern African countries.

Two specific and distinct examples of biodiversity offset implementation in the developed world are the situations in Europe and in Australia.

No Net Loss in the EU

As has been noted, biodiversity offsets are being discussed controversially in the EU, both at the EU level in the scope of the no net loss initiative and at the member state level with regard to the introduction of offset and habitat banking schemes (see Sects. 1.3 and 4.1.3).

While the results of the public consultation on the planned no net loss initiative point toward the development of a "voluntary EU framework for compensation/ offsetting including technical guidelines and benchmarking good practice" (European Commission 2015a, b), greater potential lies in the enactment of a regulatory framework for offsets at EU level in order to ensure the implementation of the principle of full coverage (Darbi et al. 2016). However, this currently appears unlikely to happen (cf. Wende et al. 2018; Tucker et al. 2018).

Australia

Australia has seen an increase in offsetting practice and debate in recent years. This is backed by legal offset requirements at both federal (in particular related to the EPBC Act[1]) and state level. Furthermore, examples of corporate biodiversity offsets preempting requirements by governmental authorities (e.g., no. 25 in the screening) have been noted.

While a clear trend toward regulatory biodiversity offsets is observed in large parts of Australia, inequalities remain among the different federal states which could best be overcome on the basis of a common regulatory framework at the federal level (similar to the federal level offsetting framework of the German Impact Mitigation Regulation set out in the Federal Nature Conservation Act).

Moreover, in some cases it was unclear whether a regulatory requirement existed (type 1) or whether offsets were driven by a corporate commitment (type 5). Therefore, the relation of these two types needs to be clarified.

Furthermore, NGOs might help in verifying offsets and making them credible.

8.1.4 Application Spectrum: What Can the Typology Be Used for?

Summarizing, the suggested typology of (voluntary) biodiversity offsets contributes to:

- Increase transparency
- Structure and differentiate
- Take into consideration different contexts, drivers, and motivations
- Enables the development of a broader evidence base (through methods and tools)
- Provides 72 cases (thereof six described in more detail) as part of this evidence base

[1]Environment Protection and Biodiversity Conservation Act.

These features or advantages of the built typology can help to analyze and further underpin a number of specific trending issues and problems associated with biodiversity offsets as described below.

Most notably, the typology accounts for the global differences in offsets and helps to identify and focus on the major characteristics and challenges in a given context. This is important because biodiversity offsets are now discussed as a global phenomenon. However, they can play out differently in different contexts. The typology clearly points out these differences in terms of contrasting tendencies worldwide (e.g., the specific situation in Europe) and visualizes them in the form of maps.

This means that for the consideration of offsets under different contexts the typology can help to focus on the prevailing aspects rather than on the whole spectrum of offsets. An example of this is corporate commitments toward no net loss/net gain that trigger the implementation of biodiversity offsets (type 5 corporate biodiversity offsets), which have found to be of minor importance in the European context. Thus, the related discourse regarding the sincerity of these corporate commitments and the associated criticism of "greenwashing" can be assumed to be largely irrelevant for the implementation of biodiversity offsets in Europe.

This example illustrates that a more differentiated and less generalized approach toward offsets is beneficial for both policy and public debate, but also for developers and practitioners, in terms of enabling clearer and more specific conclusions on the implementation of biodiversity offsets.

As has been noted, biodiversity offsets have become the subject of heated public discourses, which partly suffer from a lack of information depth and differentiation as well as arguments taken out of context. Thus, notwithstanding the contribution of the present study to an informed debate on biodiversity offsets building on specifications, context, and evidence, it also addresses emerging offset policies and schemes and may be used as a sort of textbook to gain an overview on the different shades of offset implementation and the contexts in which they occur.

For practitioners, the application of the typology underpinned and illustrated by the evidence base from the screening of biodiversity offset cases can present a valuable source of information in terms of crucial aspects and stakeholders, lessons learnt, and best practice to build upon, as well as a source to compare and contextualize their own efforts or plans.

Moreover, this evidence base (72 biodiversity offset cases compiled in the screening) can provide the starting point for an international registry of biodiversity offsets that extends beyond general system descriptions[2] and registries at national or regional level[3] and includes a large number of offset examples from around the world. This can serve as a source of information not only for practitioners, but also

[2]See www.speciesbanking.com for a rather comprehensive overview, cf. Madsen et al. (2010) and (2011).

[3]For an example, see the registry of the measures implemented by a habitat bank in Germany, the Sächsische Ökoflächen-Agentur: http://www.sls-sachsen.de/sls-oekokonto.html

for a wide range of stakeholders, most notably governments and NGOs as well as to underpin further research on biodiversity offsets.

The methodology including derived criteria and tables/forms that have been built upon these criteria, both from the general screening and from the case studies, can be used to describe and analyze further practical evidence on biodiversity offsets worldwide. This ultimately enables not only the attribution of further offset cases to the built types, but may also foster the (improved) comparison at global level while also incorporating considerations of context.

This evidence base differentiated into types is a useful prerequisite for the evaluation of the outcome of biodiversity offsets both regarding procedural aspects (governance, efficiency, etc.) and added nature conservation value (effectiveness, additionality, achievement of the goal of no net loss/net gain). It enables drawing somewhat generalizable conclusions beyond the evaluation of single cases, i.e., beyond "offsets work" vs. "offsets do not work" statements.

Furthermore, with regard to the role of different stakeholders, the present study indicates a more sophisticated understanding of the role of government that goes beyond the conventional understanding in terms of regulator and enforcing authority, which restricts the role of government to biodiversity offsets mandated by law. As a result of this study, diversified roles and responsibilities of government bodies have been observed under the different types (cf. Sect. 8.1.1). These represent not only crucial factors to success in many cases, but also offer an enormous potential for more forms of private–public partnerships, etc.

Lastly, the present study reveals where, how, and under what circumstances existing standards (e.g., the BBOP principles on biodiversity offsets) are implemented, how they are influencing and interacting with other tools at various levels (including international, national, and local).

8.2 Contextualization: Bringing Together the Theoretical Fundamentals with the Practical Evidence on Offsets

8.2.1 How Do Biodiversity Offsets Fit into the Nature Conservation Toolbox?

As ICMM IUCN (2012) finds "[b]iodiversity offsets are one of a number of environmental stewardship approaches, including traditional philanthropy, and other kinds of positive environmental contributions and compensation actions. Offsets differ from these other approaches in being more explicitly linked to project impacts."

Figure 8.4 mentions other forms of environmental stewardship and shows their relation to biodiversity offsets.

What can be seen from this and the above mentioned is that the core particularity (and also advantage) of biodiversity offsets lies in establishing a direct impact–offset relation that is also mostly determined quantitatively. While this has been criticized by some (cf. Sects. 1.3, 9.1, Appendices 6 and 10), it is clearly a more transparent

1. **Traditional philanthropy:** For example, funding support to environmental organizations with no identified link with the impacts or operations of a company.
2. **Positive environmental contributions**: For example education, training and research. These actions have been called "indirect offsets" by the Australian Government and "Additional Conservation Actions" by Rio Tinto. For example, offset investments approved by state governments in Australia include capacity building and research. The effects of such investments are not measurable as quantitative biodiversity outcomes.
3. **Compensation actions linked to the impacts of a development but not commensurate with the type and scale of impacts**: These could be as loosely linked as those of Walmart's "Acres for America" or Enbridge's "Acre for an Acre" programs. These are not No Net Loss biodiversity offset programs, and are quantified in nothing more than hectares of land rather than in terms of biodiversity value of the land.

Fig. 8.4 Other types of environmental stewardship with potential added value for biodiversity offsets. Source: ICMM IUCN (2012)

approach to businesses' negative externalities to promote biodiversity offsets as a means to counterbalance the negative residual impacts of their operations, than it is to promote traditional philanthropic actions (which may in fact be the same as the offset) in a sense of contributing proactively to nature conservation.

This is certainly not intended to replace philanthropy in general and in a business context. Instead, as Earl et al. (2010) claim, increasingly a mix of instruments is needed for best outcomes for a global natural resource management. This includes different tools of regulation, market-based instruments, and voluntary approaches.

Consequently, biodiversity offsets are one tool of an increasingly diversified nature conservation toolbox. These tools, however, are of complementary nature and highly interrelated. To this adds that "[m]any of the implementation challenges of biodiversity offsets are no different to those of biodiversity conservation more generally, for example insufficient stakeholder participation, unsustainable financing, lack of adaptive management and political constraints such as corruption." (ICMM IUCN 2012)

8.2.2 How Can the Results of This Study Inform the Theoretical Discussion on Voluntariness in Environmental Protection?

The analysis of voluntary biodiversity offsets in the scope of this study was largely fueled by the analysis of the theoretical concept of voluntariness and its application in an environmental context. Thus, at the end of this study it is necessary to put the empirical results into the broader theoretical context. This means to ask whether or how the broader debate on voluntary environmental protection, most notably with regard to the concepts of voluntary environmental approaches and corporate (environmental) responsibility can be enhanced by the specific results obtained throughout this study.

Vis Compulsiva and Influence on the Motivation

As has been noted, vis compulsiva is the situation that is most often encountered in practice. This means that ultimately most actions are driven by some sort of influence on the motivations or action goals of a person. It has been confirmed throughout this study that motivations and the different drivers leading to them are crucial triggers for biodiversity offsets. These have been found to arise mostly from context conditions, i.e., the legal and cultural context, for example, governmental bodies that favor offsets, engaged local stakeholders or promoters of global best practices in offsetting such as from the Business and Biodiversity Offsets Program and the International Finance Corporation Performance Standard.

Altruism

It has been noted in the course of the theoretical analysis that it is crucial to include the concept of altruism into any normative evaluation of voluntariness. However, the example of biodiversity offsets has shown that in practice the concept of (pure) altruism is largely intangible. It represents an ideal concept located at the very end of the voluntariness spectrum. Thus, only rare examples have been encountered in the screening of biodiversity offset cases that would at least partially fit into the group of altruistic biodiversity offsets, i.e., voluntary biodiversity offsets in the strictest sense of the word. It has been concluded that altruistic biodiversity offsets in terms of numbers (magnitude and distribution) are a fairly weak type as compared to the other built types.

Nevertheless, it has been determined that altruistic motivations can be complimentary to other motivations (however, difficult to identify). This is in line with theoretical observations that altruistic and egoistic motivations are not mutually exclusive, but rather can coexist and even be mutually reinforcing.

Voluntary Environmental Approaches

In the scope of this study, voluntary biodiversity offsets have been understood as a specific form of voluntary environmental approaches and a particular case of voluntary provision of public goods. This has been confirmed throughout the present study and many related aspects have been discussed. It refers in particular to the question as to what the role of government is, why business would participate (motivation) and whether better outcomes could be reached with a voluntary approach (efficiency).

Large firms adopt voluntary environmental approaches for solid economic reasons, i.e., businesses may gain market-driven and social benefits from adopting voluntary environmental approaches. This can also be observed with regard to voluntary biodiversity offsets (the "business case," "social license to operate").

Most importantly, the diversification of the role of government from provider and controller to facilitator and enabler, i.e., a shift from government to governance has been observed. However, this does not lead to the conclusion that voluntary biodiversity offsets—or voluntary environmental approaches—are replacing traditional environmental regulation. Rather, they are of complementary nature and in line with findings by Alberini and Segerson (2002), voluntary environmental approaches may be more successful when backed by an underlying regulatory framework.

Corporate Responsibility

Corporate responsibility and corporate biodiversity offsets raise a number of questions that are applicable to the general corporate responsibility discourse, e.g., the appropriate scale of corporate commitment and action.

As a result of globalization, corporate responsibility is increasingly subject to a transition from national economies to global ones (Castelló and Lozano 2011). This calls for global standards for the legitimacy of business conduct. A similar trend toward globalization has been observed with regard to biodiversity offsets. As corporate responsibility as a global trend now builds on a growing number of international standards, codes of conduct, and associated management systems (e.g., ISO, GRI, OECD Guidelines), biodiversity offsets are also increasingly shaped by global standards, e.g., by the Business and Biodiversity Offsets Program and the International Finance Corporation. This is particularly visible with regard to the identified type 3b enabled biodiversity offsets (by NGOs), setting "new global norms."

Moreover, corporate responsibility and biodiversity offsets both face expectations to help solve global environmental problems. Whereas the focus of corporate responsibility is broader, biodiversity offsets are specifically targeted to counterbalance ongoing biodiversity loss.

The large overlap between corporate responsibility and voluntary biodiversity offsets also refers to similar challenges, e.g., in terms of "considering nature as a resource that can be exploited forever" (Fougère and Solitander 2009), the problem of accountability (if business itself defines its goals and responsibilities) and ensuring continuous (and stable) commitment in the long term.

Furthermore, as has been noted for corporate responsibility, temporal aspects in terms of issues' maturity play an important role. Similarly, it has been found that different types of biodiversity offsets can also evolve over time and they may also be encountered consecutively, e.g., as has been noted, government enabled offset schemes are in some cases observed to be preempting regulatory offset schemes.

Finally, both corporate responsibility and biodiversity offsets have been found to be only as strong or as effective as the (diverse) factors shaping them.

Summary

While building on biodiversity offsets as a practical case, the typology that has been developed is not entirely specific or restricted to this case. In fact, it could potentially be applied to a variety of environmental (and most likely also social) aspects beyond legal compliance. Thus, the typology could add to the discussion about voluntary environmental approaches both in research and in practice. So far, these have been distinguished as business-led initiatives, government programs, and bilateral agreements (cf. Sect. 3.3).

However, this narrow focus of government on the one side and business on the other leaves out the various stakeholders that are increasingly gaining importance as has been exemplarily shown for the case of biodiversity offsets. This refers for example to the role of (nature conservation) NGOs as partner and facilitator for environmental commitments of businesses, but also to the role of large international (public and private) lenders who are increasingly taking similar responsibilities as regulators.

As has been shown in the literature on voluntary environmental approaches and corporate responsibility, context is an important criterion or driver. On this basis, the developed typology of biodiversity offsets attempts to accommodate a more refined analysis of the context of development projects.

This may also provide new impulses for the discussion on voluntary environmental approaches and motivations for corporate responsibility that account for diversified context conditions. Thus, given the similarities in terms of drivers and motivations, it is of interest whether the following types can also be transferred to voluntary environmental approaches:

1. Regulatory (induced by government)
2. Conditional (conditioned by lenders)
3. Enabled (enabled by governments/NGOs)
4. Sectoral (set sectoral standards)
5. Corporate (corporate commitments)
6. Local (induced by local communities/stakeholders)
7. Altruistic (arising from responsibility/social norms without pressures/incentives)

References

Alberini A, Segerson K (2002) Assessing voluntary programs to improve environmental quality. Environ Resour Econ 22(1):157–184

Albrecht J, Schumacher J, Wende W (2014) The German impact-mitigation regulation—a model for the EU's no-net-loss strategy and biodiversity offsets? Environ Policy Law 44(3):317–325

Castelló I, Lozano J (2011) Searching for new forms of legitimacy through corporate responsibility rhetoric. J Bus Ethics 100:11–29

Darbi M, Wende W (2013) Biodiversity offsets and German impact mitigation regulation. In: Japan-German bio web city/region symposium: new trend of landscape design 2—enhancing ecological services and biodiversity in urban, suburban and rural areas, Nagoya, S5–S17

Darbi M, Albrecht J, Schumacher J (2016) Die europäische No Net Loss-Initiative und Biodiversity Offsets. Einblicke in die aktuelle Diskussion zur Kompensation von Eingriffen in die Biodiversität aus deutscher Perspektive. In: Hebeler T, Hofmann E, Proelß A, Reiff P (eds) Jahrbuch des Umwelt- und Technikrechts 2016. Erich Schmidt, Berlin

Earl G, Curtis A, Allan C (2010) Towards a duty of care for biodiversity. Environ Manag 45 (4):682–696

eftec, IEEP et al (2010) The use of market-based instruments for biodiversity protection—the case of habitat banking. Technical report. For European Commission DG Environment

European Commission (2015a) No net loss of biodiversity public consultation. Summary overview of the responses. Available at http://ec.europa.eu/environment/nature/biodiversity/nnl/pdf/report.pdf

European Commission (2015b) Public consultation on the future EU Initiative on No Net Loss of Biodiversity and Ecosystem Services. Statistics. Available at http://ec.europa.eu/environment/nature/biodiversity/nnl/pdf/NoNetLoss%20statistics%20analysis.pdf

Fauna & Flora International (FFI) (n.d.) Approaching no net loss to biodiversity. Available at https://www.google.de/url?sa=t&rct=j&q=&esrc=s&source=web&cd=1&ved=0ahUKEwi0n6rKzrbLAhXGE5oKHfSABzEQFggcMAA&url=http%3A%2F%2Fwww.fauna-flora.org%2Fwp-content%2Fuploads%2FApproaching-NNL-to-Biodiversity.pdf&usg=AFQjCNFBBITyxYuD18i8og_HtU4gye8M0Q&sig2=bccz65ZSGHwwG24X55RN-g&bvm=bv.116573086,d.bGs&cad=rja

Fougère M, Solitander N (2009) Against corporate responsibility: critical reflections on context, thinking, practice, content and consequences. Corp Soc Responsib Environ Manag 16 (2009):217–227

Herbert M (2015) German impact mitigation regulation—national, European, international. A comparison from the viewpoint of the Federal Agency for Nature Conservation. UVP Rep 29 (3):149–151. Eingriffsregelung; Umweltverträglichkeitsprüfung; Strategische Umweltprüfung; Biologische Vielfalt

ICMM IUCN (2012) Independent report on biodiversity offsets. Prepared by The Biodiversity Consultancy. Available at www.icmm.com/biodiversity-offsets

IEEP et al (2012) Policy options for an EU No Net Loss Initiative. A tender for the European Commission. Unpublished

IFC (2012) Performance standard 6. Biodiversity conservation and sustainable management of living natural resource. Available at http://www.ifc.org/wps/wcm/connect/bff0a28049a790d6b835faa8c6a8312a/PS6_English_2012.pdf?MOD=AJPERES

Lopez Arbeláez DM, Quintero Sagre JD (2015) Compensaciones de biodiversidad: experiencias en Latinoamérica y aplicación en el contexto colombiano. Gestión y Ambiente 18(1):159

Madsen B, Moore Brands K, Carroll N (2010) State of biodiversity markets: offset and compensation programs worldwide

Madsen B, Carroll N, Kandy D, Bennett G (2011) Update: state of biodiversity markets. Forest Trends, Washington, DC

Nazari M, Proebstel D (2009) Biodiversity offsets in mining. Mining.com, environment and communities, pp 41–44

PricewaterhouseCoopers LLP (2010) Biodiversity offsets and the mitigation hierarchy: a review of current application in the banking sector. Study completed on behalf of the Business and Biodiversity Offsets Programme and the UNEP Finance Initiative). Available at http://www.unepfi.org/fileadmin/documents/biodiversity_offsets.pdf

TBC (2014) Rio Tinto Simandou Project Critical Habitat Assessment: IFC Performance Standard 6. Unpublished draft report of The Biodiversity Consultancy, January 2015

ten Kate K, von Hase A, Boucher J, Cassin J, Victurine R et al (2011) Opportunities for environmental funds in compensation and offset schemes. RedLAC capacity building project for environmental funds. RedLAC, Rio de Janeiro. Available at https://www.google.de/url?sa=t&rct=j&q=&esrc=s&source=web&cd=5&cad=rja&uact=8&ved=0CEAQFjAEahUKEwj5id_P9-nIAhXI2BoKHWeiBhU&url=http%3A%2F%2Ftoolkit.conservationfinance.org%2Fsites%2Fdefault%2Ffiles%2Fdocuments%2Fredlac-capacity-building%2F5-opportunities-environmental-funds-compensation-and-offset-schemes.pdf&usg=AFQjCNGcxk5akDMfV_11U2syzDuHxj6vsg&sig2=BG9IsPZnohSD_ww6wlDhrw&bvm=bv.106379543,d.d2s

The Biodiversity Consultancy (2015) IFC PS6. Available online http://thebiodiversityconsultancy.com/approaches/ifc-ps6/

Tucker G, Darbi M, Wende W, Quétier F, Rayment M (2018) Conclusions: lessons from biodiversity offsetting experiences in Europe. In: Wende W, Tucker GM, Quétier F, Rayment M, Darbi M (eds) Biodiversity offsets: European perspectives on no net loss of biodiversity and ecosystem services. Springer, Cham

Villarroya A, Barros AC, Kiesecker J (2014) Policy development for environmental licensing and biodiversity offsets in Latin America. PLoS One 9:e107144

Wende W, Tucker GM, Quétier F, Rayment M, Darbi M (eds) (2018) Biodiversity offsets: European perspectives on no net loss of biodiversity and ecosystem services. Springer, Cham

Outlook: Potential for Further Improvement and Research on the Developed Typology of Biodiversity Offsets

9

Abstract

The state of the scientific knowledge and the practical evidence explored encourage the analysis and investigation of the various forms of (voluntary) biodiversity offsets. However, there are a number of critical aspects associated with them. Several of these have been extensively researched, e.g., the formulation of standards and methods for loss-gain calculations. Building on these premises, the present study has focused on the different types of implementation (i.e., governance, responsibilities, and drivers) of biodiversity offsets. Notwithstanding, three major research gaps persist: (1) The overarching question of appropriateness of and limits to offsets requires research at the interface of natural science, planning, and policy design on the one hand and social sciences on the other hand. (2) So far, little or no evidence exists where offsets have been evaluated, not at the individual project level, let alone systematically. An outcome oriented critical analysis of biodiversity offsets is needed both from a nature conservation point of view and to underpin the discourse on the question of appropriateness of offsets. (3) There is clearly a need for a broader evidence base at the global level as a prerequisite for the evaluation of the outcomes and effectiveness of offsets and consequently to delineate factors for success and failure and to establish best practice. Together with previous compilations of case studies, e.g., by the Business and Biodiversity Offsets Program, the evidence base compiled in this study could add to an international registry of biodiversity offsets and a global compilation of on-the-ground examples of biodiversity.

Keywords

Appropriateness of biodiversity offsets · Commodification of nature · Outcome of biodiversity offsets · Biodiversity offsets evaluation · Biodiversity offsets case studies

M. Darbi, *Biodiversity Offsets Between Regulation and Voluntary Commitment*,
https://doi.org/10.1007/978-3-030-25594-7_9

Due to the complexity of both biodiversity and biodiversity offsets, there are a number of critical tasks associated with them. Several of these have been extensively researched, e.g., the formulation of standards and best practices as well as methods for loss–gain calculations. Building on these premises, the present study has focused on the different types of implementation (i.e., governance, responsibilities and drivers) of (voluntary) biodiversity offsets. However, further research needs remain.

In this study, specific research needs have been identified and mentioned at various points. This refers in particular to the results and open questions of the represented offset types, e.g., with regard to the question if and how government enabled offset frameworks can lead to regulatory offset schemes as well as regarding the role and perpetuity of bilateral cooperations between businesses and nature conservation NGOs.

> [R]ather than a lack of theory, it is practical issues of implementation and adaptive management (combined with external risks such as differing government expectations) that can cause offsets to fail. It is the absence of a solid track record that causes the business community to remain hesitant to invest in offsets due to uncertainty of outcomes. (ICMM IUCN 2012)

More generally, three major research gaps persist, related to the need for a critical analysis of offsets and a broad data availability to build upon (cf. Fig. 9.1):

1. The overarching question of appropriateness of and limits to offsets (cf. Sect. 1.3) requires research at the interface of natural science, planning, and policy design on the one hand and social sciences on the other hand. First scientific discourses are emerging on this topic (e.g., Spash 2015; Scharks and Masuda 2016; Benabou 2014; cf. Bekessy et al. 2010; Garrard et al. 2015; Walker et al. 2009; Sullivan 2013). This targets a more general critical analysis of biodiversity offsets, i.e., whether offsets as such or under specific conditions are appropriate or inappropriate, respectively. While not at the core of this study, it provides context and basic principles of offsets.
2. More closely related to the present study is the need for an outcome-oriented critical analysis of biodiversity offsets. As has been noted (see Sect. 1.3) little or no evidence exists where offsets have been evaluated, not at the individual project level, let alone systematically. Research on this issue is needed both from a nature conservation point of view and to underpin the discourse on the question of appropriateness of offsets (see no. 1 above). Clearly, a sound evaluation of biodiversity offsets needs to be specific and done at the project level, but the results should also allow for generalizations in terms of lessons learned and recommendation. Seen the diversity of offset approaches worldwide (Darbi et al. 2010; Poulton 2015; eftec, IEEP et al. 2010) which has been confirmed in this study (see Appendix 1), the suggested typology can help to reconcile responding to case specific conditions with a certain degree of generalization at the level of the types.

Gather baseline data!

As offsets are highly context sensitive, it is important to meticulously map and evaluate environmental baseline conditions. Research can support government agencies and developers in providing mechanisms to gather area-wide and up to date baseline data.

Focus on the governance of biodiversity offsets!

Technical aspects of biodiversity offsets have been very much in the focus of academic discussion and they are indeed the fundamentals of any single offset and even more of offset schemes. While they are the necessary preconditions for the planning and design of offsets, their implementation depends on establishing appropriate forms of governance. By analyzing the different institutions, actors as well as their specific motivations and behavior, research should identify and compare different governance regimes. This way, obstacles and success factors for the implementation of biodiversity offsets can be identified.

Advance evaluation of biodiversity offsets!

So far, most efforts are being spent on the development of standards and planning of the offset. However, what happens after that, i.e. in the long term? Little or very rough estimates exist on the outcome of offsets on the ground. Research on the evaluation of biodiversity offsets is therefore needed, e.g. in terms of their effectiveness and efficiency. This involves the identification of criteria, processes, roles and responsibilities of the different parties involved.

Improve documentation of examples of failure!

Offsets are controversially debated. One obstacle is obviously that because they are complex and complicated, probably no offset can be done a hundred percent right. It is thus crucial to learn from experience and to document examples of failure as a source for future improvement. However, showing and analyzing what can go (or went) wrong should not be used merely to expose developers, but be explained constructively.

Compile long term evidence on offsets!

The documentation of offset experience has been lacking so far, both in qualitative and in particular in quantitative terms. There are numerous tools available that need to be explored in more detail, for example compensation registers and maps. Especially the use of GIS and online tools should be fostered.

Fig. 9.1 Implications/recommendations for research. Source: Author

3. In line with previous compilations of case studies, e.g., by the Business and Biodiversity Offsets Program and the International Council on Mining and Metals (BBOP 2009a, b; ICMM 2010), the present study has attempted to gather an explorative evidence base of biodiversity offset cases in a global context. There is a need to both quantitatively (i.e., include more cases) and qualitatively (i.e., include more in-depth information) extend this database as a prerequisite for the evaluation of the outcomes and effectiveness of a broad evidence base and consequently to delineate factors for success and failure and to establish best practice.

These three aspects are further specified and described in the following subchapters.

9.1 Discussion of the Appropriateness of Biodiversity Offsets Versus the Commodification of Nature

A number of fundamental questions are inseparably related to the concept of biodiversity offsets:

- What quality criteria characterize a biodiversity offset?
- Which are the crucial and potential weak points for the implementation of biodiversity offsets?
- How do critics argue on these weak points and what are the risks of biodiversity offsets?
- What are the opportunities of biodiversity offsets? Do they outweigh the risks?

Biodiversity offsets are increasingly becoming the subject of public debate; however, in public discourses they are often framed as "a license to trash" or related to "the commodification of nature." In fact, there is a growing controversy as is illustrated in Fig. 9.2.

Despite the criticism of a rather biased argumentation and the "plea for an informed debate on biodiversity offsets" (Evans 2014), parts of the discussion on biodiversity offsets are held by highly specialized experts who know the subject and problems of biodiversity offsets very well. This has led to highly sophisticated argumentations that are touching philosophic grounds.[1]

One of the most fervent critics of biodiversity offsets as part of the "natural capital agenda" is UK environmental expert George Monbiot. He is an environmental and

[1]This includes Conferences, e.g., Natural Capital Forum and Forum on the Natural Commons and various online discussions, particularly on LinkedIn, e.g., Is biodiversity offsetting just a tool for the elite to extract more wealth from natural resources? Available at https://www.linkedin.com/grp/post/3667510-5829721851814776836.

Fig. 9.2 Love or leave? The controversy on biodiversity offsets. Blog post. Biodiversity Offsets Blog. Source: Darbi (2014a)

Love or Leave? The controversy about Biodiversity Offsets

Posted on Thursday, September 4, 2014 by Marianne Darbi

No to Biodiversity Offsetting?

As biodiversity offsets become more visible in the public they are subject to a growing controversy. Critics of biodiversity offsets have a growing lobby, especially in the UK, where the government seems to rush, wanting to push through the concept of biodiversity offsets, no matter at what expense and quality.

As a result a number of biodiversity offset pilots that are in trial are facing severe criticism and resistance, mostly by the local population and NGOs:

> *"The policy is inherently flawed: biodiversity offsetting ignores the difficulties in recreating ecosystems, it overlooks the uniqueness of different habitats, and it disregards the importance of nature for local communities." (source: 2nd Forum on the Natural Commons)*

Are biodiversity offsets a License to trash?

The "license to trash" criticism has resulted in various writings (e.g. by George Monbiot in his column in the *Guardian*), events (e.g. Forum on the Natural Commons) and campaigning (No to biodiversity offsetting), up to a recent mockumentary[a].

Controversy: What distinguishes opponents and proponents of Biodiversity Offsets?

The distinction between opponents and proponents in the current discussion comes down to a question of framing.

There are on the one side the opponents who argue from an ethical and moral (or even philosophical) perspective that biodiversity can't be measured and valued in economic terms and by its very nature is irreplaceable and thus not offsettable.

On the other side the proponents of biodiversity offsets take a more pragmatic view claiming that biodiversity offsets are inevitable whenever development takes place.

Where are we now and where to from here?

While certainly both sides are making a point, neither side has managed the issue so far: the first one is lacking to suggest solutions (while getting the bigger picture right) and the second is suggesting possibly weak solutions (that don't get the bigger picture).

A weak point is that the discussion about biodiversity offsets as a concept isn't held in academia, but more as a societal discussion – loaded with emotions and value judgments.

It is certain that biodiversity offsets concern many people. And therefore it is good that the discussion is held in public. But the "hustle and bustle of quickly picked arguments" that can be frequently observed doesn't help very much.

So, biodiversity offsets — whether you love or leave them — should become subject of an informed debate instead.

Arguments against the Natural Capital and Biodiversity Offsets

Monbiot gives the following four arguments (with increasing significance) against what he calls the "Natural Capital Agenda" underpinned with different examples, one of them being biodiversity offsets.

1. Pricing nature is nonsense because the numbers are not reliable.
2. Unbundling ecosystem services and trading them separately means damaging the holistic system of nature.
3. The power of the economic system as a whole is pitted against the power of nature.
4. The problem of values and framing: taking the position of the opponent (the monetisation of nature) means losing the own values (intrinsic value of nature).

Fig. 9.3 George Monbiot's criticism on Biodiversity Offsets and the Natural Capital Agenda: "The pricing of Everything." Source: extracted from Darbi (2014b)

PRO	CON
• Biodiversity offsets are part of the "toolbox". • Biodiversity offsets are an inevitable solution to biodiversity loss from development. • Biodiversity offsets are a business case.	• Biodiversity offsets justify the commodification of nature for development. • Biodiversity offsets are impossible to achieve (because biodiversity is too complex to be measures and restored).
• Biodiversity offsets are an interim situation. • Biodiversity offsets may only be used in specific contexts.	

Fig. 9.4 PRO and CON attitudes toward biodiversity offsets. Source: Author

political activist who writes a weekly column for The Guardian and is the author of a number of books.

George Monbiot has presented a detailed (however mostly negative) analysis of his criticism on the Natural Capital Agenda (in the United Kingdom) including biodiversity offsets as one part of it at his annual lecture at the Sheffield Political Economy Research Institute as shown in Fig. 9.3.

The example of George Monbiot represents only one of a number of different attitudes toward biodiversity offsets that can be observed in practice (both in favor and against them) as shown in Fig. 9.4 (for more information see Appendix 9).

These different attitudes as well as the related reservations toward biodiversity offsets and the discourses around biodiversity offsets need to be further analyzed.

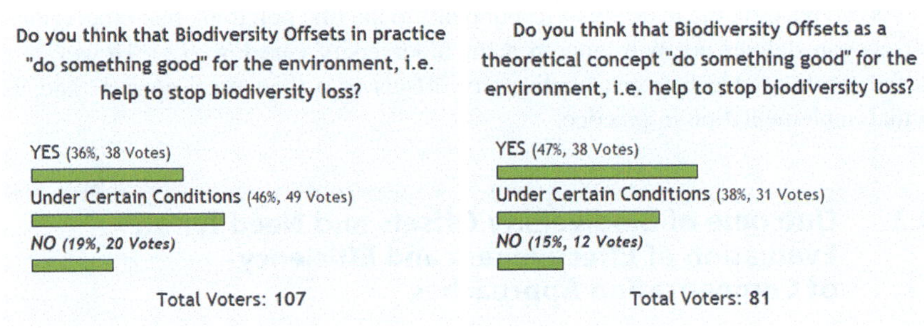

Fig. 9.5 Polls on the appropriateness of biodiversity offsets to deliver positive outcomes for biodiversity. Source: Darbi (n.d.)

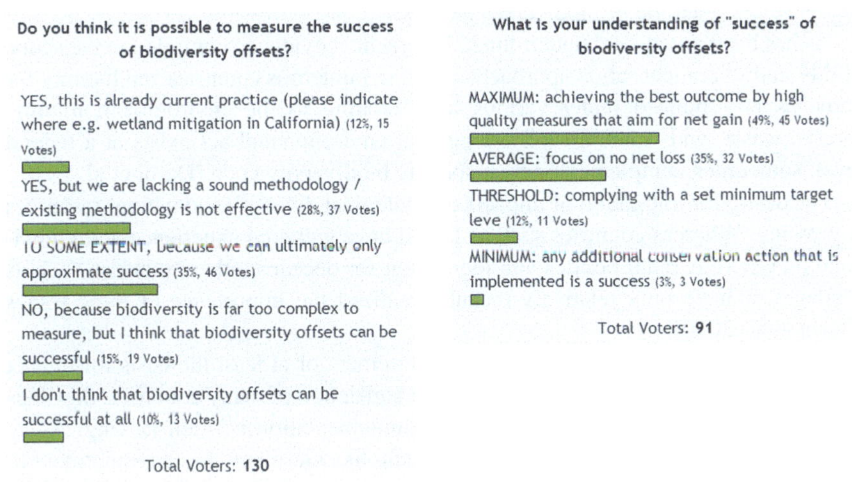

Fig. 9.6 Polls on how to measure the success of biodiversity offsets. Source: Darbi (2015)

First attempts in this regard have been made in the course of this study (though they are not the specific focus). A number of posts on the Biodiversity Offsets Blog have addressed the controversy of biodiversity offsets through:

- Own opinion posts (see Fig. 9.2)
- Guest posts (see Appendixes 6 and 7)
- Posts related to emerging literature/sources (see Appendix 8)
- Simple opinion polls (see Figs. 9.5 and 9.6)

A series of simple (two-question) polls was conducted through the Biodiversity Offsets Blog. The questions were intentionally formulated normatively—as any evaluation of the concept of biodiversity offset is highly normative.

As can be seen, the majority of respondents to the first poll think that biodiversity offsets can deliver positive outcomes for biodiversity (see Fig. 9.5). However, a difference is visible between biodiversity offsets as a theoretical concept and its actual implementation in practice.

9.2 Outcome of Biodiversity Offsets and Need for an Evaluation of Effectiveness and Efficiency of Compensation Approaches

Altorfer (2004) suggests that voluntariness needs to refer to its effects. Consequently, voluntary biodiversity offsets ultimately need to be judged by their outcome. This is a complex task because there is a variety of different issues that influence the outcome of biodiversity offsets.

As has been noted (and underpinned by practical evidence compiled in the scope of this study), compensation approaches exist in numerous countries and liability for damages is stipulated under various laws relating to the environment, mining, forests, waste, and water. Usually, a general environmental act exists at a federal level, sometimes complemented by a specific biodiversity code (Darbi et al. 2010).

The desire for mitigation of impacts on biological diversity and natural resources is growing. Whereas countries such as Brazil have gathered experience and knowledge on environmental politics and legislation for decades, other countries such as Madagascar have only relatively recently realized the importance of these issues (Darbi et al. 2010).

Despite the fact that a compensation requirement or at least the basis for impact mitigation is explicated in the environmental legislation of many countries, this does not represent the situation in practice and implementation is often lacking. As an example the case of France can be cited, which, like Germany, has a requirement to compensate for negative environmental impacts stipulated in the environmental law since 1976. However, while in Germany an extensive and relatively well functioning compensation scheme has evolved over the past four decades (German Impact Mitigation Regulation) the French compensation requirement has not been properly implemented. Only recently has the French ministry for the environment in cooperation with the private "CDC Biodiversité" begun to implement a first pilot on biodiversity offsets (personal communication with Delphine Morandeau, 2010).

In a global context, impact mitigation mostly focuses on EIA as the main implementation tool (Darbi et al. 2009). However, this does not always prove to be effective with regard to the actual conservation outcomes needed to compensate for the impacts. A weakness is that in many countries there is no general methodology available to determine the compensation demand and design compensation measures. Usually a case-by-case approach is taken. This implies a lack of general predefined, comparable, and transparent criteria which complicates the valuation and makes it less transparent and more difficult to compare between projects. Yet this problem has been widely identified in several countries, evidenced by a lack of standardized valuation schemes for biological diversity and impacts. Taking a case-by-case approach means that one of the central questions, how to determine the compensation ratio, cannot be completely answered (Darbi et al. 2009).

Another aspect that mainly influences the outcome of biodiversity offsets is the goal that it targets. In the scope of this study, four different varieties have been distinguished:

- Positive balance (Net Gain)
- Neutral balance (No Net Loss)
- Minimum balance (reduced loss)
- Not effective (loss)

The goal of no net loss is most commonly encountered in practice. However, commitments toward net gain are increasing.

Despite these improving biodiversity offset standards, the question remains in each particular case what no net loss or net gain means and how it can be quantified.

More generally, one should also consider that no net loss is only technically feasible because we have already reached a certain level of deterioration, i.e., for every new impact and the associated loss a part from this already deteriorated (in the past) environment will be restored.

What can be seen from this is that while offsets may be planned with best intentions, it is crucial to evaluate the ultimate outcome in terms of both effectiveness and efficiency. However, these are hardly monitored and evaluated. This is probably the most important component to increasing the success (and possibly also the acceptability) of biodiversity offsets.

As can be seen from Fig. 9.6, 75% of respondents to an opinion poll conducted on the Biodiversity Offsets Bog think that it is possible to measure the success of biodiversity offsets at least to some extent. Nevertheless, 28% think that a sound methodology is lacking or existing methodology is not effective.

This highlights that further research is needed to underpin and operationalize the evaluation of biodiversity offsets. For this purpose, a number of different tools may be explored. One simple, yet promising, example is the use of a SWOT analysis as shown in Fig. 9.7. This includes the concise representation of Strengths, Weaknesses, Opportunities and Threats (SWOT) in the form of a table.

This can by no means be exhaustive and replace a detailed evaluation. Nevertheless, it can give a first impression on the specifications and context conditions with regard to the likely outcomes of the specific type or case of biodiversity offsets and eventually factors that are affecting the success of the offset.

9.3 The Need for a Broader Empirical Base of Implemented Biodiversity Offset Cases

As has been noted in Sect. 8.1.4, the evidence base compiled in this study could provide the starting point for an international registry of biodiversity offsets. There is clearly a need for such an information base at the global level, supported by many

Strengths	Weaknesses
• Biodiversity harm is addressed directly by the responsible party (polluter pays principle). • Etc.	• Due to its complexity, biodiversity offsets cannot accurately measure biodiversity losses and gains. • Etc.
Opportunities	**Threats**
• Can add to strategic goals. • Etc.	• Economic framing in terms of "commodification" can lead to a "license to trash". • Etc.

Fig. 9.7 SWOT analysis of biodiversity offsets. Source: Author

experts worldwide. For example ICMM IUCN (2012) argues for a "simple online source of offset case studies":

> Offset practitioners are not in contact with each other except through word of mouth and forums such as BBOP, and offset case studies are insufficiently documented. A simple online gazetteer of offset case studies, practitioners, potential partners, etc could be created that is searchable by relevant criteria (eg country). This could also cover those with experience in major types of ecosystem (reefs, rainforest, desert). Design and implementation of offsets should be covered. (ICMM IUCN 2012)

To some extent, registries at national or regional level already exist, e.g., in the scope of US Wetland Mitigation and German Impact Mitigation Regulation.[2] The first attempt to go beyond these national boundaries has been made by Speciesbanking.com, hosted by the US based Ecosystem Marketplace Network (see Fig. 9.8).

> Speciesbanking.com is a global information clearinghouse for a segment of biodiversity markets focusing on biodiversity offsetting, compensation and banking. Until now, there has been no centralized information resource to serve researchers, policy makers, buyers, sellers, and other market participants. This site aims to provide a succinct answer to the question "What is happening in biodiversity offset and compensation programs around the world?" SpeciesBanking.com provides information at two levels: at the Program level and at the level of individual biodiversity banks. (Speciesbanking.com)

Despite this comprehensive overview of compensation programs and habitat banking worldwide, a global compilation of on-the-ground examples of biodiversity

[2]For an example see the registry of the measures implemented by a habitat bank in Germany, the Sächsische Ökoflächen-Agentur: http://www.sls-sachsen.de/sls-oekokonto.html

Fig. 9.8 Offset and compensation programs and banks by region. Source: Screenshot from www.speciesbanking.com

offsets in the form of case studies is still lacking and requires both further research and documentation (cf. Appendix 5).

9.4 Concluding Remarks

Not only is biological diversity "diverse" in the literal sense of the word, the same holds to be true for biodiversity offsets which arise in different contexts, under different conditions and are influenced and implemented by a variety of stakeholders using diverse modes of governance.

Taking into account this complex and varied situation, the present study has attempted to contribute to drawing a more differentiated picture of biodiversity offsets going beyond "black vs. white," "good vs. bad," or "mandatory vs. voluntary" classifications of biodiversity offsets that are simply no longer fit for purpose.

As has been noted, biodiversity offsets have become part of the nature conservation toolbox and are rooted in both the scientific discourse and in practice.

Given that background, the state of the scientific knowledge and the practical evidence explored throughout this study encourage the analysis and exploration of the various forms of (voluntary) biodiversity offsets.

However, a lot still needs to be done, especially in terms of a critical process and outcome oriented documentation, as well as evaluation of biodiversity offsets. This needs to be done on a sound and broad theoretical and empirical scientific base. In this regard, the present study provides a starting point for further in-depth research on voluntary biodiversity offsets.

References

Altorfer H (2004) Zusammenfassende Thesen. In: Ammann H (ed) Freiwilligkeit zwischen liberaler und sozialer Demokratie. Seismo, Zürich

BBOP (2009a) Biodiversity offset cost-benefit handbook. BBOP, Washington, DC. Available at: www.forest-trends.org/biodiversityoffsetprogram/guidelines/cbh.pdf

BBOP (2009b) Biodiversity offset implementation handbook. BBOP, Washington, DC. Available at: www.forest-trends.org/biodiversityoffsetprogram/guidelines/oih.pdf

Bekessy SA, Wintle BA, Lindenmayer DB, McCarthy MA, Colyvan M, Burgman MA, Possingham HP (2010) The biodiversity bank cannot be a lending bank. Conserv Lett 3:151–158

Benabou S (2014) Making up for lost nature?: a critical review of the international development of voluntary biodiversity offsets. Environ Soc 5:103–123

Darbi M (2014a) Love or leave? The controversy about biodiversity offsets. Biodiversity Offsets Blog. Available at http://www.biodiversityoffsets.net/love-leave-controversy-biodiversity-offsets/

Darbi M (2014b) George Monbiot's criticism on Biodiversity Offsets and the Natural Capital Agenda: "The pricing of everything". Biodiversity Offsets Blog. Available at http://www.biodiversityoffsets.net/george-monbiots-criticism-biodiversity-offsets-natural-capital-agenda-pricing-everything/

Darbi M (2015) New poll: do you think it is possible to measure the success of biodiversity offsets? Biodiversity Offsets Blog. Available at http://www.biodiversityoffsets.net/new-poll-think-pos%C2%ADsi%C2%ADble-mea%C2%ADsure-suc%C2%ADcess-bio%C2%ADdi%C2%ADver%C2%ADsity-offsets/

Darbi M (n.d.) Open discussion platform. Biodiversity Offsets Blog. Available at http://www.biodiversityoffsets.net/open-discussion-platform/

Darbi M, Ohlenburg H, Herberg A, Wende W, Skambracks D, Herbert M (2009) International approaches to compensation for impacts on biological diversity. Final Report. Available at http://www.forest-trends.org/biodiversityoffsetprogram/library/new/Biokom_Final%20Report_IOER_TUB.pdf

Darbi M, Ohlenburg H, Herberg A, Wende W (2010) Impact mitigation and biodiversity offsets-compensation approaches from around the world: a study on the application of Article 14 of the CBD (Convention on Biological Diversity). BfN-Schr.-Vertrieb im Landwirtschaftsverl.

eftec, IEEP et al (2010) The use of market-based instruments for biodiversity protection—the case of habitat banking. Technical report. For European Commission DG Environment

Evans M (2014) Gina vs. the reef? A plea for informed debate on biodiversity offsets. Available at https://mcevansresearch.wordpress.com/2014/08/21/gina-vs-the-reef-a-plea-for-informed-debate-on-biodiversity-offsets/comment-page-1/#comment-1943

Garrard G et al (2015) Offset policies don't work. So maybe we should be weighing up the alternatives. Decision Point Online, 91/August 2015

ICMM (2010) Mining and biodiversity. A collection of case studies—2010 edition. ICMM, London

ICMM IUCN (2012) Independent report on biodiversity offsets. Prepared by The Biodiversity Consultancy. Available at www.icmm.com/biodiversity-offsets

Poulton D (2015) Key issues in biodiversity offset law and policy: a comparison of six jurisdictions. Report for Ontario Nature

Scharks T, Masuda YJ (2016) Don't discount economic valuation for conservation. Conserv Lett 9 (1):3–4

Spash CL (2015) Bulldozing biodiversity: the economics of offsets and trading-in nature. Biol Conserv 192:541–551

Sullivan S (2013) After the green rush? Biodiversity offsets, uranium power and the 'calculus of casualties' in greening growth. Hum Geogr 6(1):80–101

Walker S, Brower AL, Stephens RTT, Lee WG (2009) Why bartering biodiversity fails. Conserv Lett 2:149–157

Appendix 10

Abstract

The screening of offset cases was at the core of the netnographic research approach and evolved over the years. In the scope of the screening 90 possible biodiversity offset cases were gathered to underpin the typology as shown in Appendix 1. This exploratory list of worldwide biodiversity offset cases (which by no means intends to be exhaustive, nearly exhaustive, or representative) was then checked for adequate and simple accessible (online) information quality (using mainly the internet search engines Google and Google scholar). A number of these have been rejected due to poor information quality or other reasons (e.g. project failure/abandonment). The remaining 72 cases are displayed on a world map and described in a table, notably regarding the following aspects: Name/brief description (and impact type), Country/location, Sources and notes, and Information quality. Furthermore, the two core criteria (Influence and initiators; Motivation and ultimate goal) are displayed with their respective values to tick and possibly some short explanation (e.g. government influence, US Wetland Mitigation following Clean Water Act). From the latter, building on the information available, a normative judgment is made to which type the specific case is assigned. One case can be characteristic for one type or represent a combination of two or more types

In addition, Appendix 2 presents an introduction to the Biodiversity Offsets Blog. This is complemented and implemented by a number of blog posts (Appendices 5–10).

Appendix 3 summarizes definitions of Corporate (Social) Responsibility and Appendix 4 provides list of Members of the Advisory Group of the Business and Biodiversity Offsets Programme

M. Darbi, *Biodiversity Offsets Between Regulation and Voluntary Commitment*,
https://doi.org/10.1007/978-3-030-25594-7_10

Keywords

Biodiversity offsets case studies · Regulatory biodiversity offsets · Conditional biodiversity offsets · Enabled biodiversity offsets · Sectoral biodiversity offsets · Corporate biodiversity offsets · Local biodiversity offsets · Altruistic biodiversity offsets · Biodiversity Offsets Blog · Corporate social responsibility · Business and Biodiversity Offsets Programme · BBOP

10.1 Appendix 1: Screening of Biodiversity Offset Cases Worldwide

The results of the screening are presented in a world map below (caption overleaf) and further explained in the following table.

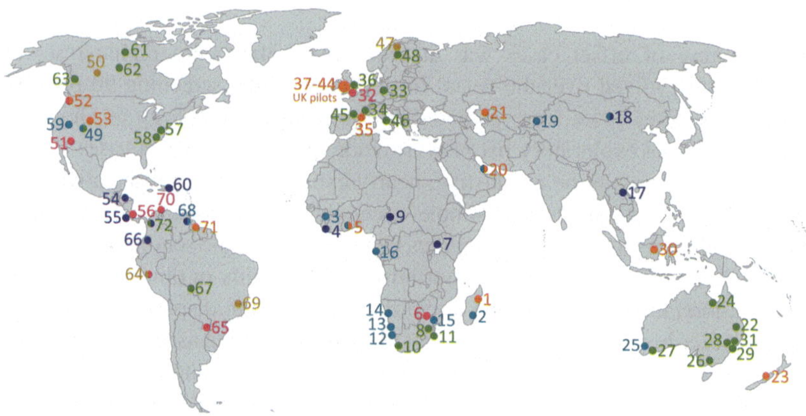

Green: type 1 regulatory biodiversity offsets, dark blue: type 2 conditional biodiversity offsets, orange: type 3 enabled biodiversity offsets, turquoise: type 5 corporate biodiversity offsets, red: type 6 local biodiversity offsets, ochre: type 7 altruistic biodiversity offsets

Africa

1. Ambatovy BBOP Pilot
2. QIT Mining Madagascar QMM
3. Simandou
4. Piloting aggregated offsets and mining in Liberia
5. BBOP Pilot Akyem Gold Mining Project
6. Potgietersrust platinums limited (PPRust) BBOP Pilot
7. Bujagali Energy Limited
8. Anglo Coal Isibonelo colliery
9. Petroleum Development and Pipeline Project
10. Mount Royal Golf Estate
11. Pulp United Pulp Mill
12. De Beers Marine Kleinzee / Alexander Bay Sea Areas
13. Namdeb Diamond Corporation
14. Rossing Uranium Mine
15. Palabora
16. Shell Foundation offset on Gamba Complex

Asia

17. Nam Theun 2 Hydropower
18. Oyu Tolgoi LLC
19. Kumtor Gold Mine
20. PEARL GTL QATAR BBOP Pilot
21. Petroleum and Natural Gas Development in the Ustyurt

Australia

22. Koala Offset in South East Queensland
23. Strongman Coal BBOP Pilot
24. Mc Arthur River Mine
25. Newmont Boddington mine
26. Basslink Undersea Power Cable
27. Raventhorpe Nickel Project
28. Karuah bypass offset
29. Cumber and Plain Wood-land / Western Sydney
30. IndometCoal Project
31. Warkworth Coal Mine

Europe

32. Thames ink Railway offsets
33. Burgberg Zschaitz habitat bank
34. CNM high speed rail bypass
35. Operation Cossure
36. Bardon Hill Quarry
37. South Devon UK pilot
38. North Devon UK pilot
39. Essex UK pilot
40. Doncaster UK pilot
41. Development in Dudley
42. Project Pinewood
43. Nottinghamshire UK pilot
44. Warwickshire UK pilot
45. A65 Langon-Pau motorway
46. Apennine Wind Farms
47. Openpit Mining Mertainen
48. Ume Delta River Railway

North- & Central America

49. Kennecott Utah Copper Mine
50. Suncor Biodiversity Offset
51. Resolution Copper Mining
52. Bainbridge Island BBOP pilot
53. Jonah Natural Gas Field
54. Via San Lorenzo – Olanchito
55. Hidroelectrica Reventazon
56. Braulio Carrillo National Park / construction of a highway
57. Tygart Valley River Stream and Wetland Bank
58. South Fork Hoppers Creek Mitigation
59. Newmont Long Canyon Project
60. Pueblo Viejo gold mine
61. NGTL pipeline proposals in Horne River area
62. Total Joslyn oil-sands mine
63. Enbridge Northern Gateway pipelines

South America

64. Antamina Mine
65. Hidroelectrica Yacyreta
66. Pasto Mocoa road
67. Gasoducto Gasbol
68. Brisas Gold & Copper Project
69. AngloGold Ashanti Brasil Mineracao
70. Cerrejón mine
71. Amaila Fall Hydro Project
72. Gramalote Mining Project

no	Name / brief description (and impact type)	Country / location	Influence and initiators							Motivation and ultimate goal			Type	Sources and notes	Information quality
			government	Financial institute	Sector	Corporation	Local community	NGO(s)	without	Pressure / compliance	Incentive / cost-benefit	Altruism / responsib.			
Africa															
1.	Ambatovy BBOP Pilot Project Case Study (Nickel mine)	Atsinanana, Madagascar		X	X ICMM)		X (license to operate)	**X** (BBOP)			X		**Type 3 b enabled biodiversity offsets**, (type 6b local biodiversity offsets, type 2 conditional biodiversity offsets, type 4 sectoral biodiversity offsets)	Ambatovy.com, Dickinson and Berner (2010), Hayes et al (n.d.), Waeber (2013), Burgin (2008), Berner et al (2010), von Hase et al (2014), The Ambatovy Project (2009), BBOP (n.d.a), ten Kate (2007), Kormos and Kormos (2011), Hayes et al (2013a and b) *Active BBOP pilot*	++
2.	QIT Mining Madagascar QMM (ilmenite mining)	Fort Dauphin, Madagascar				X (Rio Tinto pilot)		X (IUCN)					**Type 5 corporate biodiversity offsets**, (type 3b enabled biodiversity offsets)	Darbi et al (2010), QMM (n.d, 2007a, b and c), BBOP (2009b), Rio Tinto (2004 and 2006), Temple et al (2011), ten Kate et al (2011), Rio Tinto (2015)	++
3.	Simandou (open-pit iron-ore mine)	Guinea	X	X (IFC)		X (Rio Tinto)	X			X			**Type 5 corporate biodiversity offsets, (type** 2 conditional biodiversity offsets, type 6b local biodiversity offsets)	Rio Tinto (2008a and b), Kormos et al (2014), TBC (2014) Rio Tinto (2012), Boyes (2014a and b) *Rio Tinto pilot site, creation of the Pic de Fon Foret Classe protected area in a process of effective collaboration between government, the private sector and local communities*	++

no	Name / brief description (and impact type)	Country / location	Influence and initiators							Motivation and ultimate goal			Type	Sources and notes	Information quality
			government	Financial institute	Sector	Corporation	Local community	NGO(s)	without	Pressure / compliance	Incentive / cost-benefit	Altruism / responsib.			
4.	Piloting aggregated offsets and mining in Liberia	Liberia		**X** (World Bank)									**Type 2 conditional biodiversity offsets**	Johnson and Hund (2015), World Bank Group (2015), World Rainforest Movement (2015), PROFOR (2015)	++
5.	BBOP Pilot Project Case Study Akyem Gold Mining Project (gold mine)	Eastern Region, Ghana	X	X		**X** (Newmont)	X	**X** (BBOP)		X	X		**Type 3 b enabled biodiversity offsets, type 5 corporate biodiversity offsets**, (type 6b local biodiversity offsets, type 2 conditional biodiversity offsets)	Burgin (2008), Newmont Golden Ridge Limited (2009), Newmont (2014 and 2015), OECD (2013) *No longer on BBOP's website, but by Newmont promoted as a BBOP pilot and ongoing*	+
6.	Potgietersrust platinums limited (PPRust) BBOP Pilot (Platinum mine)	South Africa					**X**	X (BBOP)			X		**Type 6b local biodiversity offsets**, (type 3b enabled biodiversity offsets)	Burgin (2008), BBOP (n.d.a), ten Kate (2007), Department of Environmental Affairs and Development Planning (2007)	+
7.	Bujagali Energy Limited (Hydropower plant and transmission line)	Uganda		X (World Bank, African Development Bank)						X			**Type 2 Conditional biodiversity offsets**	Burgin (2008), BBOP (2009b), Independent Review mechanism (2012), ten Kate et al (2011), R. J. Burnside International Limited (2010)	+

(continued)

no	Name / brief description (and impact type)	Country / location	Influence and initiators							Motivation and ultimate goal			Type	Sources and notes	Information quality
			government	Financial institute	Sector	Corporation	Local community	NGO(s)	without	Pressure / compliance	Incentive / cost-benefit	Altruism / responsib.			
8.	Anglo Coal Isibonelo colliery (open cast coal mining)	Mpumalanga, South Africa	**X**							X			**Type 1 regulatory biodiversity offsets**	Department of Environmental Affairs and Development Planning (2007), Copans (2007), AngloAmerica South Africa (2015)	+
9.	Petroleum Development and Pipeline Project	Chad and Cameroon		X (World Bank, IFC)						X			**Type 2 Conditional biodiversity offsets**	BBOP (2009b), The World Bank and International Finance Corporation (2000), ten Kate et al (2004 and 2011), Kormos and Kormos (2011), OECD (2013)	-
10.	Mount Royal Golf Estate (extansion of golf course with residential and commercial development)	Swartland, South Africa	**X**							X			**Type 1 regulatory biodiversity offsets**	BBOP (2009b), ten Kate et al (2011), Department of Environmental Affairs and Development Planning (2007) *Little project specific information*	-
11.	Pulp United Pulp Mill (bleached chemi thermos mechanical pulp mill)	KwaZulu-Natal, South Africa	**X** (Record of Decision incl. offset)							X			**Type 1 regulatory biodiversity offsets**	BBOP (2009b), ten Kate et al (2011), Department of Environmental Affairs and Development Planning (2007), OECD (2013)	-
12.	De Beers Marine Kleinzee and Alexander Bay Sea Areas (offshore diamond mining)	Kleinzee, South Africa			ICMM	X (De Beers Net Neutral Impact, NNI)		WWF					**Type 5 corporate biodiversity offsets**, (type 3b enabled biodiversity offsets, type 4 sectoral	ICMM (2010) *Collaborates with WWF and the South African National Biodiversity Institute to develop an Offshore Marine Protected Areas Network, Mitigation hierarchy is part of De Beers Environmental Standard*	-

no	Name / brief description (and impact type)	Country / location	Influence and initiators							Motivation and ultimate goal			Type	Sources and notes	Information quality
			government	Financial institute	Sector	Corporation	Local community	NGO(s)	without	Pressure / compliance	Incentive / cost-benefit	Altruism / responsib.			
													biodiversity offsets)		
13.	Namdeb Diamond Corporation (diamond mine, De Beers)	Sperr-gebiet, Namibia				X (De Beers NMI)							**Type 5 corporate biodiversity offsets**	*ICMM (2010)* *Jointly owned by the government of Namibia and De Beers, Mitigation hierarchy is part of De Beers Environmental Standard*	-
14.	Rossing Uranium Mine (Rio Tinto)	Namibia				X (Rio Tinto)		X (Fauna Flora International)					**Type 5 corporate biodiversity offsets,** (type 3b enabled biodiversity offsets)	Rio Tinto (n.d.), Marsh (n.d.) *Rio Tinto pilot site*	-
15.	Palabora	South Africa				X (Rio Tinto)		X (Fauna Flora International)					**Type 5 corporate biodiversity offsets,** (Type 3b enabled biodiversity offsets)	Rio Tinto (n.d.), Marsh (n.d.) *Rio Tinto pilot site*	-
16.	Shell Foundation offset to support Smithsonian Institution's Monitoring and Assessment of Biodiversity	Gabon				X (Shell)	X	X (Smithonian Institution)					**Type 5 corporate biodiversity offsets**, (type 6b local biodiversity offsets	Burgin (2008), WBSCD (n.d.)	-

(continued)

no	Name / brief description (and impact type)	Country / location	Influence and initiators							Motivation and ultimate goal			Type	Sources and notes	Information quality
			government	Financial institute	Sector	Corporation	Local community	NGO(s)	without	Pressure / compliance	Incentive / cost-benefit	Altruism / responsib.			
	Program on Gamba Complex														
Asia															
17.	Nam Theun 2 Hydropower Project	Central Lao PDR		X (World Bank, Asian Development Bank)						X			**Type 2 conditional biodiversity offsets**	BBOP (2009b), ten Kate et al (2011), Kormos and Kormos (2011), OECD (2013), International Advisory Group to the World Bank (2005), Singh (2009), http://www.eia.nl/fr/publications/rapports-consultatifs-et-projets/027-i *World Bank has promoted Nam Theun 2 as a model of social and environmental good practice*	++
18.	Oyu Tolgoi LLC (copper/gold mine)	Southern Gobi, Mongolia		**X** (IFC)		**X** (Rio Tinto)							**Type 2 conditional biodiversity offsets, Type 5 corporate biodiversity offsets**	Pollard (2013), Pilgrim and Murphy (n.d.), EBRD (n.d.), Oyu tolgoi LLC Biodiversity Strategy (2011), TBC & FFI (2012a, b, c and d)	++
19.	Kumtor Gold Mine (open pit gold mine)	Issyk Kul, Kyrgyzstan				X (Centerra)		X (FFI)					**Type 5 corporate biodiversity offsets**, (type 3b enabled biodiversity offsets)	BBOP (2009b), ten Kate et al (2011), KUMTOR (2016), Nazari (2013) *Follows Centerra's corporate policy and standards*	+
20.	PEARL GTL PROJECT, QATAR BBOP Pilot (GTL)	Qatar				**X** (Shell)	X (regulatory)	**X** (BBOP)		X	X		**Type 3 b enabled biodiversity offsets, type 5**	Burgin (2008), BBOP (n.d.a), ten Kate (2007) *No follow-up information*	+

no	Name / brief description (and impact type)	Country / location	Influence and initiators: government	Financial institute	Sector	Corporation	Local community	NGO(s)	without	Motivation and ultimate goal: Pressure / compliance	Incentive / cost-benefit	Altruism / responsib.	Type	Sources and notes	Information quality
							good-will)						**corporate biodiversity offsets**, (Type 6 b local biodiversity offsets)		
21.	Offsets for Petroleum and Natural Gas Development in the Ustyurt	Ustyurt Plateau, Uzbekistan	**X** (Uzbek government, UNDP)								X		**Type 3a enabled biodiversity offsets**	Bull et al. (2013b), UNDP (2010) *The Uzbek Government is collaborating with the United Nations Development Programme to investigate mitigation and compensation measures, including offsets, for reducing impacts to biodiversity from the extractive sector.*	+
Australia and Oceania															
22.	Koala Offset in South East Queensland	Queensland, Australia	X							X			**Type 1 regulatory biodiversity offsets**	Key and Darbi (2015)	++
23.	Strongman Coal Mine, Solid Energy New Zealand BBOP Pilot (coal mine)	New Zealand	X			X (Solid Energy New Zealand)	X (public scrutiny, permit)	**X** (BBOP)					**Type 3b enabled biodiversity offsets**, (type 5 corporate biodiversity offsets, type 6 a and b local biodiversity offsets)	BBOP (n.d.a), Solid Energy New Zealand Limited (2009), von Hase (2104), OECD (2013) *Active BBOP pilot*	++

(continued)

no	Name / brief description (and impact type)	Country / location	Influence and initiators							Motivation and ultimate goal			Type	Sources and notes	Information quality
			government	Financial institute	Sector	Corporation	Local community	NGO(s)	without	Pressure / compliance	Incentive / cost-benefit	Altruism / responsib.			
24.	Mc Arthur River Mine Open Cut Project	Northern Territory, Australia	**X** (Nor-thern Terri-tory Govern-ment)					X (Conser vation Interna-tional)					**Type 1 regulatory biodiversity offsets,** (type 3b enabled biodiversity offsets)	Burgin (2008), Rodricks (2010), URS (2005a and b, 2006), Erskine (2006a and b), Environment Protection Authority (2006)	++
25.	Newmont Boddington mine (Gold mine)	Western Australia				**X** (New-mont)							**Type 5 corporate biodiversity offsets**	Newmont Boddington Gold PTY Ltd (2014a and b), Newmont (2014 and 2015)	+
26.	Basslink Under-sea Power Cable	Victoria, Australia	**X** (Victo-ria's Native Vegeta-tion Manage ment)							X			**Type 1 regulatory biodiversity offsets**	BBOP (2009b), Bull et al (2013a), Westerweller and Price (2006), Duncan and Hay (2007), ten Kate et al (2011)	+
27.	Raventhorpe Nickel Project	Bandalup corrior, Western Australia	**X**			X (BHP Biliton)							**Type 1 regulatory biodiversity offsets**, (type 5 corporate biodiversity offsets)	Burgin (2008), ICMM (2010), Rodricks (2010), Environmental Protection Authority (2013)	+
28.	Karuah bypass offset	Hunter Valey, New South	**X** (NSW Bioban-king										**Type 1 regulatory biodiversity offsets**	Burgin (2008), Rowe and Phibbs (2005), Department of Environment and Conservation NSW (2006)	+

no	Name / brief description (and impact type)	Country / location	Influence and initiators							Motivation and ultimate goal			Type	Sources and notes	Information quality
			government	Financial institute	Sector	Corporation	Local community	NGO(s)	without	Pressure / compliance	Incentive / cost-benefit	Altruism / responsib.			
		Wales, Australia	scheme), DEC[a]											*In the type of the Karuah bypass, the Roads and Traffic Authority (RTA) acknowledged that it could not avoid, minimise and mitigate all the impacts on biodiversity on-site. The RTA provided 89 hectares of compensatory habitat to offset the loss.*	
29.	Cumberland Plain Woodland of Western Sydney (clearing of natural vegetation)	New South Wales, Australia	**X** (NSW Biobanking scheme)										**Type 1 regulatory biodiversity offsets**	Burgin (2008), Office of Environment & Heritage (2011 and 2015), Department of Infrastructure and Regional Development (2015), Department of Environment, Climate Change and Water (NSW) (2010) *Growth Centres Biodiversity Offset Program*	+
30.	IndometCoal Project (metallurgical coal resource, BHP Billiton)	Kalimantan, Indonesia			ICMM			X (cooperation with FFI)					**Type 3b enabled biodiversity offsets**, (type 4 sectoral biodiversity offsets)	ICMM (2010), Kormos and Kormos (2011)	+
31.	Open cut extension at the Warkworth Coal	Upper Hunter Valley, New	**X**			X (Rio Tinto)							**Type 1 regulatory biodiversity offsets**, (Type	Hannam (2014), Fish et al. (n.d.), Nature Conservation Council (2015)	-

[a] Department of Environment and Conservation NSW (DEC)

(continued)

no	Name / brief description (and impact type)	Country / location	Influence and initiators: government	Financial institute	Sector	Corporation	Local community	NGO(s)	without	Motivation and ultimate goal: Pressure / compliance	Incentive / cost-benefit	Altruism / responsib.	Type	Sources and notes	Information quality
	Mine (Rio Tinto Coal Australia)	South Wales, Australia	(EPBC and TSC Act)[b]										5 corporate biodiversity offsets)		
Europe															
32.	Thameslink Railway offsets	London Borough of Lambeth, UK	X (Defra demonstration project)				X Grass-roots collaboration of industry, Conservation & local government				X	X	**Type 6 local biodiversity offsets**, (type 3 enabled biodiversity offsets)	Parsons Brinckerhoff (2013), Woodley and Baker (2014), Woodley (2015), Kemp (2014), Department for Environment Food and Rural Affairs (2013), CIEEM (2013) *As part of the programme's commitment to sustainable development, Thameslink has an objective to 'maintain and enhance biodiversity' and a target to 'enhance our habitat footprint to achieve a net biodiversity gain via the mitigation hierarchy,' in addition to the requirements conditioned on programme through the Transport and Works Act Order.*	++
33.	Burgberg Zschaitz habitat bank (offset for windpower development)	Saxony, Germany	**X**				X			X			**Type 1 regulatory biodiversity offsets,** (type 6b local	Darbi et al. (2014) *Stands exemplarily for the multitude of offset projects that have been carried out in 40 years practice with the German impact mitigation regulation*	++

[b] Environment Protection and Biodiversity Conservation (EPBC) Act and Threatened Species Conservation (TSC) Act

no	Name / brief description (and impact type)	Country / location	Influence and initiators							Motivation and ultimate goal			Type	Sources and notes	Information quality
			government	Financial institute	Sector	Corporation	Local community	NGO(s)	without	Pressure / compliance	Incentive / cost-benefit	Altruism / responsib.			
													biodiversity offsets)		
34.	Contournement Nîmes Montpellier - A new high speed rail bypass	Southern France	**X** (Environmental Code)										**Type 1 regulatory biodiversity offsets**	Quétier et al. (2015b), Département de l'Herault (2013), DREAL Languedoc-Roussillon (2014) *Under Art. L.411 Environmental code (protected species)*	+
35.	Operation Cossure	Crau Plain, Southern France	**X**				X	X			X		**Type 3a enabled biodiversity offsets**, (type 6b local biodiversity offsets)	CDC Biodiversité (n.d.b), Morandeau and Vilaysack (2012) *This pilot scheme is being conducted in liaison with the Ministry of Ecology to experiment ways of creating "natural asset reserves". Planners can use it as a source of pro-biodiversity initiatives, especially for offset arrangements linked to the residual impacts of their projects as part of ecologically neutral strategies.* *Operation Cossure is a local project put together by key actors in the Crau region themselves: local authorities, farmers, naturalists, scientists and government departments.*	+
36.	Bardon Hill Quarry	Leiceister shire, UK	X (Natural England)							X			**Type 1 regulatory biodiversity offsets**	Aggregate Industries UK Ltd (2010), Temple et al (2010), Bardon Aggregates (2009), SLR (2014) *to address concerns raised by Natural England in its planning consultation response to provide "an agreed programme of mitigation and*	+

(continued)

no	Name / brief description (and impact type)	Country / location	Influence and initiators							Motivation and ultimate goal			Type	Sources and notes	Information quality
			government	Financial institute	Sector	Corporation	Local community	NGO(s)	without	Pressure / compliance	Incentive / cost-benefit	Altruism / responsib.			
														compensation that will be delivered as part of this proposed development"	
37.	South Devon UK pilot	Devon, UK	X								X		**Type 3 a enabled biodiversity offsets**	South Devon Biodiversity Offsetting Guidance (2014), DNLP (n.d.) *Prioritises offset design to benefit priority habitats and two key species, the Greater Horseshoe Bat and Cirl Bunting. Illustrates that offsetting does not change existing protection and processes for protected species (Cirl Buntings) and habitats (South Hams SAC) – but that it can be used to benefit protected species.*	+
38.	The North Devon UNESCO Biosphere Reserve UK pilot	Devon, UK	X								X		**Type 3 a enabled biodiversity offsets**	North Devon Biosphere (2012), DNLP (n.d.), North Devon Biosphere (n.d.) *Focuses on priority habitats within the Biosphere Reserve and aims to also offset for loss of the benefits provided by these habitats (ecosystem services such as flood attenuation, flood storage, access, pollution control and carbon sequestration).*	+
39.	Essex UK pilot (infrastructure development, housing scheme)	Essex, UK	X								X		**Type 3 a enabled biodiversity offsets**	Defra (2013b), OpenNESS (2016), King (2014), Environment Bank (2013) *in cooperation with the Environment Bank*	+
40.	Doncaster UK pilot	Doncaster, UK	X								X		**Type 3 a enabled biodiversity offsets**	Biodiversity Offsetting in Doncaster Guidance on the Process (n.d.), Doncaster Metropolitan Borough Council (2015)	-

no	Name / brief description (and impact type)	Country / location	Influence and initiators							Motivation and ultimate goal			Type	Sources and notes	Information quality
			government	Financial institute	Sector	Corporation	Local community	NGO(s)	without	Pressure / compliance	Incentive / cost-benefit	Altruism / responsib.			
41.	Development Project in Dudley, West Midlands	West Midlands, UK	X								X		**Type 3a enabled biodiversity offsets**	FIEEM, J. B. C. (2012) *retrospective application of the Defra biodiversity offsetting guidance currently being trialled in pilot schemes in England to a development project granted planning permission in 2009*	-
42.	Project Pinewood (ARUP)	Bucking-hamshire, UK	X					X (BBOP)			X		**Type 3 a/b enabled biodiversity offsets**	The Natural Capital Initiative (2010) *Project Pinewood represents one of the first development proposals in the UK to formally pilot the concept of biodiversity offsetting and to apply the methodology for calculating and implementing offsetting requirements developed by the Business and Biodiversity Offsets Program.*	-
43.	Nottinghamshire UK pilot	Notting-hamshire, UK	X								X		**Type 3 a enabled biodiversity offsets**	Yorkshire Wildlife Trust (2014), Nottinghamshire County Council (2012a, b, c, d and e) *Sowerby Gateway (925 houses, school & neighbourhood centre), Olympia Park (mixed use housing), Cragg Wood (34 houses, access roads & public open space), Wildford Lane (commercial development)*	-
44.	Warwickshire, Coventry and Solihull UK pilot	Warwick-shire, UK	X								X		**Type 3 a enabled biodiversity offsets**	Warwickshire County Council (2014), Lowe (2014)	-
45.	Offsetting the ecological	Southern France	**X**							X			**Type 1 regulatory**	CDC Biodiversité (n.d.a)	-

(continued)

no	Name / brief description (and impact type)	Country / location	Influence and initiators							Motivation and ultimate goal			Type	Sources and notes	Information quality
			government	Financial institute	Sector	Corporation	Local community	NGO(s)	without	Pressure / compliance	Incentive / cost-benefit	Altruism / responsib.			
	impact of the A65 Langon-Pau motorway												**biodiversity offsets**		
46.	Apennine Wind Farms	Italy	**X** (EU BHD, national law)							X			**Type 1 regulatory biodiversity offsets**	BBOP (2009b), ten Kate et al (2011), OECD (2013) *Little information*	-
47.	Compensation Outline Openpit Mining Mertainen	Mertai-nen, North Sweden							X			X	**Type 7 altruistic biodiversity offsets**	Eriksson (2014), Enetjärn et al. (2015) *Voluntary initiative by developer, later provisioned by Environmental Code (Species Protection Ordinance)*	-
48.	Ume Delta River Railway	Ume River, Sweden	**X** (EU BHD, national law)							X			**Type 1 regulatory biodiversity offsets**	Enetjärn et al. (2015) *Impact on the Ume River delta Natura 2000 site (SCI and SPA) and Ramsar site, Regulatory driver: Habitats directive via Swedish Environmental Code on Natura 2000*	-
North- and Central America and the Caribbean															
49.	Kennecott Utah Copper Mine (expansion of the tailings of a copper mine)	Utah, US	**X** (US Wet-land Mitiga-tion, Clean Water Act)			X (Rio Tinto)				X			**Type 1 regulatory biodiversity offsets, Type 5 Corporate biodiversity offsets**	BBOP (2009b), TEEB (2011), Rodricks (2010), Rio Tinto (2008c), ten Kate et al (2004 and 2011), Biodiversity Offsetting Type Study (n.d.) *The offset went beyond regulatory requirements*	++

no	Name / brief description (and impact type)	Country / location	Influence and initiators							Motivation and ultimate goal			Type	Sources and notes	Information quality
			government	Financial institute	Sector	Corporation	Local community	NGO(s)	without	Pressure / compliance	Incentive / cost-benefit	Altruism / responsib.			
50.	Suncor Biodiversity Offset project in Canada (oil sands extraction)	Alberta, Canada			X Canada's Oil Sands Innovation Alliance (COSIA)	X Suncor			X		X	X	**Type 7 altruistic biodiversity offsets**, (Type 5 corporate biodiversity offsets, type 4 sectoral biodiversity offsets)	Suncor (2016), Darbi (2015a), Chittenden and Kolodychuk (2005) *the COSIA was formed after the project*	+
51.	Resolution Copper Mining	Arizona, US			ICMM	X (Rio Tinto NPI)	X (requires land exchange with gov.)			X	X		**Type 6b Local biodiversity offsets**, (Type 5 Corporate biodiversity offsets, type 4 sectoral biodiversity offsets)	ICMM (2010), Resolution Copper Company (2013), Resolution Copper Mining (n.d.)	+
52.	Type study of the BBOP pilot biodiversity offset work on Bainbridge Island (real estate and infrastructure)	Washington, US	X				X smooth progress	**X** (BBOP)			X	X	**Type 3 a/b enabled biodiversity offsets, Type 6b local biodiversity offsets,** (type 7 altruisitc offsets)	Burgin (2008), City of Bainbridge Island (2009), ten Kate (2007), Bainbridge Island Land Trust (2015), OECD (2013) *Former BBOP pilot (2009) continued without BBOP (local level), real estate and infrastructure on small scale by a private developer*	+
53.	Jonah Natural Gas Field	Wyoming, US						X (TNC)			X	X	**Type 3b enabled biodiversity**	BBOP (2009b), Kiesecker et al (2009), ten Kate et al (2011)	+

(continued)

no	Name / brief description (and impact type)	Country / location	Influence and initiators							Motivation and ultimate goal			Type	Sources and notes	Information quality
			government	Financial institute	Sector	Corporation	Local community	NGO(s)	without	Pressure / compliance	Incentive / cost-benefit	Altruism / responsib.			
													offsets, (type 7 altruistic biodiversity offsets)		
54.	Via San Lorenzo – Olanchito (road infrastructure)	Northern Honduras		X (World Bank)						X			**Type 2 conditional biodiversity offsets**	Quintero (2013), Lopez and Quintero (2015), Quintero (2007), Mandle et al (2013) *2008 World Bank provided funding for road improvement conditional on protection of 1400 ha hummingbird habitat.*	+
55.	Hidroelectrica Reventazon (hydroelectric power station)	Costa Rica	X (Costa Rican Institute of Electricity ICE)	X (Inter-American Development Bank IDB)									**Type 2 conditional biodiversity offsets**	Lopez and Quintero (2015)	-
56.	Establishment of the Braulio Carrillo National Park for the construction of a highway	Costa Rica					X			X	X		**Type 6 local biodiversity offsets**	Mandle et al (2013) *A real-world example of the coordinated approach comes from Costa Rica in the 1970s where the construction of a new highway from San José to Puerto Limón was coordinated with the establishment of the Braulio Carrillo National Park. The planned route would be impacting areas of high conservation value, which raised concerns among environmental groups.*	-

no	Name / brief description (and impact type)	Country / location	Influence and initiators							Motivation and ultimate goal			Type	Sources and notes	Information quality
			government	Financial institute	Sector	Corporation	Local community	NGO(s)	without	Pressure / compliance	Incentive / cost-benefit	Altruism / responsib.			
57.	Tygart Valley River Stream and Wetland Bank	Elkins, West Virginia, US	**X** (US Wetland Mitigation)										**Type 1 regulatory biodiversity offsets**	Kelly (2014)	-
58.	South Fork Hoppers Creek Mitigation	North Carolina, US	**X** (US Wetland Mitigation)										**Type 1 regulatory biodiversity offsets**	Kelly (2014)	-
59.	Newmont Long Canyon Project (Gold mine)	Nevada, US				**X** (Newmont)							**Type 5 corporate biodiversity offsets**	Newmont (2014 and 2015)	-
60.	Pueblo Viejo gold mine expansion	Dominican Republic		**X** (IFC PS6)		X (Barrick Gold)				X			**Type 2 conditional biodiversity offsets,** (type 5 corporate biodiversity offsets)	Hardner & Gullison (n.d.)	-
61.	NGTL pipeline proposals in Horne River area: caribou habitat offsets	Alberta, Canada	**X**										**Type 1 regulatory biodiversity offsets**	Poulton (2015a, b and c)	-

(continued)

no	Name / brief description (and impact type)	Country / location	Influence and initiators							Motivation and ultimate goal			Type	Sources and notes	Information quality
			government	Financial institute	Sector	Corporation	Local community	NGO(s)	without	Pressure / compliance	Incentive / cost-benefit	Altruism / responsib.			
62.	Total Joslyn oilsands mine: offsets for species at risk	Alberta, Canada	X										**Type 1 regulatory biodiversity offsets**	Poulton (2015a, b and c)	-
63.	Enbridge Northern Gateway pipelines	Alberta, Canada	X										**Type 1 regulatory biodiversity offsets**	Poulton (2015a, b and c) *Enbridge Northern Gateway pipeline conditions: 10 conditions for 4 different types of offsets (wetlands, freshwater fish habitat, marine habitat, caribou habitat).*	-
South America															
64.	Antamina Copper and Zinc Mine	Ancash, Peru			X (ICMM)			X				X	**Type 7 altruistic biodiversity offsets, type 6 local biodiversity offsets,** (Type 3b Enabled biodiversity offsets, type 4 sectoral biodiversity offsets)	BBOP (2009b), ten Kate et al (2011), ICMM (2010), BNI (2006), Lopez and Quintero (2015) *Has been awarded the Sustainable Development Prize by Ministry of Energy and Mines for its Polylepsis Program*	++
65.	Hidroelectrica Yacyreta (hydroelectric power station)	Argentina, Paraguay		X (World Bank, IDB)						X			**Type 2 conditional biodiversity offsets**	Darbi et al (2010), Lopez and Quintero (2015), Quintero (2007) *Fulfilling requirements of lenders such as the World Bank and IDB, Environmental Assessment was*	++

no	Name / brief description (and impact type)	Country / location	Influence and initiators							Motivation and ultimate goal			Type	Sources and notes	Information quality
			government	Financial institute	Sector	Corporation	Local community	NGO(s)	without	Pressure / compliance	Incentive / cost-benefit	Altruism / responsib.			
														updated in 1992 and compensation measures were established.	
66.	Pasto Mocoa in the Amazon (Mocoa-Saanciso road)	Colombia		X (Inter-American Development Bank IDB)				X (WWF)		X	X		**Type 2 conditional biodiversity offsets**, (Type 3b enabled biodiversity offsets)	Lopez and Quintero (2015), Rincon (2014), Quintero (2013) *IDB engagement was an opportunitiy to influence policies*	+
67.	Gasoducto Gasbol (pipeline)	Bolivia, Brazil	X (SNUC-Act)	X (Inter-American Development Bank IDB)				X (WCS)		X			**Type 1 regulatory biodiversity offsets,** (type 2 conditional biodiversity offsets, type 3b enabled biodiversity offsets)	Lopez and Quintero (2015), Quintero (2007), Castillo (n.d.) *quality screening prompted new, revised EAs. Taking the time to revise this initial step proved fruitful: the project was subsequently awarded the World Bank Green Award for outstanding environmental management (by IAIA)*	+
68.	Brisas Gold and Copper Project (Open pit mine, Gold Reserves Inc)	Eastern Venezuela		X (IFC PS 6, equator principles)		X (GRI's corporate commitment)							**Type 5 corporate biodiversity offsets, type 2 conditional biodiversity offsets**, (type 6 local	BBOP (2009b), ten Kate et al (2011), OECD (2013), Nazari et al (2009), Doswald et al (2012) *The ESIA was designed to meet GRI's corporate commitments to best practice and sustainability and the IFC Performance Standards and the Equator Principles.*	+

(continued)

no	Name / brief description (and impact type)	Country / location	Influence and initiators							Motivation and ultimate goal			Type	Sources and notes	Information quality
			government	Financial institute	Sector	Corporation	Local community	NGO(s)	without	Pressure / compliance	Incentive / cost-benefit	Altruism / responsib.			
													biodiversity offsets)		
69.	AngloGold Ashanti Brasil Mineracao (gold mine)	Minas Gerais, Brazil			ICMM				X			X	**Type 7 altruistic biodiversity offsets**, (type 4 sectoral biodiversity offsets)	ICMM (2010)	-
70.	Cerrejón Coal (coal mining)	La Guajira, Colombia						CI			X	X	**Type 6b local biodiversity offsets**, (type 3b enabled biodiversity offsets, type 7 altruistic biodiversity offsets)	ICMM (2010), Conservation Internatioal (2016a) *Cerrejón is engaging with public and private organizations through cooperative agreements with the regional environmental authority and with CI Colombia.*	-
71.	Amaila Fall Hydro Project	Guyana											**Type 3 b enabled biodiversity offsets**	Conservation Internatioal (2016b), EPA Guyana (2014)	-
72.	Gramalote Mining Project	Antioquia region, Colombia	**X**	**X**				X					**Type 1 regulatory biodiversity offsets, type 2 conditional biodiversity offsets**, (type 3b enabled	Arango and Franco (2014), Vorhies (2014) *New regulations by the Colombian Government: no net loss of biodiversity from major development projects, compliance with IFC PS and BBOP standard*	-

no	Name / brief description (and impact type)	Country / location	Influence and initiators							Motivation and ultimate goal			Type	Sources and notes	Information quality
			government	Financial institute	Sector	Corporation	Local community	NGO(s)	without	Pressure / compliance	Incentive / cost-benefit	Altruism / responsib.			
													biodiversity offsets)		
Rejected types with poor information quality															
73.	Malua Biobank	Sabah, Malaysia												Malua Biobank (2012), Brock (2015) *controversial*	-
74.	BBOP offset in Mexico	Mexico												Burgin (2008) *No information*	--
75.	Road and Massai Tourism lodges BBOP pilot	Kenia						X (BBOP)						Ten Kate (2007) *No information*	--
76.	Manaus Energia, Balbina Hydropower plant	Brazil												BBOP (2009b), ten Kate (2011) *Controversial and little information*	--
77.	Heavy Crude Pipeline (OCP)	Ecuador												Ten Kate et al (2004), Goodland (2005) *Project Fails to Comply with World Bank Operational Policies*	--
78.	Ncandu (dam for drinking water)	KwaZulu-Natal, South Africa												Department of Environmental Affairs and Development Planning (2007) *Controversial and little information*	--
79.	Cerro Vanguardia (open pit gold mine, AngloGold Ashanti)	Santa Cruz, Argentina			ICMM									ICMM (2010) *little information*	--

(continued)

no	Name / brief description (and impact type)	Country / location	Influence and initiators							Motivation and ultimate goal			Type	Sources and notes	Information quality
			government	Financial institute	Sector	Corporation	Local community	NGO(s)	without	Pressure / compliance	Incentive / cost-benefit	Altruism / responsib.			
80.	Greater Norwich UK pilot	Greater Norwich, UK	X								X		Type 3 a enabled biodiversity offsets	Defra (2013b) *little information*	-
81.	Belize Roads and Drainage	Belize												Quintero (2007) *little information*	–
82.	Tocantins Development	Brazil												Quintero (2007) *little information*	–
83.	Cartagena Water Supply	Colombia												Quintero (2007) *little information*	–
84.	Santa Fe Water Supply	Colombia												Quintero (2007) *little information*	–
85.	Bogotá water supply	Colombia												Quintero (2007) *little information*	–
86.	Argentina Flood Protection	Argentina												Quintero (2007) *little information*	–
87.	Iron Ore, Labrador City	Canada												Burgin (2008) *little information*	–
88.	Weda Bay Nickel	Indonesia												Environment Deprtment (n.d.)	–
89.	Netherlands: Compensation Bank Pilot Projects													Morandeau & Vilaysack (2012)	–
90.	Exeter and East Devon Growth Point UK pilot	Devon, UK	X								X		Type 3 a enabled biodiversity offsets	DNLP (n.d.) *Focuses on how offsetting can be used to meet the objectives of the Green Infrastructure Strategy.*	–

10.2 Appendix 2: Biodiversity Offsets Blog

Introduction and Welcome to the Biodiversity Offsets Blog
Welcome to the Biodiversity Offsets Blog! This site aims to provide an interdisciplinary platform for the information and exchange on Biodiversity Offsets, the Mitigation Hierarchy and the wider context. *The About-section in the menu above gives you a brief introduction to what biodiversity offsets are and summarizes the objectives of this blog. You may also want to get to know about me and my PhD (which is the basis for this blog).* *The Biodiversity Offsets Blog has only recently started (it went online in August 2014) and thus will continuously be updated with any upcoming information. The whole bunch of information that has risen in the past few years will also be incorporated bit by bit.* *As the focus of this platform is to bring people and their expertise together, please get in contact if you have anything to share. It is highly welcomed if you would like to write posts or reviews or share photographs (Please request author rights). You can also comment to any post or start a discussion on the open discussion platform.* *PLEASE NOTE THAT THIS IS NOT A PLATFORM FOR THE PROMOTION OF BIODIVERSITY OFFSETS: STRONG OPINIONS FROM BOTH PROPONENTS AND OPPONENTS (AND ALL THAT LIES IN BETWEEN) ARE VERY MUCH WELCOME!* *In this sense, you are also kindly requested to answer the current polls (see in the sidebar on the right or below if you are using mobile devices). Thank you.* *Please find below a list of all posts or choose from the menu above for a specific focus.* *FOR NETWORKERS* *Have a look at the Experts and Exchange section if you want to get to know experts in the field of biodiversity offsets from all over the world with their respective focus of expertise or explore past and present PhD topics.* *FOR INFORMATION SEEKERS* *Have a look at the sources section to explore relevant literature as well as a growing list of links to websites.* FOR PRACTITIONERS *It is greatly appreciated if you'd contribute to and exchange on biodiversity offsets on the ground. See information on policies and governance and also worldwide examples of biodiversity offsets.*

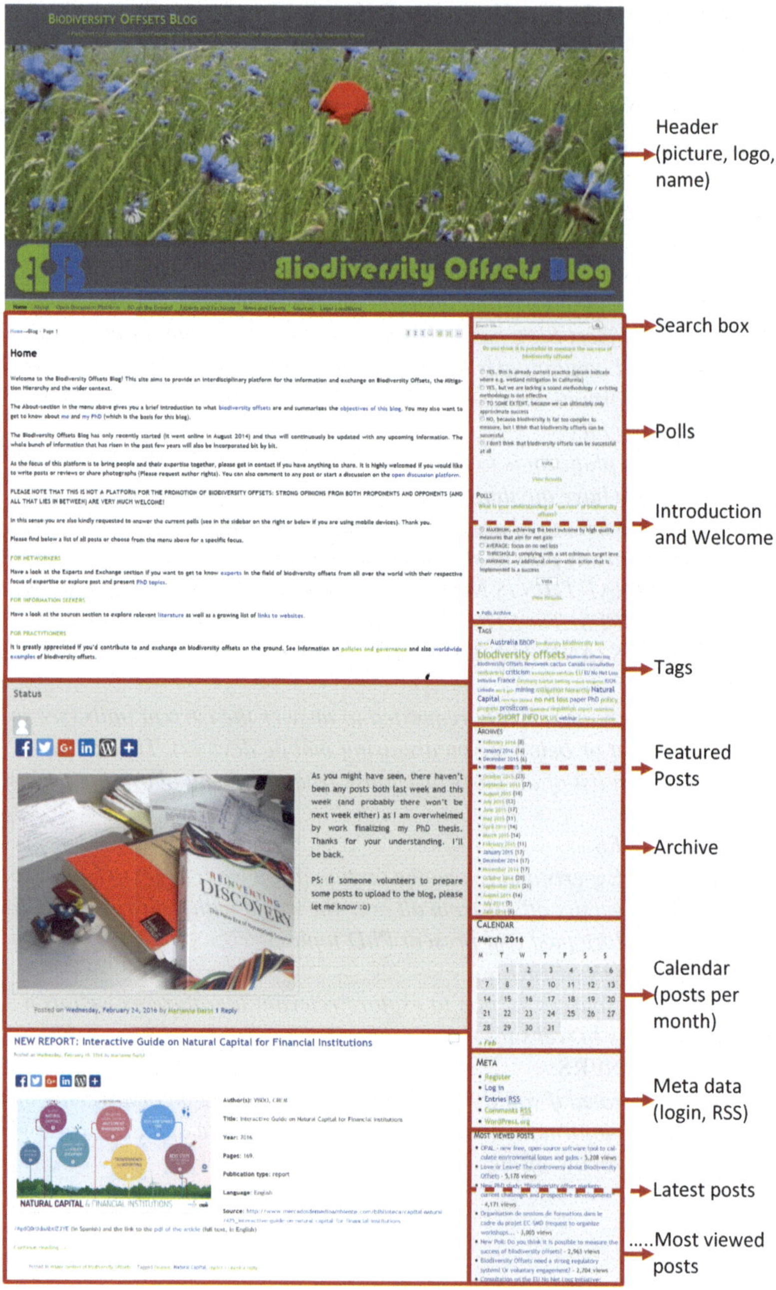
Biodiversity Offsets Blog
Home
Status
NEW REPORT: Interactive Guide on Natural Capital for Financial Institutions
Header (picture, logo, name)
Search box
Polls
Introduction and Welcome
Tags
Featured Posts
Archive
Calendar (posts per month)
Meta data (login, RSS)
Latest posts
Most viewed posts

10.3 Appendix 3: Definitions of Corporate (Social) Responsibility (Table 10.1)

Table 10.1 Definitions of corporate (social) responsibility

Definition source	Definition
Steiner (1971)	CSR is "enlightened self-interest"
Davis (1973)	"good corporate citizenship": what goes beyond obeying the law in terms of "what every good citizen does..."
Carroll (1983)	CSR involves the conduct of a business so that it is economically profitable, law abiding, ethical, and socially responsible ... Thus, CSR is composed of four parts: economic, legal, ethical, and voluntary or philanthropic.
Kilcullen and Kooistra (1999)	CSR is the degree of moral obligation that may be ascribed to corporations beyond simple obedience to the laws of the state.
Woodward-Clyde (1999)	CSR has been defined as a "contract" between society and business wherein a community grants a company a license to operate and in return the latter meets certain obligations and behaves in an acceptable manner.
Khoury et al. (1999)	Corporate social responsibility is the overall relationship of the corporation with all of its stakeholders. These include customers, employees, communities, owners/ investors, government, suppliers, and competitors. Elements of social responsibility include investment in community outreach, employee relations, creation and maintenance of employment, environmental stewardship, and financial performance.
Piacentini et al. (2000)	CSR is the voluntary assumption by companies of responsibilities beyond purely economic and legal responsibilities.
UK Government (2001)	Corporate social responsibility recognizes that the private sector's wider commercial interests require it to manage its impact on society and the environment in the widest sense [...] CSR goes beyond legal obligations, involving voluntary, private sector-led engagement, which reflects the priorities and characteristics of each business, as well as sectoral and local factors.
Marsden (2001)	A socially responsible corporation is one that runs a profitable business that takes account of all the positive and negative environmental, social, and economic effects it has on society.
McWilliams and Siegel (2001)	Actions that appear to further some social good, beyond the interests of the firm and that is required by law.
Commission of the European Communities (2002)	Corporate social responsibility is about companies having responsibilities and taking actions beyond their legal obligations and economic/business aims. [...] This can be summed up as the triple bottom line approach: i.e., economic, social, and environmental.

(continued)

Table 10.1 (continued)

Definition source	Definition
Business for Social Responsibility (2003)	Corporate social responsibility is achieving commercial success in ways that honor ethical values and respect people, communities, and the natural environment.
Ethical Performance (2003)	At its best, CSR is defined as the responsibility of a company for the totality of its impacts, with a need to embed society's values into its core operations as well as into its treatment of its social and physical environment.
Global Corporate Social Responsibility Policies Project (2003)	Global corporate social responsibility can be defined as business practices based on ethical values and respect for workers, communities, and the environment.
EU Commission (2006)	[CR] is a concept whereby companies integrate social and environmental concerns in their business operations and in their interaction with their stakeholders on a voluntary basis. It is about enterprises deciding to go beyond minimum legal requirements and obligations stemming from collective agreements in order to address societal needs.
Castelló and Lozano (2011)	CSR refers to the voluntary initiative, it integrates social and environmental concerns in business operations, as well as responsibilities.
Blowfield and Murray (2014)	Corporate responsibility comprises (a) the responsibilities of business in the context of wider society, (b) how those responsibilities are defined and negotiated, and (c) how they are managed and organized.

Source: Dahlsrud, A. (2008): How Corporate Social Responsibility is defined: an Analysis of 37 Definitions. In: Corporate Social Responsibility and Environmental Management

10.4 Appendix 4: List of Members of the Advisory Group of the Business and Biodiversity Offsets Programme

(as of November 2015)

Companies with Biodiversity Footprint to Offset

- Ambatovy Project
- Cemex
- Eiffage
- ERAMET - PT WEDABAY Nickel Project
- New Britain Palm Oil Ltd.
- Newcrest Mining Limited
- Solid Energy Coals of New Zealand
- Sveaskog
- Tahi Estate
- TOTAL SA
- Winstone Aggregates

Service Providers

- Arup
- Biodiversity Works
- Biotope
- Daemeter Consulting
- Earthmind
- Ecoacsa
- EcoDecisión
- Environ Corporation
- Environmental Resources Management
- Golder Associates
- Hardner & Gullison Associates
- Hogan Lovells
- International Conservation Services CC
- Markit Environmental Registry
- New Forests
- Nollen Group
- Proforest
- Resource Environmental Solutions, LLC
- Response Ability, Inc.
- Scientific Certification Systems
- SLR Consulting
- Treweek Environmental Consultants
- The Biodiversity Consultancy
- The Environment Bank
- Tonkin and Taylor
- Wildlands Inc.

Financial Institutions

- CDC Biodiversité
- Citi
- European Bank for Reconstruction and Development
- Global Environment Fund
- International Finance Corporation
- Inter-American Development Bank
- Mizuho Bank, Ltd.

Governments and Intergovernmental Organizations

- Department of Conservation, New Zealand
- Department for Environment and Rural Affairs—Defra (UK)
- Department of Sustainability & Environment, Government of Victoria, Australia
- Forestry Department, Sabah, Malaysia

- Government of Catalonia, Spain
- International Union for Conservation of Nature (IUCN)
- Ministry of Agricultural and Rural Development (MARD), Vietnam Directorate of Forestry
- Ministry of Ecology, Energy, Sustainable Development, and Spatial Planning, France
- Ministry of Environment and Tourism, government of Namibia
- Ministry of Infrastructure and the Environment, The Netherlands
- Ministry of Mines and Energy, Namibia
- Ministry of Natural Resources and the Environment (MONRE), government of Vietnam
- Ministry of Nature, Environment and Tourism, government of Mongolia
- National Institute of Ecology, Mexico
- National Environment Management Authority, Uganda
- Ramsar Convention on Wetlands
- South African National Biodiversity Institute
- United Nations Development Programme (Environment and Energy Group)
- United Nations Environment Programme—World Conservation Monitoring Centre
- Wildlife Division, Forestry Commission, Government of Ghana

Conservation and Civil Society Groups

- BirdLife International
- Brazilian Biodiversity Fund (Funbio)
- Centre for Research-Information-Action for Development in Africa
- Conservation International
- Earthwatch Institute
- Ecoagriculture Partners
- Fauna & Flora International
- Forest Trends
- Grupo Ecológico Sierra Gorda, I.A.P., México
- International Institute for Environment and Development
- Leibniz Institute of Ecological and Regional Development (IOER)
- Nature Conservation Resource Center, Ghana
- Rainforest Alliance
- Royal Botanic Gardens, Kew
- The Nature Conservancy
- Tulalip Tribes, US
- Wildlife Conservation Society
- WWF-UK
- Zoological Society of London

Individuals

- Steve Botts
- Susie Brownlie
- Marc Christensen
- Michael Crowe
- Kevin Erwin
- Toby Gardner
- Martin Hollands
- Louise Johnson
- Daniela Lerda
- Paul Mitchell
- Dave Poulton
- Dave Richards
- Shelagh Rosenthal
- Marian Weber

10.5 Appendix 5: Worldwide On-the-Ground Examples of Biodiversity Offsets—Your Help Is Needed! Post on the Biodiversity Offsets Blog

Worldwide On-the-Ground Examples of Biodiversity Offsets—Your Help Is Needed!
Posted on Thursday, August 21, 2014, by Marianne Darbi

On-the-ground examples of biodiversity offsets are needed as evidence and illustration for the discussion on biodiversity offsets

I am planning to compile and present real on-the-ground examples for biodiversity offsets (that should go to here).

(continued)

List and individual cases

To do so, I have envisaged two things: a very simple list with worldwide on-the-ground examples of biodiversity offsets (like the one I have started here) and individual cases that will be presented in single posts.

Now, I know that while in some countries this is a pretty simple task (at least in quantitative terms, do n't get me wrong) because only a few cases exist, in others like the US, Germany and Australia because of the multitude of cases, it would be neither feasible nor make any sense to try to give an overview on all of them.

Can you help to understand what biodiversity offsets mean in practice?

Still, no matter if the system as such is quite well established and known/ represented or if it's a pilot, I am sure that real world examples increasingly help to understand what biodiversity offsets mean in practice:

to generally understand what biodiversity offsets really are and what is out there worldwide

- what are biodiversity offsets?
- where and when are they implemented?
- who cares about them?
- to put theory and practice into relation and to look at the results
- is there a theory-practice-gap?
- what are the tricky points?
- what is really crucial?
- and most importantly, what are the actual outcomes?

What kind of projects are best suitable?

So, I wonder, if from your experience you could present a real world biodiversity offset example in a post?! That would be tremendous. The more examples will be included here, the better!

The choice of the example(s) is of course up to you: a project that you've been involved in and that you know well, a very positive or negative example, a controversial one, something that is well known or something hardly ever anyone has heard of, a huge one or a small — its up to you!

How can you contribute on-the-ground examples of biodiversity offsets?

If you would like to start right away please subscribe. I will then change your role from (read only) subscriber to (read and write) contributor. Once this is done, you can log in and create your post(s). The Blog works with wordpress. By clicking on the little house in the top left corner you can (after logging in) directly access the dashboard. Then click on "posts" and "add new". Enter your text and don't forget to save (on the right).

(continued)

Or if you prefer you can also send me something on LinkedIn or via email (marianne.darbi[at]gmx.de).

If (much to my regret) you don't find the time for this (at least not now), please get in touch and simply name some cases for the list (name, country location, responsible and maybe a source or short description).

Looking very much forward to hear from you!

10.6 Appendix 6: Biodiversity Offsetting—A Fool's Errand?—A Comment by James Brown. Post on the Biodiversity Offsets Blog

Biodiversity Offsetting—A Fool's Errand?—A Comment by James Brown

Posted on Thursday, April 23, 2015, by guest

A picture I took in Banagher Glen, an ancient woodland, on my second day of sampling

This is a guest post by James Brown, PhD student at Queen's University Belfast.
This comment has previously been published on QuBio blog of the School of Biological Sciences. It is the expression of the author's thoughts and experiences and as such is acknowledged as a fruitful contribution to the discussion on biodiversity offsets. If you want to react or clarify your own position (underpin or disprove), please leave a reply below!

As someone who is currently studying the genetic variation in native Irish trees I require good material for sampling as such I plan to go to natural ancient woodland to collect my samples. As part of my background reading I recently came across a number of news articles about the process of biodiversity offsetting. In basic terms it involves the destruction of a woodland or other ecosystem for property development or other reasons with the creation of a new woodland elsewhere. The idea is that no diversity will be lost due to the creation of this new woodland and in some cases it may be justified by the

(continued)

promise of building a much larger woodland. There are currently six pilot schemes testing offsetting in England.

So what's the problem then you might ask? Well the problem is that it doesn't work like that. It is not a simple black and white process wherein you remove one woodland, create another and this doesn't have any negative impact. Living trees and shrubs are supported by and they themselves support a wide range of other organisms including those in the soil such as fungi as well as organisms which feed on them, bacteria, yeast, nematodes and mites which are associated with both live and dead roots, those which inhabit dead areas of wood including fungi and invertebrates, epiphyte communities which exploit all surfaces of trees such as lichens and mosses, animals which feed on plant products for example nectar along with plants themselves as well as animals which feed on fungi and animals which live in the plants. It would be impossible to replicate the exact same ecosystem elsewhere. Not to mention the loss and inability to replicate the genetic diversity contained within ancient woodlands.

There is also the issue that woodlands themselves are part of a larger ecosystem so removing one from an area could have negative impacts on the surrounding area. Woodlands provide a whopping 17 different ecosystem services according to the Millennium Ecosystem Assessment. These include provisioning services such as timber, regulating services such as air and water regulation and cultural services such as recreation and tourism. By removing a woodland these services may be lost from an area completely or only partially fulfilled by a newly created woodland. In some cases woodlands may help prevent flooding due to their structure and the greater uptake of water by trees. Thus their removal may put areas at risk.

In terms of the creation of new woodlands there doesn't seem to be any clarity as to how this would be carried out. For one example where would they get the seeds from? Would they get seeds from the local area or would they import because it might be cheaper? There is the issue that they might inadvertently import infected material to create woodlands which would cause even more damage by spreading disease to other areas. There is also the question of enforcement, who will make sure that developers keep their promises? Also when exactly does the new woodland get created, is it after the destruction of the old one, is it after the development of whatever is replacing the old one? And finally what factors determine whether or not a woodland is eligible for destruction? Already in England there is controversy over plans to destroy an ancient woodland in order for a petrol station to be built.

Biodiversity is a key part of life and is very valuable but by placing monetary values on things such as woodlands it gives the false impression that they can be traded along with the idea that bigger is better which is not

(continued)

necessarily the case. The problem is with a growing global population more land is needed for development and so solutions need to be found in terms of finding areas for new properties. Biodiversity offsetting should not be that solution in the case of veteran or ancient woodlands. If biodiversity offsetting is to be used then clearly enforceable rules are required to ensure as little damage as possible is done to our already dwindling ecosystems.

10.7 Appendix 7: Behold the Power of Fungus . . . and Biodiversity Offsets—A Guest Post by Naazia Ebrahim. Post on the Biodiversity Offsets Blog

Behold the Power of Fungus . . . and Biodiversity Offsets—A Guest Post by Naazia Ebrahim

Posted on Wednesday, May 27, 2015, by guest

This is a guest post by Naazia Ebrahim of the OECD Environment Directorate.
This comment has previously been published on OECD Insights. It is the expression of the author's thoughts and experiences and as such is acknowledged as a fruitful contribution to the discussion on biodiversity offsets. If you want to react or clarify your own position (underpin or disprove), please leave a reply below!

When you think of biodiversity conservation, you probably think of the classic images: the polar bear, the lion, the elephant, the giraffe. The ecological community likes to call them charismatic megafauna, with only a hint of satire.

But did you know that the only thing that can neutralise the deadliest, antibiotic-resistant superbug on this planet is a fungus? Now, note that it was discovered in the soil of a Canadian national park, and it rather makes the argument (well, the anthropogenic argument) for conservation of biodiversity in all its shapes and forms by itself. Behold the power of a fungus!

Unfortunately, most biodiversity has been having a rough time of it lately.

As we have all heard recently, WWF's 2014 Living Planet Report has reported a 52% decline in mammals, birds, reptiles, amphibians and fish overall from 1970 to 2010, while IUCN's Red List indicates that a quarter of mammals, over a tenth of birds, and 41% of amphibians are at risk of

(continued)

extinction. The decline is worse in the tropics, and particularly in Latin America, where species populations have dropped by 83% since 1970. Significantly scaled up efforts will be needed if we are to reach the 2011–2020 Aichi Biodiversity Targets—agreed upon at the 2010 conference of the Convention on Biological Diversity—in time. And this is true not only for conservation, but also for the sustainable use of biodiversity and natural resources.

Here at the OECD, we're looking at how to scale up finance for biodiversity, and how instruments for conservation and sustainable use of biodiversity can be designed and implemented in more effective ways. One instrument, biodiversity offsets, has recently been gaining much attention from government and business alike. Based on the polluter pays approach, biodiversity offsets operate under a "no-net-loss" principle, and have the potential to reduce the costs of achieving environmental objectives.

To be sure, there are difficulties. The most obvious one is that biodiversity is not like carbon: one unit emitted here does not equal one unit saved there. Biodiversity is highly specific, and often highly localised; there are many ecosystems that wouldn't necessarily exist if ecological conditions changed slightly. So project developers need to be particularly careful at building in safeguards; offsets must be a last resort after avoidance and mitigation; offset design must adhere to strict requirements for ecological equivalence; and monitoring and verification must be extremely robust.

As we work through establishing good practice insights, we hope that biodiversity offsets will be able to provide developers with an additional tool to reduce their adverse impacts on biodiversity in a cost-effective way. That really would be a win-win—and one that would make all superbug-fighting fungi happy.

Useful links

The OECD held a workshop on biodiversity offsets in November 2013, with representation from governments, industry, and civil society. Keep an eye out for our publication, forthcoming in early 2015.

Preliminary OECD Policy Highlights on Biodiversity Offsets

Chapter on Biodiversity Offsets from Scaling up Finance Mechanisms for Biodiversity

10.8 Appendix 8: The Commodification of Nature? Post on the Biodiversity Offsets Blog

The Commodification of Nature?—Article in BIODIV'2050
Posted on Saturday, May 23, 2015, by Marianne Darbi

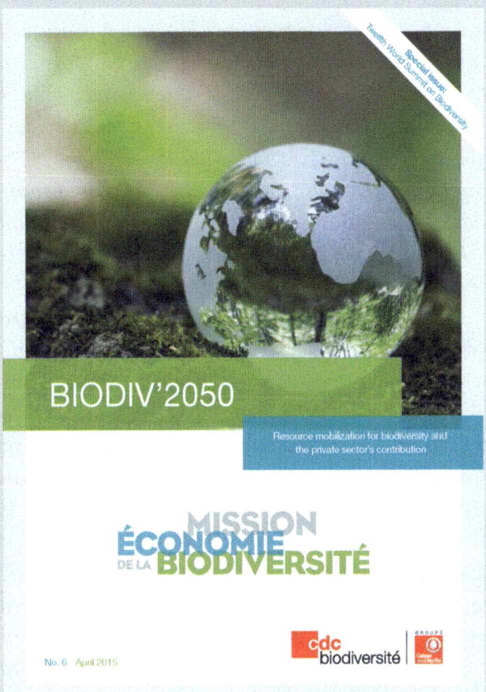

As I have already announced in my latest post, the latest issue of Mission Économie de la Biodiversité's "BIODIV'2050" (No. 6 — April 2015) is out. Beside the previously mentioned article entitled "Thinking out the appropriate frameworks: biodiversity offsets and safeguards" another article asks "HOW CAN THE PRIVATE SECTOR CONTRIBUTE TO RESOURCE MOBILIZATION TO REACH THE AICHI TARGETS?" It is highlighted that there is a lack of international consensus on the participation of the private sector and on the financing mechanisms, e.g. with regard to the "commodification of nature":

The commodification of nature?

The economic approach to biodiversity, the mobilization of innovative financial mechanisms and the involvement of businesses in natural capital conservation are controversial issues. One argument is that these approaches may lead to the commodification of nature, the creation of biodiversity markets and/or the appropriation of nature by the private sector. In other words, species, habitats and ecosystem services would be assimilated to merchandise, like any other, and would be monetized at the daily rate to be "bought" or "sold" by companies or financial institutions speculating for profit. But this vision is far from reality on the ground.

Paying a physician does not imply commodifying health. Paying compensation in case of a life-threatening accident does not mean putting a price tag

(continued)

on human life. Similarly, calculating the economic value of biodiversity does not mean pricing its value.

Economic valuation of biodiversity aims at recognising that biodiversity is useful and rare, and that decisions, (public or private) concerning territorial development or production have been taken up to now without giving sufficient heed to the value of biodiversity and hence has been detrimental to the ecosystems on which the economic activities depend. The vision of nature adopted in this approach, thus, is very anthropocentric. However it doesn't not imply the "pricing" of species for sale on any market driven by supply and demand. And were such "markets" ever to exist, standard economic theory tells us that for public goods such as biodiversity or ecosystem services, markets of this type would not be effective.

With regard to economic instruments, often assimilated to so-called market-based instruments, experience, here again, is far different from prejudicial opinions, one way or the other. In the case of biodiversity offsets or payments for environmental services (PES) for instance, it is not the species or the ecosystem services, which by definition cannot be appropriated, that are bought or sold, but rather the conservation or restoration actions that bring about changes in practices on lands where people have user rights. Quite contrary to the idea of commodification, when these actions are connected to the introduction of conservation easements, privately-owned lands, in a way, could been seen as being part again of the public domain serving general interest. Furthermore, the context of implementation of those instruments is strictly regulated by public authorities, guarantors of the robustness of defined conservation and restoration actions, as defined and implemented.

Last, businesses that engage in biodiversity action are mostly encouraged to incorporate the value of ecosystem services in their decision-making process in order to recognise the impact and dependence of their activities on biodiversity. The primary goal here is to limit the cost of these activities to society and construct business models that combine development and biodiversity conservation.

You can read the article and download the full issue here (open access). For more information see also the abstract and table of contents of the issue below.

10.9 Appendix 9: List with Further Blog Posts on the Controversy of Biodiversity Offsets

http://www.biodiversityoffsets.net/manifesto-for-conservation-banks-in-spain-a-guest-post-by-mercados-de-medio-ambiente/

http://www.biodiversityoffsets.net/a-la-espera-de-los-bancos-de-conservacion-de-la-naturaleza-waiting-for-conservation-banks-in-spain-a-guest-post-by-alfonso-carretero/

http://www.biodiversityoffsets.net/behold-the-power-of-fungus-and-biodiversity-offsets-a-guest-post-by-naazia-ebrahim/
http://www.biodiversityoffsets.net/are-biodiversity-offsets-still-relevant-for-india-a-comment-by-divya-narain/
http://www.biodiversityoffsets.net/compensation-for-biodiversity-losses-can-varied-approaches-deliver-a-comment-by-carlos-ferreira/
http://www.biodiversityoffsets.net/opinion-biodiversity-offsets-corporate-responsibility-opportunity-paradox-comment-carlos-ferreira/
http://www.biodiversityoffsets.net/will-offsets-save-sink-protected-areas-new-paper-pilgrim-bennun/
http://www.biodiversityoffsets.net/biodiversity-offsetting-can-certain-circumstances-improve-nature-biodiversity-conservation-comment-heidi-wittmer-consultation-eu net-loss-initiative-2/
http://www.biodiversityoffsets.net/theory-practice-gap-biodiversity-offsets-comment-alan-key/
http://www.biodiversityoffsets.net/nature-sale-2nd-forum-natural-commons-2-june-2014-regents-park-hub-london/
http://www.biodiversityoffsets.net/biodiversity-offsetting-a-fools-errand-a-comment-by-james-brown/
http://www.biodiversityoffsets.net/does-biodiversity-offsetting-make-nature-a-commodity-a-comment-by-agata-szafraniuk/
http://www.biodiversityoffsets.net/case-studies-biodiversity-offsetting-voices-ground-critical-briefing-note-friends-earth/
http://www.biodiversityoffsets.net/no-net-loss-initiative-risk-chance-nature-political-session-german-green-party-left-issue-biased/

10.10 Appendix 10: George Monbiot's Criticism on Biodiversity Offsets and the Natural Capital Agenda: "The Pricing of Everything"

The Natural Capital Agenda: The Pricing, Valuation, Monetization, Financialization of Nature in the Name of Saving It

Monbiot starts off with giving arguments from supporters of the Natural Capital Agenda. He refers to two points: giving a value to nature and raising money to save it:

> Those who support this agenda say, "Look, we are failing spectacularly to protect the natural world—and we are failing because people aren't valuing it enough. Companies will create a road scheme or a supermarket—or a motorway service station in an ancient woodland on the edge of Sheffield—and they see the value of what is going to be destroyed as effectively zero. They weigh that against the money

(continued)

to be made from the development with which they want to replace it. So if we were to price the natural world, and to point out that it is really worth something because it delivers ecosystems services to us in the form of green infrastructure and asset classes within an ecosystems market (i.e. water, air, soil, pollination and the rest of it), then perhaps we will be able to persuade people who are otherwise unpersuadable that this is really worth preserving."

They also point out that through this agenda you can raise a lot of money, which is not otherwise available for conservation projects.

He concludes:

These are plausible and respectable arguments. But I think they are the road to ruin—to an even greater ruin than we have at the moment. Let me try to explain why with an escalating series of arguments. I say escalating because they rise in significance, starting with the relatively trivial and becoming more serious as we go.

Pricing nature is nonsense because the numbers are not reliable

Perhaps the most trivial argument against the Natural Capital Agenda is that, in the majority of cases, efforts to price the natural world are complete and utter gobbledygook. And the reason why they are complete and utter gobbledygook is that they are dealing with values which are non-commensurable.

They are trying to compare things which cannot be directly compared. The result is the kind of nonsense to be found in the Natural Capital Committee's latest report, published a couple of weeks ago (4). The Natural Capital Committee was set up by this Government, supposedly in pursuit of better means of protecting the natural world.

It claimed, for example, that if fresh water ecosystems in this country were better protected, the additional aesthetic value arising from that protection would be £700 million. That's the aesthetic value: in other words, what it looks like [...] It's just not possible to have meaningful figures for benefits which cannot in any sensible way be measured in financial terms.

Rare examples where pricing actually works to the benefit of nature

Now there are some things that you can do. They are pretty limited, but there are some genuinely commensurable pay-offs that can be assessed. So, for instance, a friend of mine asked me the other day, "What's the most lucrative investment a land owner can make?". I didn't know. "An osprey! Look at Bassenthwaite in the Lake District where there's a pair of ospreys breeding and the owners of the land have 300,000 people visiting them every year. They charge them for car parking and they probably make a million pounds a year."

You can look at that and compare it to what you were doing before, such as rearing sheep, which is only viable because of farm subsidies: you actually lose money by keeping sheep on the land. So you can make a direct comparison because you've got two land uses which are both generating revenue (or losing revenue) that is already directly costed in pounds. I've got no problem with that. You can come out and say there is a powerful economic argument for having ospreys rather than sheep.

(continued)

There are a few others I can think of. You can, for instance, look at watersheds. There is an insurance company which costed Pumlumon, the highest mountain in the Cambrian mountains, and worked out that it would be cheaper to buy Pumlumon and reforest it in order to slow down the flow of water into the lowlands than to keep paying out every year for carpets in Gloucester.

There were quite a few assumptions in there, as we don't yet have all the hydrological data we need, but in principle you can unearth some directly commensurable values—the cost of insurance pay-outs, in pounds, versus the cost of buying the land, in pounds—and produce a rough ballpark comparison. But in the majority of cases you are not looking at anything remotely resembling financial commensurability.

Unbundling ecosystem services and trading them separately means damaging the holistic system of nature

Problem Two is that you are effectively pushing the natural world even further into the system that is eating it alive. Dieter Helm, the Chairman of the Natural Capital Committee, said the following in the same report I quoted from just a moment ago. "The environment is part of the economy and needs to be properly integrated into it so that growth opportunities will not be missed."

There, ladies and gentlemen, you have what seems to me the Government's real agenda. This is not to protect the natural world from the depredations of the economy. It is to harness the natural world to the economic growth that has been destroying it. All the things which have been so damaging to the living planet are now being sold to us as its salvation; commodification, economic growth, financialisation, abstraction. Now, we are told, these devastating processes will protect it.

[...]

Now nature is to be captured and placed in the care of the financial sector, [...] we need to "unbundle" ecosystem services so they can be individually traded.

That's the only way in which it can work—this financialisation and securitisation and bond issuing and everything else they are talking about. Nature has to be unbundled. If there is one thing we know about ecosystems, and we know it more the more we discover about them, it's that you cannot safely disaggregate their functions without destroying the whole thing. Ecosystems function as coherent holistic systems, in which the different elements depend upon each other. The moment you start to unbundle them and to trade them separately you create a formula for disaster.

The power of the economic system as a whole against the power of nature

Problem Three [...] is, of course, power.

Power is the issue which seems to get left out of the Natural Capital Agenda. And because it gets left out, because it it is, I think, deliberately overlooked, what we are effectively seeing is the invocation of money as a kind of fairy dust, that you sprinkle over all the unresolved problems of power in the hope that they will magically resolve themselves. But because they are unresolved, because they are unaddressed, because they aren't even acknowledged; the natural capital agenda cannot possibly work.

[...]

You haven't changed anything by sprinkling money over the problem, you have merely called it something new. You have called it a market as opposed to a political

(continued)

system. But you still need the regulatory involvement of the state to make that market work. Because we persuade ourselves that we don't need it any more because we have a shiny new market mechanism, we end up fudging the issue of power and not addressing those underlying problems.

[. . .]

You do not solve the problem without confronting power. But what we are doing here is reinforcing power, is strengthening the power of the people with the money, the power of the economic system as a whole against the power of nature.

Criticism on Biodiversity Offsets: the example of Smithy Woods

Let's start on the outskirts of Sheffield with Smithy Wood. This is an ancient woodland, which eight hundred years ago was recorded as providing charcoal for the monks who were making iron there. It is an important part of Sheffield's history and culture. It is full of stories and a sense of place and a sense of being able to lose yourself in something different. Someone wants to turn centre of Smithy Wood into a motorway service station.

This might have been unthinkable until recently. But it is thinkable now because the government is introducing something called biodiversity offsets. If you trash a piece of land here you can replace its value by creating some habitat elsewhere. This is another outcome of the idea that nature is fungible and tradeable, that it can be turned into something else: swapped either for money or for another place, which is said to have similar value.

What they've said is, "We're going to plant 60,000 saplings, with rabbit guards around them, in some other place, and this will make up for trashing Smithy Wood." It seems to me unlikely that anyone would have proposed trashing this ancient woodland to build a service station in the middle of it, were it not for the possibility of biodiversity offsets. Something the Government has tried to sell to us as protecting nature greatly threatens nature.

Protest against plans to build a motorway through Smithy Wood and use biodiversity offsets (photo: Sandra Bell)

(continued)

If nature is valued at £x, development will be valued at £x+1

Say we decide that we're going to value nature in terms of pounds or dollars or euros and that this is going to be our primary metric for deciding what should be saved and what should not be saved. This, we are told, is an empowering tool to protect the natural world from destruction and degradation. Well you go to the public enquiry and you find that, miraculously, while the wood you are trying to save has been valued at £x, the road, which they want to build through the wood, has been valued at £x+1. And let me tell you, it will always be valued at £x+1 because cost benefit analyses for such issues are always rigged.

The barrister will then be able to say, "Well there you are, it is x+1 for the road and x for the wood. End of argument." All those knotty issues to do with values and love and desire and wonder and delight and enchantment, all the issues which are actually at the centre of democratic politics, are suddenly ruled out. They are outside the box, they are outside the envelope of discussion, they no longer count. We've been totally disempowered by that process.

Problem of values and framing: taking the position of the opponent (the monetization of nature) means losing the own values (intrinsic value of nature)

But the real problem, and this comes to the nub of the argument for me, is over the issues which I will describe as values and framing. [...] In other words, we are trying to make a case to people who just don't care about the natural world. How do we convince them, when they don't share those values, to change their minds? To me the answer is simple. We don't.

[...]

This, in effect, is what we are being asked to do through the natural capital agenda. We are saying "because our opponents don't share our values and they are the people wrecking the environment, we have to go over to them and insist that we're really in their camp. All we care about is money. We don't really care about nature for its own sake. We don't really believe in any of this intrinsic stuff. We don't believe in wonder and delight and enchantment. We just want to show that it's going to make money."

The alternative to the Natural Capital Agenda is mobilization

George Monbiot concludes with his alternative to the Natural Capital Agenda — action and activism:

So you say to me, "Well what do we do instead? You produce these arguments against trying to save nature by pricing it, by financialisation, by monetisation. What do you do instead?"

Well, ladies and gentlemen, it is no mystery. It is the same answer that it has always been. The same answer that it always will be. The one thing we just cannot be bothered to get off our bottoms to do, which is the only thing that works. Mobilisation.

The manufacturer's authorised representative in the EU is Springer Nature Customer Service Centre GmbH, Europaplatz 3, 69115 Heidelberg, Germany. If you have any concerns regarding our products, please contact ProductSafety@springernature.com

Printed and bound by CPI Group (UK) Ltd, Croydon, CR0 4YY
22/09/2026
02220948-0017